Chemistry of Molecules of Life

Chemistry of Molecules of Life

Dr. Darshan Singh
Associate Professor
Department of Chemistry
Daulat Ram College
University of Delhi, Delhi.

****

Dr. Anuradha
Assistant Professor
Department of Chemistry
Daulat Ram College
University of Delhi, Delhi.

****

Dr. Divya Mathur
Assistant Professor
Department of Chemistry
Daulat Ram College
University of Delhi, Delhi.

Manakin PRESS

CRC Press
Taylor & Francis Group
Boca Raton London New York

CRC Press is an imprint of the
Taylor & Francis Group, an **Informa** business

First published 2026
by CRC Press
4 Park Square, Milton Park, Abingdon, Oxon, OX14 4RN

and by CRC Press
2385 NW Executive Center Drive, Suite 320, Boca Raton FL 33431

CRC Press is an imprint of Informa UK Limited

British Library Cataloguing-in-Publication Data
A catalogue record for this book is available from the British Library

Print edition not for sale in South Asia (India, Sri Lanka, Nepal, Bangladesh, Pakistan and Bhutan)

ISBN13: 9781041206828 (hbk)
ISBN13: 9781041214830 (pbk)
ISBN13: 9781003727453 (ebk)

DOI: 10.1201/9781003727453

Typeset in Times New Roman
by Manakin Press, Delhi

Manakin
PRESS

Preface

The present book **"Chemistry of Molecules of Life"** has been written for the students of B.Sc (Hons) Chemistry and B.Sc Programme courses as per the Choice based credit system and National Education Policy (NEP 2020) syllabus. Life on Earth is built from different types of molecules popularly known as the Molecules of Life. Each one of them is vital for every single organism on Earth. Without any of these molecules, a cell and organism would not be able to survive. These molecules are important either structurally or functionally for cells and, in most cases, they are important in both ways.

This book provides insight into the chemistry of biomolecules in detail. The book comprises 6 chapters on these biomolecules.

Chapter 1 is on carbohydrates chemistry. Carbohydrates and classifications, stereochemistry, interconversions of monosaccharides, structure determination of glucose and fructose including Fischer's proof, Haworth projections, mutarotation and properties. Disaccharides and polysaccharides along with their structure and glycosidic linkage.

Chapter 2 is on nucleic acid. Compositions of nucleic acids, structure, synthesis and reaction of purine and pyrimidine bases, structure determination of nucleoside and nucleotides, structure of nucleic acids, RNA and DNA. Importance of biological processes like replication, genetic code, transcription, translation and denaturation/renaturation of DNA.

Chapter 3 is on amino acids, peptides and proteins. Structure and classification of amino acids, synthesis of amino acids, chemical & physical properties, biological importance of amino acids. Electrophoresis, isoelectric point and structure determination of peptides including N-terminal and C-terminal analysis and synthesis using N-protecting and C-protecting groups and various other methods for peptide synthesis. Proteins and levels of protein structure, importance of denaturation and renaturation in biological processes.

Chapter 4 is on enzymes. Enzymes and their types, properties, nomenclature, classification. Inhibitors and their applications, mechanism of enzyme actions, lock & key mechanism, Induced fit mechanism, features of active sites, cofactors and coenzymes, factors affecting enzyme activity, Michaelis-Menten equation and drug receptor interactions

Chapter 5 is on lipids, types of lipids and their classification. Triglycerides, classification, nomenclature, waxes, complex lipids and their functions, cholesterol, hydrogenation of oil, saponification value, iodine number, acid value and rancidity.

Chapter 6 is on concepts of energy, ATP, metabolism and its types, glycolysis and its applications, Krebs cycle and its significance, fermentation and its types and calorific value.

First of all, we express our sincere thanks to the almighty. We would like to express our heartfelt gratitude to family members and faculty members of Daulat Ram College, University of Delhi for their invaluable guidance, constant encouragement, and generous cooperation at all stages of our work. We would like to thank our Principal, Prof. Savita Roy, Daulat Ram College, University of Delhi for motivation and providing necessary facilities.

Authors have made a sincere attempt to present the content in a simple and lucid manner. We believe that the present book will fulfill the expectation of the readers. We would like to receive your valuable suggestions and inputs regarding the subject matter and will be incorporated in our next assignment.

Authors

Detail Contents

1

Carbohydrates

1.1 Introduction

Carbohydrates, the most abundant biomolecules on Earth, constitutes around 75% of the dry mass of plants. Carbohydrates are produced in green plants *via* photosynthesis using chlorophyll. In this process, CO_2 from the air and H_2O from the soil are used, and the sunlight is the energy source that is absorbed by chlorophyll.

Photosynthesis

$$6\ CO_2 + 6\ H_2O \xrightarrow[\text{sunlight}]{\text{chlorophyll}} \underset{(+)\text{-glucose}}{C_6H_{12}O_6} + 6\ O_2$$

Respiration

$$C_6H_{12}O_6 + 6\ O_2 \longrightarrow 6\ CO_2 + 6\ H_2O$$

Carbohydrates have The Following Functions:

1. Plants use carbohydrates in the form of cellulose and starch, which serves as structural element and energy reserve, respectively.
2. Carbohydrate provides energy to humans and animals through diet.
3. Carbohydrate is stored in the form of glycogen which is a short-term energy reserve.
4. Carbohydrates form a structural framework for the synthesis of other biochemical substances such as proteins, lipids, and nucleic acids.

DOI: 10.1201/9781003727453-1

5. Carbohydrates linked to lipids are glycolipids that are structural components of cell membranes.
6. Carbohydrates linked to proteins are glycoproteins which function in a variety of cell recognition processes.

The word carbohydrate originated from "carbon hydrates" and simple and common carbohydrates have the generic formula = $C_n(H_2O)_m$. For example, glucose has the molecular formula, $C_6H_{12}O_6$, and sucrose, $C_{12}H_{22}O_{11}$, which can also be written as $C_6(H_2O)_6$ and $C_{12}(H_2O)_{11}$, respectively. However, this definition is not recognized now as only a few carbohydrates possess the formula $C_n(H_2O)_m$.

A **Carbohydrate** is a *polyhydroxy aldehyde*, a *polyhydroxy ketone*, or a compound that yields polyhydroxy aldehydes or polyhydroxy ketones upon hydrolysis. Saccharide is a word derived from the Latin word which means "sweet sand". Carbohydrates are often classified according to the number of saccharide units they contain.

Carbohydrates can be classified based on the molecular size as monosaccharides, oligosaccharides, and polysaccharides.

Monosaccharides are carbohydrates containing single polyhydroxy aldehyde or polyhydroxy ketone unit. Monosaccharides cannot be broken into simpler units by hydrolysis. For example, glucose and fructose are monosaccharides. Mostly they are water-soluble, white, and crystalline solids.

Oligosaccharides are carbohydrates with two to ten monosaccharide units covalently bonded to each other. For example, in a disaccharide, there are two monosaccharide units covalently bonded to each other. Sucrose and lactose are disaccharides. Complete hydrolysis of an oligosaccharide produces monosaccharides. Upon hydrolysis, a disaccharide produces two monosaccharides, a trisaccharide three monosaccharides, a hexasaccharide six monosaccharides, and so on.

A polysaccharide is a polymeric carbohydrate that contains many monosaccharide units covalently bonded to each other. Polysaccharides often contain several thousand monosaccharide units. The paper we use is mainly cellulose, cotton in our clothes, and the wood in our houses is also made up of polysaccharides. Starch which is a component of many types of foods is also a polysaccharide.

1.2 Stereoisomerism in Carbohydrates

Compounds identical in composition and the structural formula but different in the spatial configuration are known as stereoisomers. In the case of carbohydrates also various stereoisomers are possible because they contain more than one chiral carbon. For example, sucrose has nine chiral centres, and glucose has four chiral carbons. Thus, because of a multitude of chiral centers, carbohydrates exhibit stereoisomerism.

1.2.1 Enantiomers and Diastereomers

Fischer projection formula is usually used to represent an open chain structure of carbohydrates. In the case of the monosaccharides, carbon chain is positioned vertically with the carbonyl group (aldehyde or ketone) at or near the top. The smallest monosaccharide known with a chiral center is glyceraldehyde (2,3-dihydroxypropanal) which exists as two enantiomers. The enantiomer with the -OH group on the right is D-glyceraldehyde, and the enantiomer with -OH group on the left is L-glyceraldehyde (Figure 1.1).

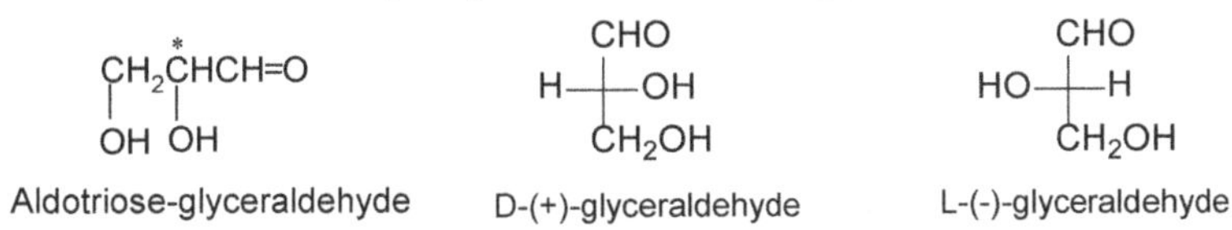

Figure 1.1: Enantiomers of glyceraldehyde

Aldotetroses have two chiral centers, and each chiral carbon (C-2 and C-3) can exist in either the R or S configuration. Thus, the four possible stereoisomers will be R,R; S,S; S,R; and R,S (Figure 1.2). Here, the aldose with 2R,3R configuration is the mirror image of 2S,3S and 2R,3S is the mirror image of 2S,3R. The mirror images have identical physical properties and rotate the plane of polarized light by equal amounts in opposite directions and are known as enantiomers. However, aldotetrose isomer, 2R,3R is not a mirror image of either the 2R,3S- or 2S,3R-isomer, and it differs from them in chemical and physical properties and is known as diastereomers. The isomers 2R,3R and 2S,3S are known as erythrose and 2R,3S and 2S,3R are known as threose. Thus, each erythrose is a diastereomer of both threose isomers and aldotetroses exists as enantiomers (+)-erythrose and (-)-erythrose, (+)-threose and (-)-threose.

The D, L system used to assign the position of –OH in glyceraldehyde enantiomers can be extended to other monosaccharides also with more than one chiral center.

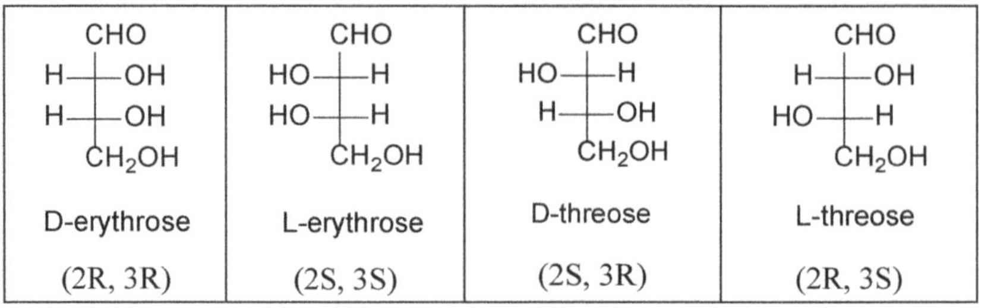

Figure 1.2: Stereoisomers of aldotetroses

The *erythro* and *threo* nomenclature is used for various other molecules where, if identical or similar groups are on the same side of the vertical chain in the Fischer projection, the isomer is termed as erythro isomer. If the substituents are opposite to each other, then it is called threo isomer.

For *n* number of chiral centers in a molecule there can be 2^n possible stereoisomers. Aldohexoses have four chiral carbons and so there are 16 possible stereoisomers, one of which is (+)-glucose as represented below (Figure 1.3). Here also, the sugars can be divided into two classes, D- or L-, if the hydroxyl of the penultimate carbon is on the right it is known as Dextro or D-sugar or if it is towards the left then it is Levo or L-sugar. The D- and L-isomer nomenclature is different from the d-, l-, (+)-, or (-)- representation which is about the rotation of the plane of the plane-polarized light by the molecule.

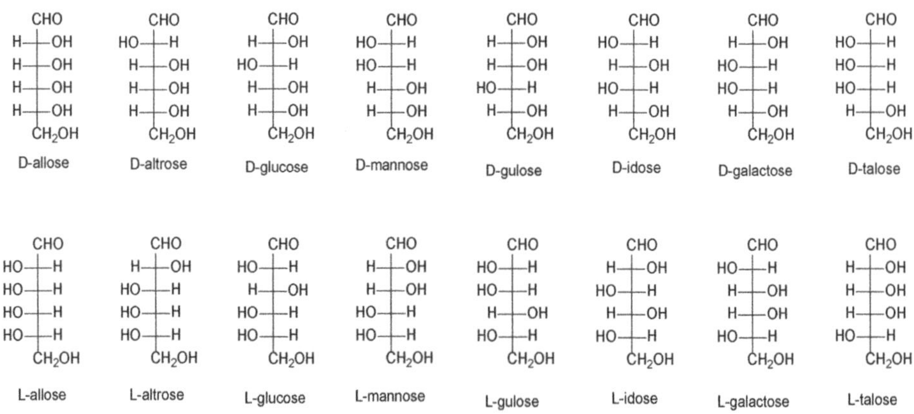

Figure 1.3: Stereoisomers of aldohexoses

1.3 Monosaccharides

Monosaccharides can be classified by both their number of carbon atoms and their functional group. The most common monosaccharides are with three to seven carbon atoms. A monosaccharide with three-carbon is called a triose, with four, a tetrose, five carbon atoms, pentoses, and with six carbon atoms it is known as hexoses.

Further, monosaccharides are classified as aldoses or ketoses based on the type of carbonyl group. An aldose is a monosaccharide with an aldehyde functional group. A ketose is a monosaccharide that contains a ketone functional group. A six-carbon monosaccharide with an aldehyde functional group will be named as aldohexose; a six-carbon monosaccharide with a ketone functional group is a ketohexose.

1.3.1 Cyclic Forms of Monosaccharides

Experimental evidence indicates that monosaccharides containing five or more carbon atoms exist not only as open-chain structures but are also in equilibrium with two cyclic structures that are in dominant forms at equilibrium. The linear and cyclic forms are structural isomers that exist in equilibrium with each other displaying ring-chain tautomerism.

The cyclic forms result from the ability of carbonyl group to react intramolecularly with a hydroxyl group, resulting in a cyclic hemiacetal. This addition of hydroxyl group across the carbon-oxygen double bond resulting in ring formation also produces a chiral center at carbon 1, leading to two stereoisomers *α-isomer and β-isomer known as* ***anomers*** that are diastereomers.

In case of glucose, the alcohol on C-5 reacts with the aldehyde (C-1) to form a six-membered ring called the "pyranose" form in reference to the cyclic ether pyran. In this process a new chiral centre is formed at C-1 leading to the formation of two diastereomers, α and β forms (Figure 1.4).

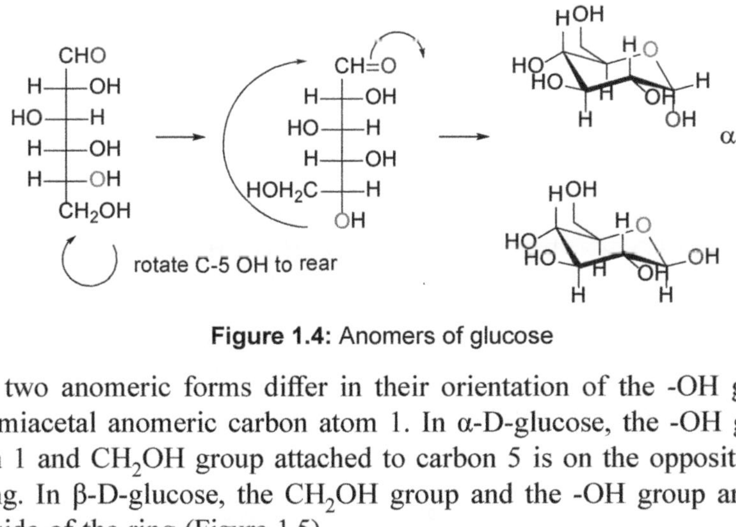

Figure 1.4: Anomers of glucose

These two anomeric forms differ in their orientation of the -OH group on the hemiacetal anomeric carbon atom 1. In α-D-glucose, the -OH group on carbon 1 and CH_2OH group attached to carbon 5 is on the opposite side of the ring. In β-D-glucose, the CH_2OH group and the -OH group are on the same side of the ring (Figure 1.5).

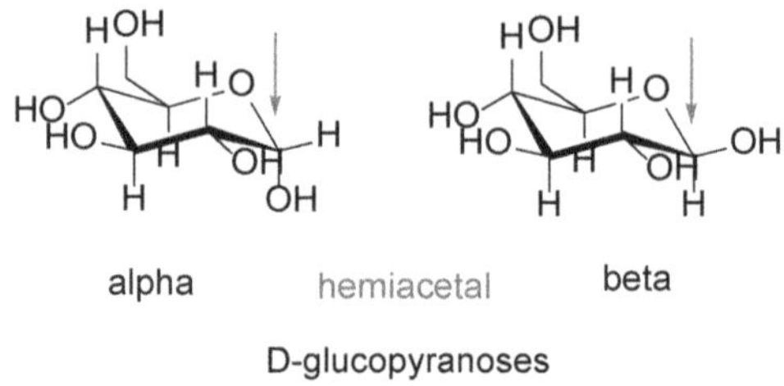

α-D-glucose	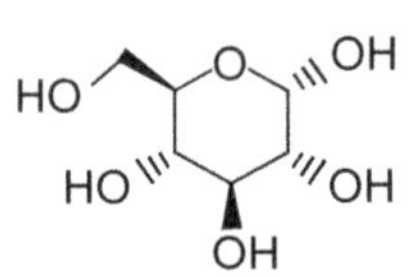	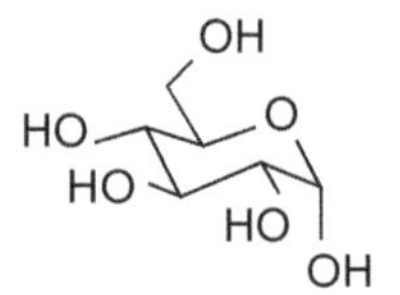	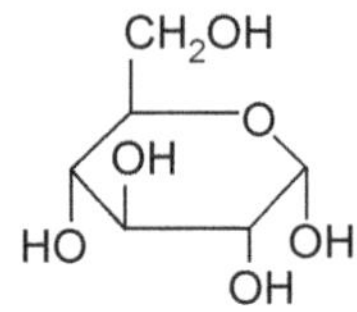
β-D-glucose	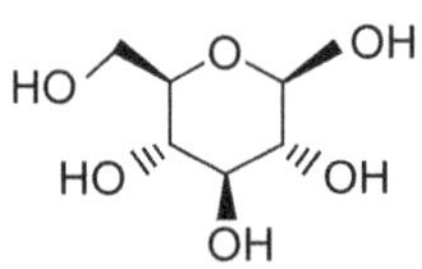	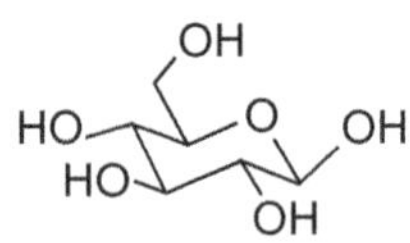	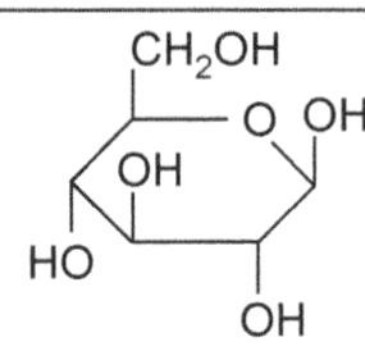

Figure 1.5: D-glucopyranoses

The anomeric carbon is important to the reactivity of carbohydrates because it is the site where the ring opening occurs in the cyclic structure leading to the formation of the carbonyl and hydroxyl group.

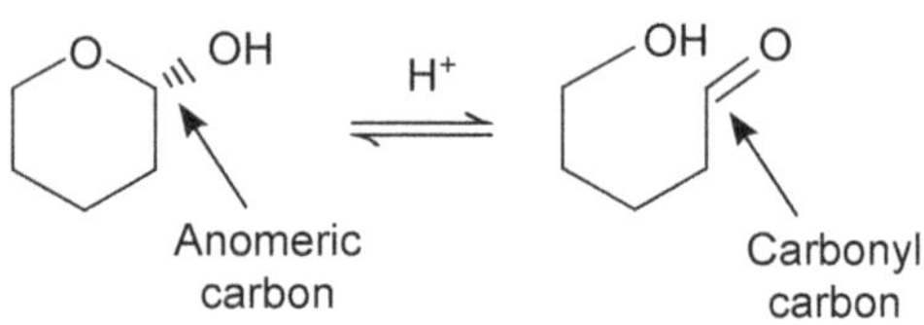

It's also possible for the hydroxyl group on C-4 of glucose to attack the aldehyde group which leads to the formation of a five-membered ring (Figure 1.6). This is known as the furanose form, in reference to the cyclic 5-membered ether furan.

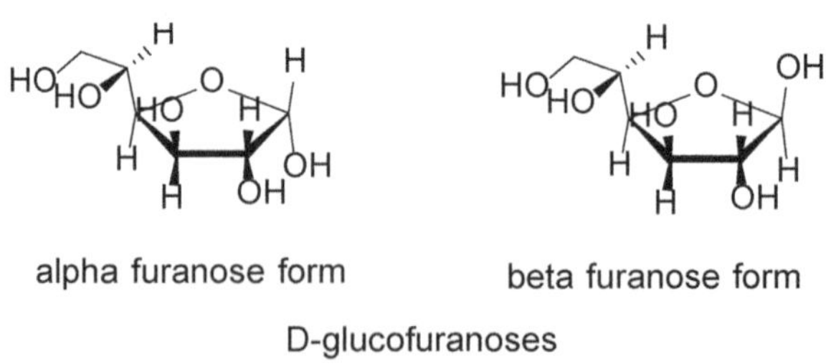

Figure 1.6: D-glucofuranoses

1.4 Mutarotation

Each anomer of D-glucose can be synthesized and isolated as pure compounds. The α-anomer of D-glucose has a specific rotation of +112° in water and the β-anomer has a specific rotation of +18.7°. However, when either of the anomer is dissolved in water, the value of the specific rotation changes over time, eventually reaching the value of +52.5° for both. Thus, specific rotation of α-D-glucopyranose decreases from +112° to +52.5° and for β-D-glucopyranose it increases from +18.7° to +52.5°. This can be explained by analysing the composition of the solution with this specific rotation of +52.5°. It was found to contain 63% of β-D-glucose, 37% α-D-glucose, and less than 0.01% open-chain form.

Thus, in an aqueous solution of D-glucose, a dynamic equilibrium exists with continual interconversion among α, β and open-chain forms (Figure 1.7). This behaviour is known as "mutarotation" which is due to the change in the optical rotation of the pure α- and β- anomers of glucose upon dissolution in water. Due to ring-chain tautomerism, the α- and β-forms slowly interconvert *via* open chain structure until equilibrium is established.

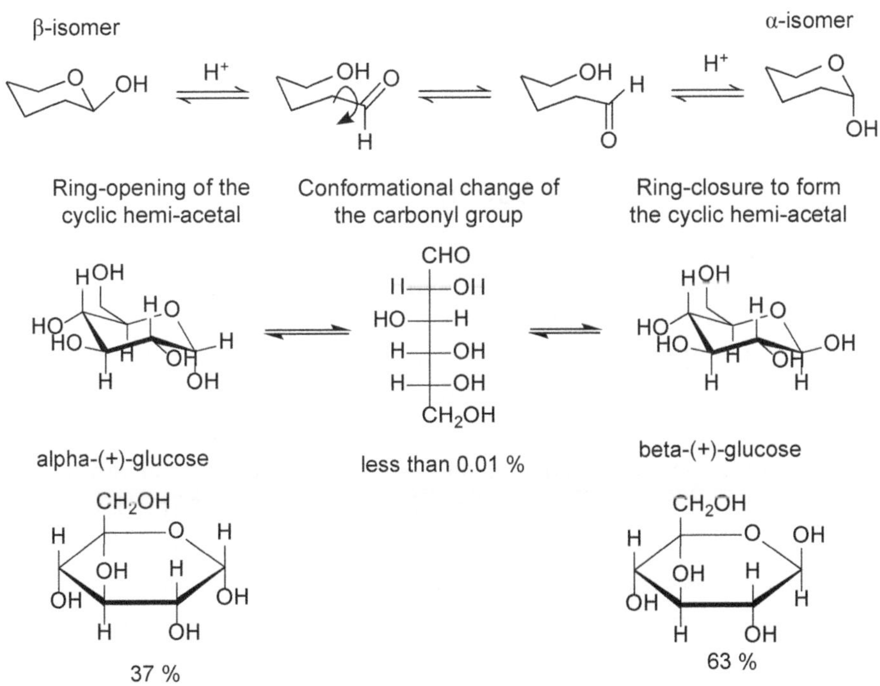

Figure 1.7: Mutarotation in D-glucose

Intramolecular cyclic hemiacetal formation and the equilibrium between forms are not restricted to only glucose. All aldoses bearing five or more carbon atoms show similar equilibria, but with different percentages of the alpha, beta, and open-chain forms. Ketoses like fructose also cyclize but form a five-membered ring.

1.5 Epimers

Epimers are diastereomers differing only in the configuration at one chiral center. For example, D-mannose and D-galactose differ in the configuration of –OH group at C-2 and C-4, respectively, and are thus, epimers of glucose (Figure 1.8).

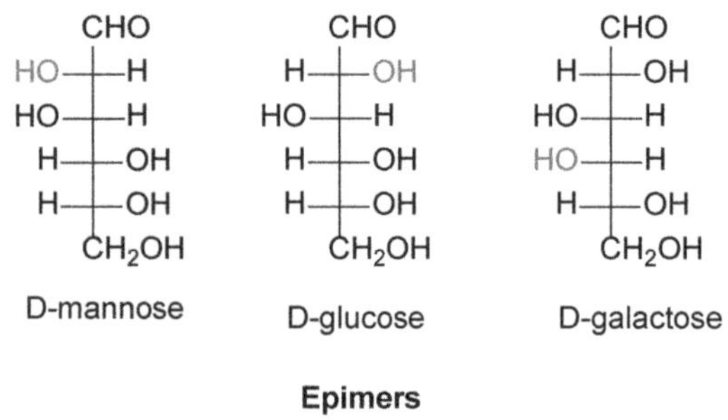

Figure 1.8: Epimers

1.6 Reactions of Monosaccharides

1.6.1 Oxidation Reaction

Oxidation reaction in monosaccharide yields three different types of acidic sugars, depending upon the oxidizing agent. Tollen's, Fehling's, Benedict's reagents, bromine water are weak oxidizing agents that can oxidize the aldehyde group in the aldose to an aldonic acid. Since aldoses act as reducing agents and get oxidized, they are called reducing sugars. In turn, Ag^+ ion in Tollen's solution get reduced to Ag resulting in formation of silver mirror. Cu^{2+} ion in Benedict's and Fehling's solution get reduced to Cu^+ ion in the form of red precipitate of cuprous oxide. Thus reducing sugar is a carbohydrate that gives a positive test with Tollen's, Fehling's and Benedict's solutions. Ketoses also give this positive test because these three solutions are basic in nature.

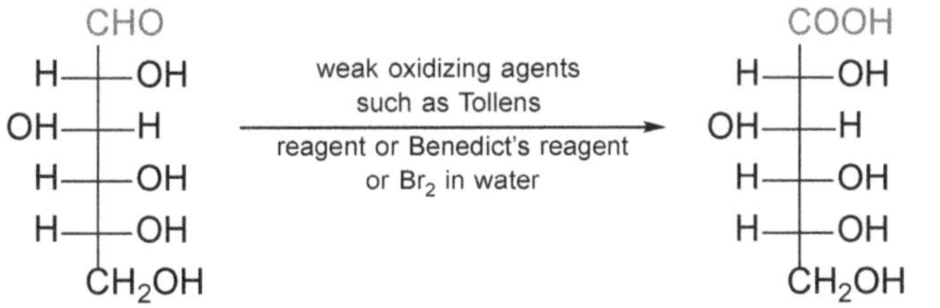

D-Glucose D-Gluconic acid

Ketose undergoes a structural rearrangement through tautomerization when they isomerize to aldoses *via* an enediol in basic medium. The aldose then reacts with Tollens and Benedict's solution giving a positive test. Thus all monosaccharides, both aldoses and ketoses, are reducing sugars.

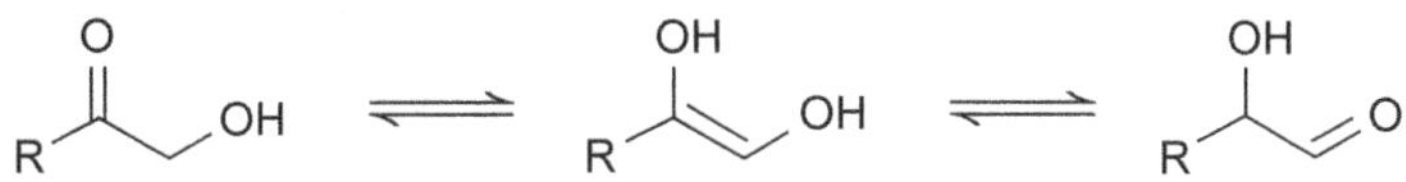

Ketose Enediol Aldose

Strong oxidizing agents such as dilute nitric acid can oxidize both the carbonyl group and the terminal primary alcohol in a monosaccharide leading to the formation of a dicarboxylic acid known as aldaric acids.

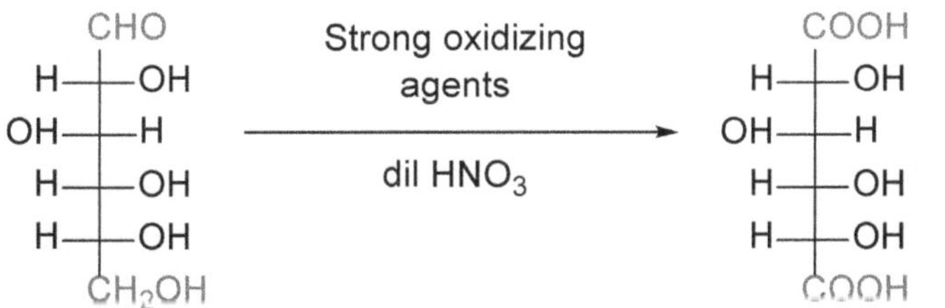

D-Glucose D-Glucaric acid

1.6.2 Reduction Reaction

The carbonyl group in an aldose or a ketose can be reduced to a hydroxyl group by using hydrogen or $NaBH_4$ as the reducing agent eading to the synthesis of sugar alcohol. For example, the reduction of D-glucose gives D-glucitol, also known as D-sorbitol.

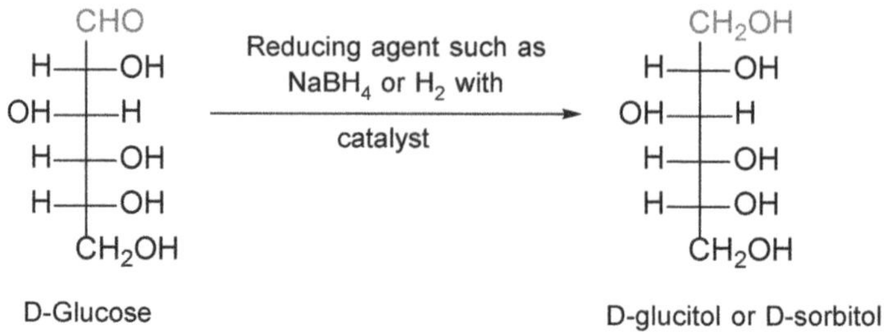

1.6.3 Glycoside Formation

Cyclic forms of monosaccharides are basically hemiacetals that can react with alcohols in the presence of acid catalyst to form anomeric acetals. The monosaccharide acetals are known as glycosides. Acetals of glucose are glucosides, of mannose are mannosides and fructose are fructosides. The reaction involves the loss of an equivalent of water and the product is stable to base and alkaline oxidants such as Tollen's reagent. In acidic solutions, glycosides undergo hydrolysis to produce the sugar and alcohol again. For example, methyl-α-D-glucopyranoside undergoes hydrolysis under acidic medium to produce anomeric mixture of D-glucose together with small amount of open chain form of glucose and methanol.

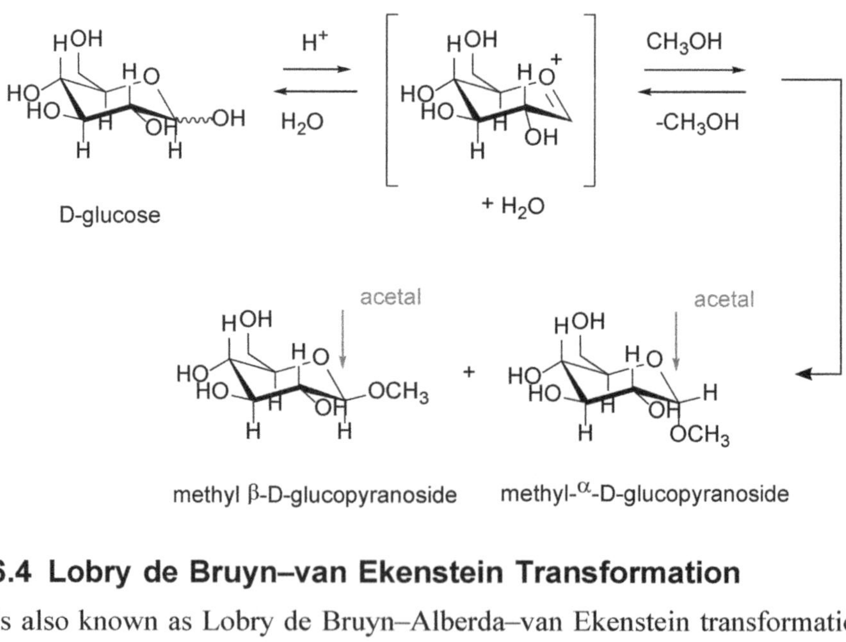

1.6.4 Lobry de Bruyn–van Ekenstein Transformation

It is also known as Lobry de Bruyn–Alberda–van Ekenstein transformation. Monosaccharides in aqueous basic solutions get converted into isomers due to the formation of tautomeric enediol as reaction intermediate. Thus in this catalyzed transformation, an aldose get converted into the ketose isomer or *vice versa*. For example, when D-glucose is dissolved in an aqueous basic solution it rearranges to D-fructose as well as D-mannose. This is because of the formation of enolate ion and also since the stereochemical configuration is lost in the enol form, it can be protonated from two faces, resulting in the formation of glucose and epimer D-mannose. Therefore, final product will be a mix of D-glucose, D-fructose and D-mannose.

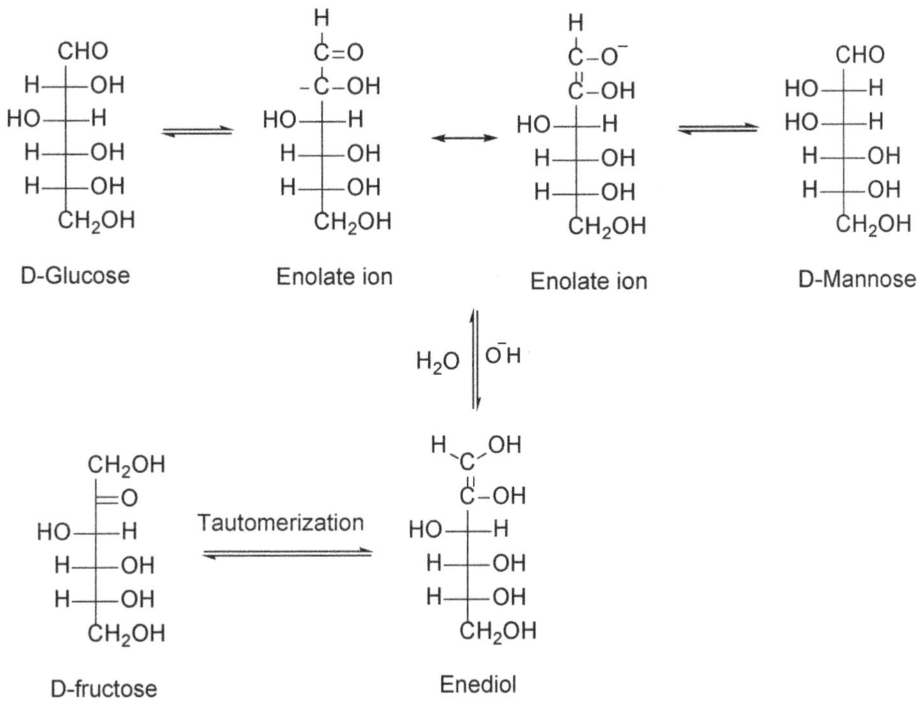

1.6.5 Conversion to Esters

Monosaccharides with excess acetic anhydride and weak base such as pyridine at room temperature convert all the hydroxyl groups into esters and forms glucose pentaacetate. Here if a mixture of α- and β-anomers of glucose is used as a starting material then the product will be corresponding to the ratio of the starting anomeric mixture. The acetylation reaction will be faster than ring opening reaction and subsequent mutarotation.

In case of the use of lewis acids such as zinc chloride and iodine as catalyst with acetic anhydride, glucose-α-pentacetate is formed as it is a thermodynamically favoured product. This is because when an anomeric mixture of D-glucose is reacted with acetic anhydride, it forms an anomeric mixture of glucose pentaacetate. But due to the anomeric effect, glucose-α-pentacetate is predominantly formed.

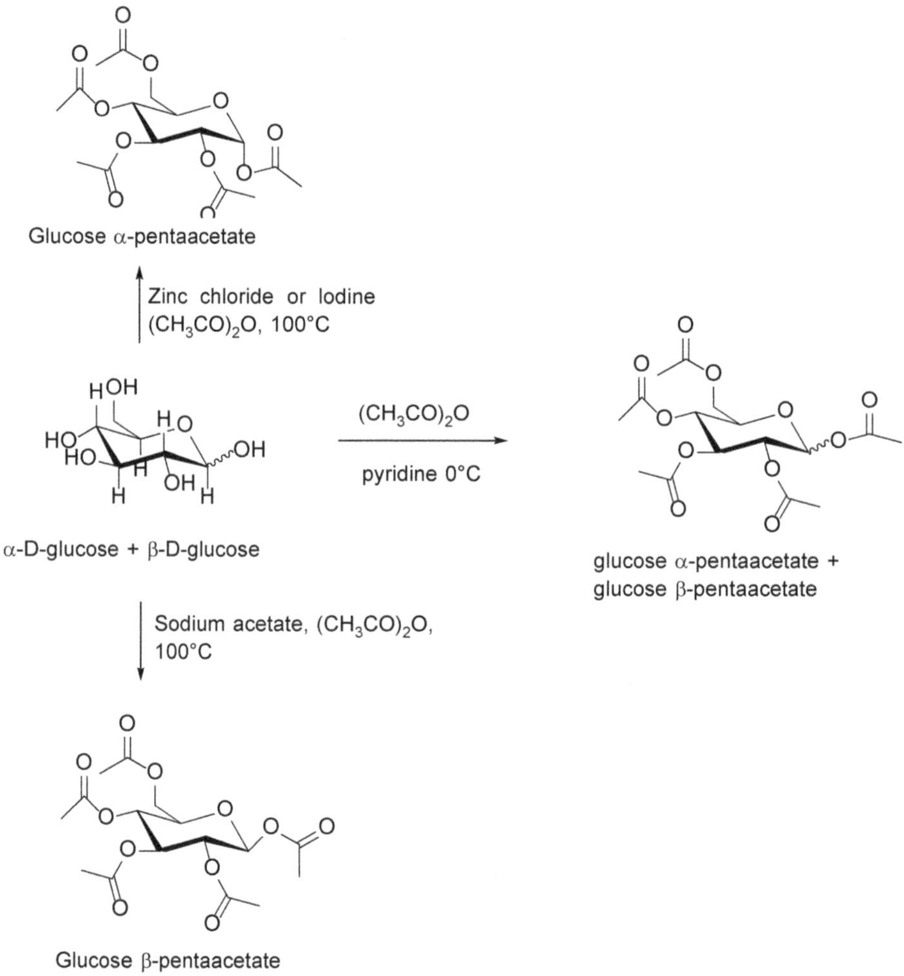

Sodium acetate is a weak base and when it is used with acetic anhydride, the reaction proceeds sluggishly at room temperature. Therefore, the reaction is carried out at high temperature and in those conditions β product is formed preferentially. This is because at high temperature mutarotation is faster than the acetylation reaction. Since the β-anomer of glucose is more stable than its α-anomer, β component reacts with acetic anhydride, giving glucose β-pentaacetate as the major product.

1.6.6 Periodate Oxidations

Vicinal-glycols have adjacent hydroxyl groups and can be cleaved to corresponding aldehydes and/or ketones through oxidative cleavage by the action of aqueous periodic acid (HIO_4). Oxidative cleavage also takes place when the –OH group is adjacent to the carbony group. This reaction is also known as the Malaprade reaction. It provides valuable information and can be successfully applied for structural analysis by measuring the number of molar equivalents of periodic acid consumed. Lead tetraacetate ($Pb(OAc)_4$) brings about similar cleavage and give good results in organic solvents. The reaction is called *Criegee reaction.*

Periodate oxidative cleavage occurs through a cyclic intermediate. Primary alcohols oxidize to formaldehyde after cleavage, whereas secondary alcohols oxidize to aldehydes after cleavage and then subsequently to formic acid. The aldehyde group gets oxidize to formic acid and keto group into carboxylic acids after cleavage which further gets oxidized to CO_2. Both cis- and trans-glycols undergo the oxidation with HIO_4 with cis-glycols reacting more rapidly than trans-glycols.

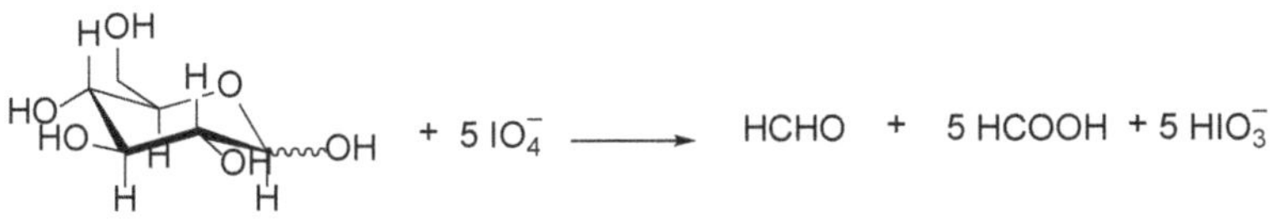

D-glucose

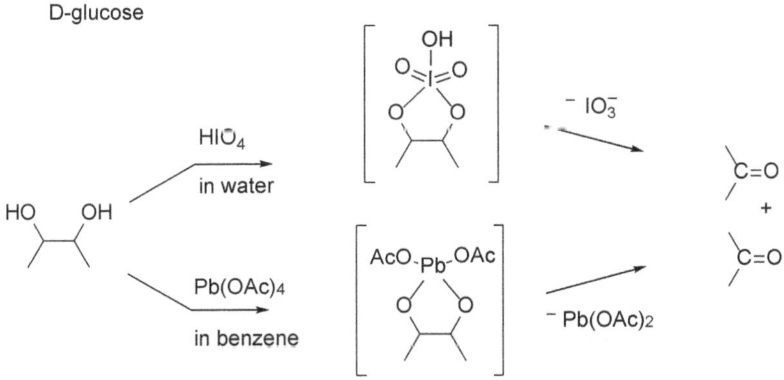

D-glucose consumes five HIO_4 molecules to yield one molecule of formaldehyde and five molecules of formic acid.

1.6.7 Formation of Osazones

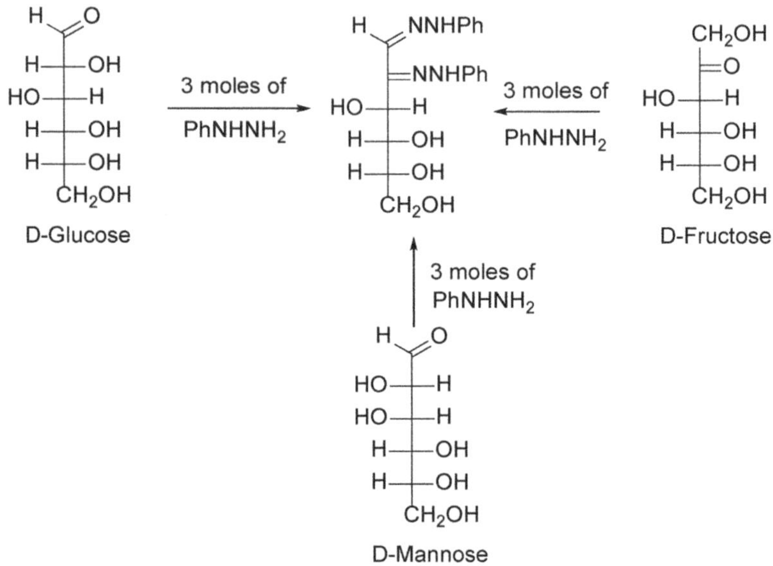

Glucose, mannose and fructose react with phenylhydrazine to obtain the same osazone derivative. In this reaction, the carbonyl group of the monosaccharide reacts with three equivalents of phenylhydrazine, and two phenylhydrazone groups are introduced in the structure, one at C-1 and the other at C-2 making a osazone. In the first step, phenylhydrazine derivative formed due to the carbonyl group undergoes rearrangement due to which a second carbonyl group is generated at C-2. This is known as Amadori **rearrangement**. Again the second molecule of phenylhydrazine reacts with the newly generated

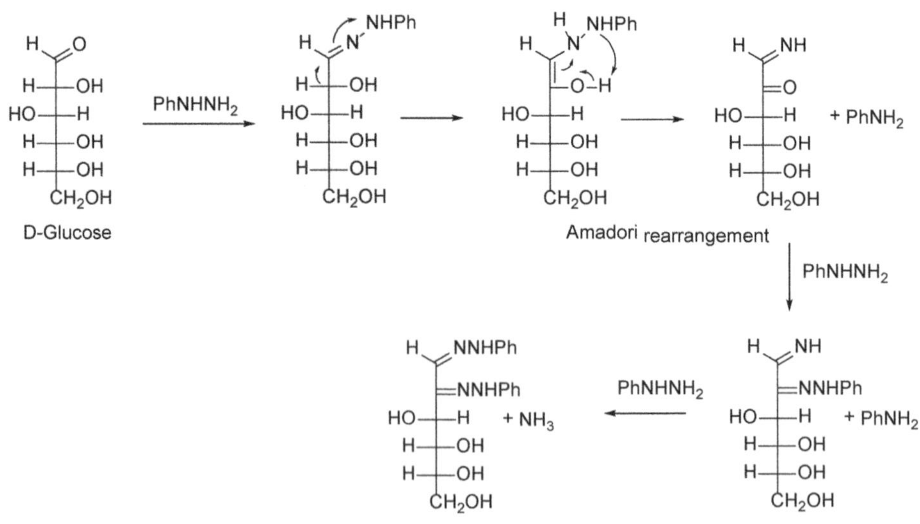

carbonyl group and finally we get the osazone derivative together with aniline and ammonia. Amadori rearrangement does not occur beyond C-2 because the osazone formed is stabilized through intra-molecular hydrogen bonding and forms a six membered stable cyclic structure.

1.7 Synthesis and Degradation of Monosaccharides

1.7.1 Kiliani-Fischer Synthesis

Nucleophilic addition of hydrogen cyanide to an aldose leads to attack of cyanide ion on the aldehyde group which results in a cyanohydrin. This extends the length of the carbon chain by one carbon. It also creates a new stereocenter, resulting in two diastereomers. Hydrolysis using barium hydroxide and with further acidification, a diastereomeric mixture of aldonic acid is obtained. It quickly reacts itself to form a more stable diastereomeric mixture of lactones. They can now be separated by chromatography or by

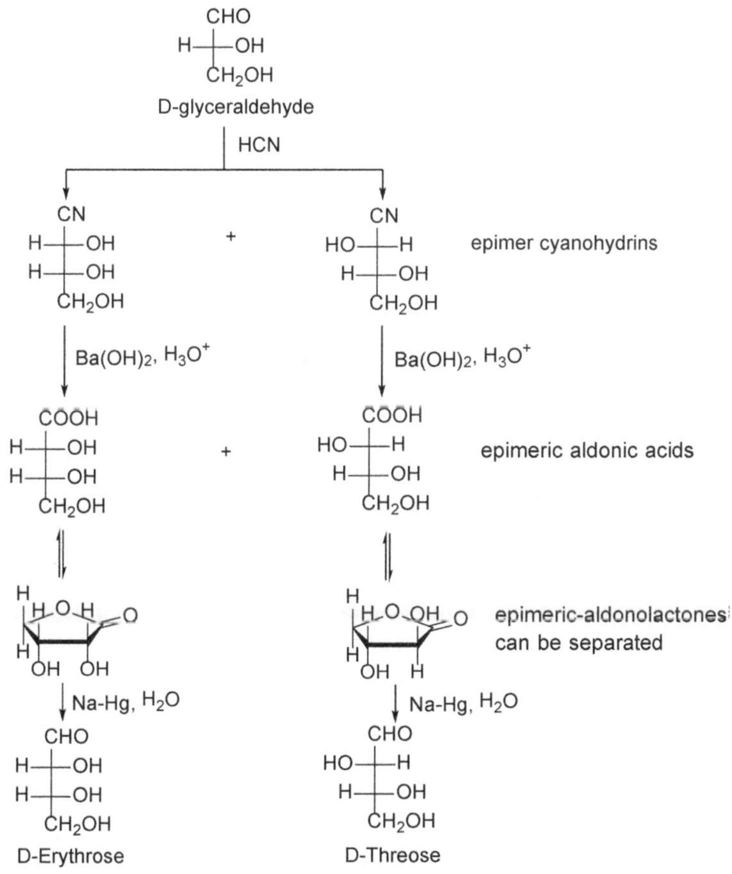

numerous other separation methods. The desired lactone is reduced with sodium amalgam. Thus, glyceraldehyde can be converted to erythrose and threose. Arabinose can be converted to glucose and mannose.

1.7.2 Ruff Degradation

Ruff degradation is used to reduce the number of carbon atoms in aldose by one carbon, thus it shortens the chain length. It involves selectively oxidizing the aldehyde to a carboxylic acid by bromine water. In the next step an iron (III) salt [$Fe_2(SO_4)_3$] with hydrogen peroxide is added which leads to oxidative decarboxylation which means loss of carbon dioxide with concomitant oxidation of the adjacent OH at C-2 to an aldehyde group. Thus, ribose can be converted to erythrose and glucose can be converted to arabinose.

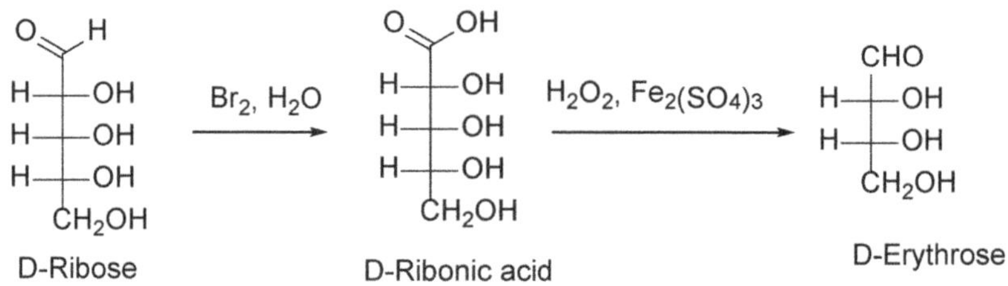

1.8 Glucose

The most common carbohydrate known and studied is glucose with the molecular formula $C_6H_{12}O_6$. It is a monosaccharide, an aldohexose with six carbons and bearing a aldehyde group and a reducing sugar. To deduce its structure some simple yet confirmatory reactions were employed by scientists.

Hydriodic acid (HI) reductively removes oxygen functional groups from a molecule. In glucose treatment of HI yielded hexane which concluded that there are six carbons in an unbranched manner in glucose. Reaction of glucose with HCN gave cyanohydrin which indicated the presence of an aldehyde carbonyl group (Figure 1.9). Further it was confirmed by the reduction of glucose as it yielded a hexa-alcohol called sorbitol or glucitol, which on further treatment with acetic anhydride gave hexaacetate. It's mild oxidation with bromine water gave mono-carboxylic acid, gluconic acid, while strong oxidizing agents such as dilute nitric acid gave a diacid named glucaric acid. This further supports the theory of the presence of six-carbon chain in glucose. Now only five oxygens remained which were to be established after the confirmation of the presence of aldehyde group. Further, glucose with acetic anhydride yielded penta-acetate derivative, thus glucose was thought to have hydroxyl groups.

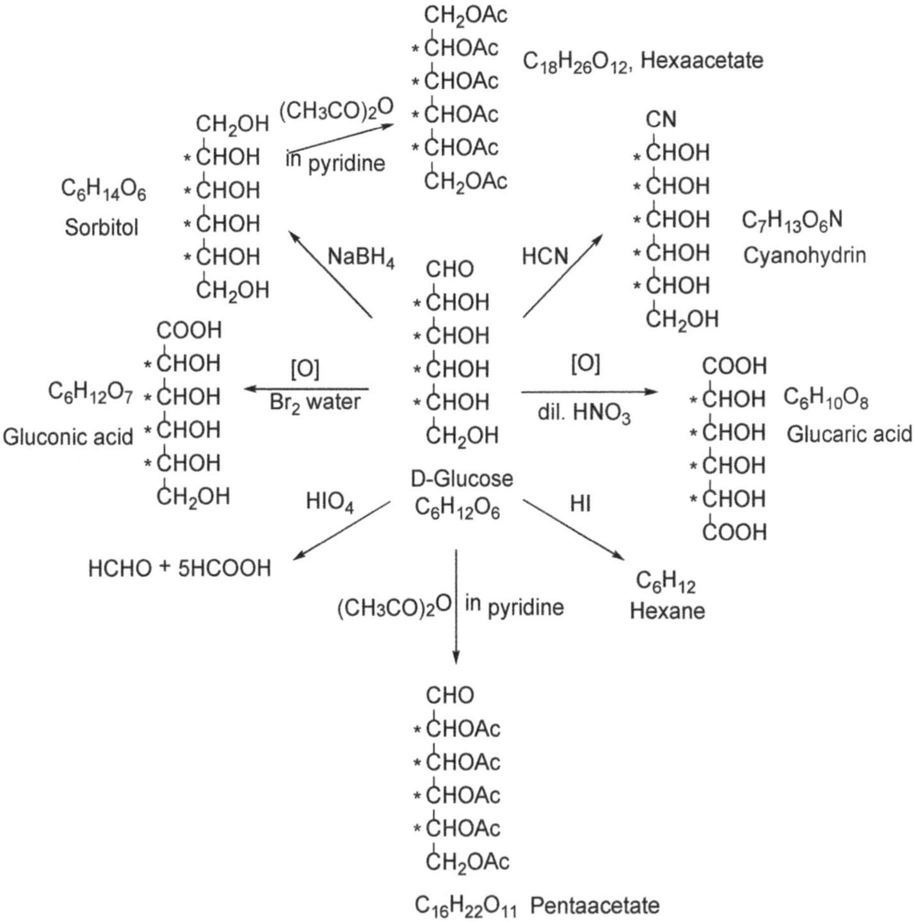

Figure 1.9: Chemical reactions for the confirmation of the structure of glucose

Now these hydroxyl groups cannot be geminal since normally it would be unstable taking into consideration the presence of adjacent carbonyl group and also germinal hydroxyls will lead to loss of water. Thus, one hydroxyl group was assigned to each last five carbon atoms. Glucose was cleaved by periodic acid due to vicinal diol moieties in its structure into one molecule of formaldehyde and five molecules of formic acid.

Now all these reactions gave a rough idea of the structure of glucose. But with four chiral centres, there can be as many as sixteen stereoisomers having this constitution, which means eight diastereomeric pairs of enantiomers, eight D-sugars and eight L-sugars. German chemist Emil Fischer in 1891 successfully found out the structure of glucose and received the Nobel Prize for chemistry in 1902 for this accomplishment.

1.9 Fischer's Proof of the Configuration of D-(+)-Glucose

To establish the structure of (+)-glucose, Fischer arbitrarily restricted his attention to only eight D-sugars.

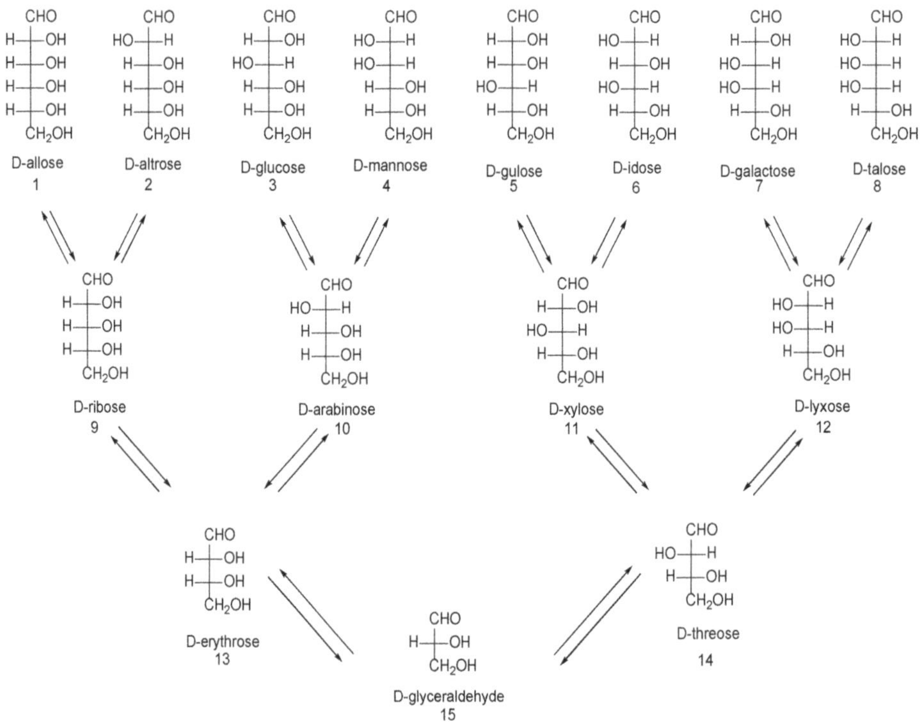

He assigned structure 3 to (+)-glucose and it was based on the following assumptions:

Glucose on oxidation with dilute nitric acid gives an optically active aldaric acid. Therefore glucose cannot have structures 1 or 7, which would give optically inactive aldaric acids.

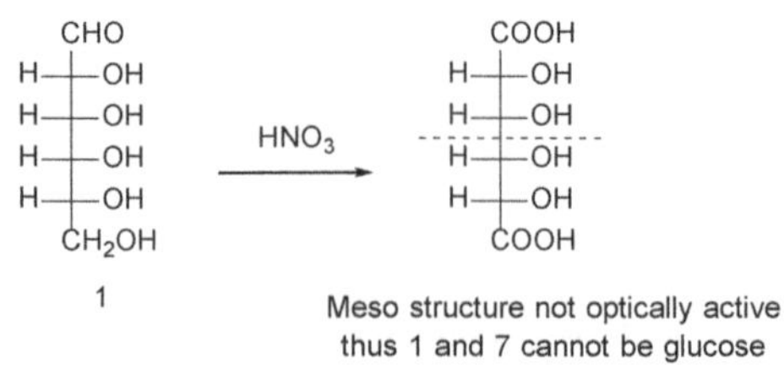

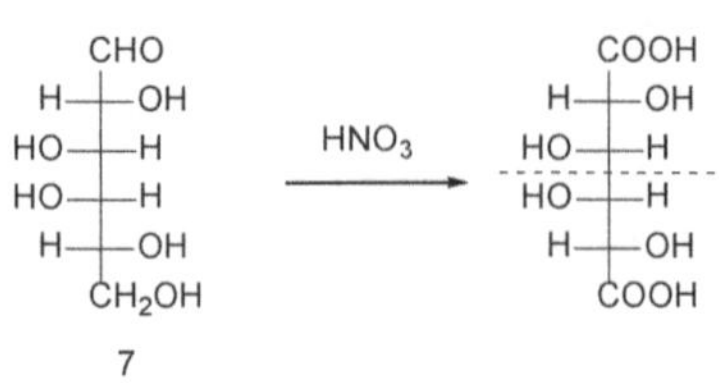

Glucose gives arabinose through Ruff degradation. Oxidation of arabinose with nitric acid gives optically active arabinaric acid. Therefore, by establishing structure of arabinose we can reach to the structure of glucose. Now, arabinose cannot have structures 9 or 11 because they will give optically inactive aldaric acids. If arabinose cannot be 9 or 11, then glucose cannot be 1, 2, 5 or 6.

$$\text{(+)-Glucose} \xrightarrow{\text{Ruff degradation}} \text{(+)-Arabinose} \xrightarrow{HNO_3} \text{(+)-Arabinaric acid}$$

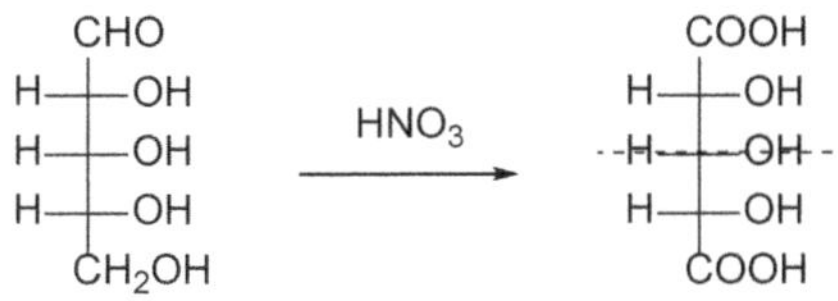

9

Meso structure not optically active
thus 2, 5 and 6 cannot be glucose

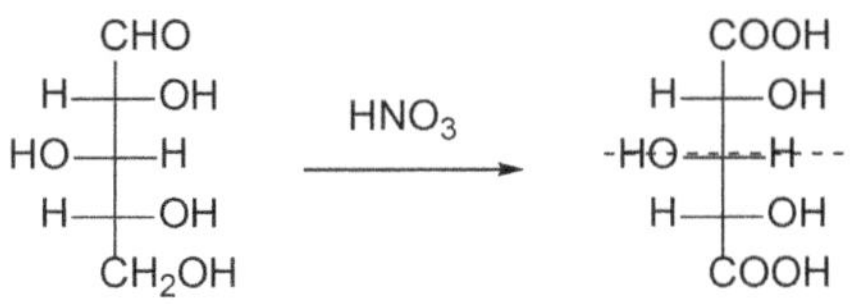

11

For arabinose, we are left with structure 10 or 12. Killiani-Fischer chain extension reaction of arabinose gives two structures, both of which on treatment with nitric acid give optically active aldaric acids. Therefore,

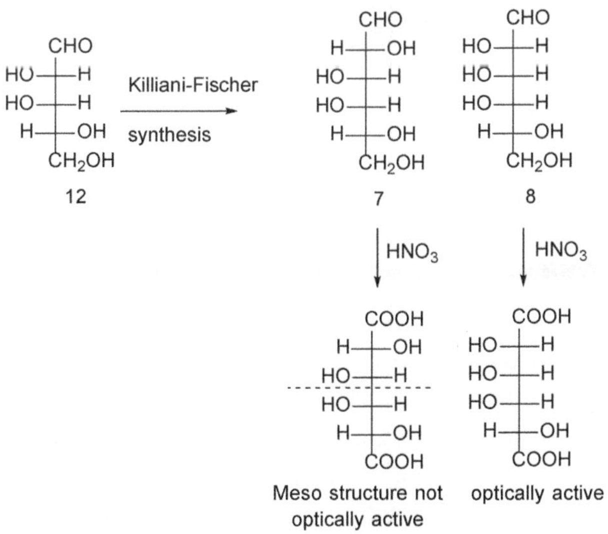

arabinose cannot be structure 12, since it would give 7 and 8 in a Killiani-Fischer chain extension. Now, 8 would give an optically active aldaric acid, but 7 would give an optically inactive aldaric acid. Thus, this eliminates 12 as structure of arabinose, and in turn eliminates 7 and 8 as structure of glucose.

Now since structure of arabinose is 10, therefore the structure of glucose would be either 3 or 4.

Fischer developed a method to interchange the ends of a sugar that is the aldehyde can be converted to a CH_2OH and the CH_2OH to an aldehyde. Fischer rationalized that if ends of 3 were interchanged, a new L-aldohexose would be obtained. On the other hand, if the ends of 4 were interchanged, it would lead to the same structure. Practically on glucose when the CH_2OH and CHO groups were interchanged a new sugar was obtained, named L-gulose. This concludes that structure of glucose is 3 and that of mannose is 4.

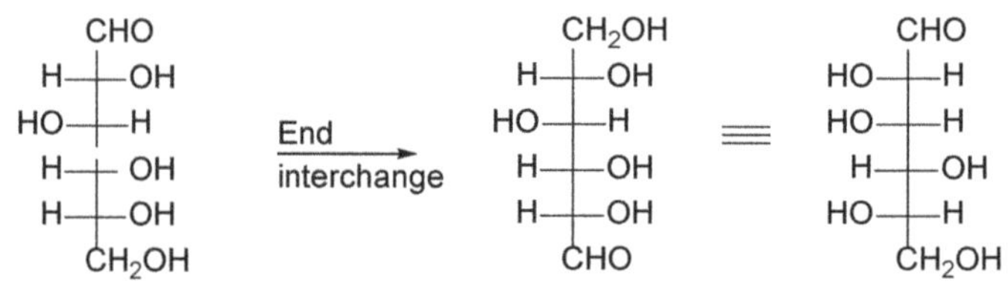

3

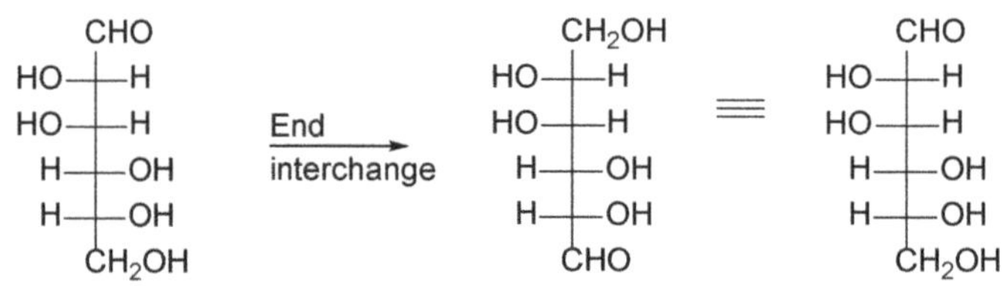

4

Same structure as 4

1.10 Fructose

Fructose is mostly found together with glucose and sucrose in honey and fruit juices. Bees use an enzyme to hydrolyze or break the sucrose in the nectar of flowers into its components, glucose and fructose. Disaccharide, sucrose consists of fructose along with glucose as the monosaccharides. It is the most important ketose and is also a reducing sugar. Another name of fructose is levulose, because of its levorotatory property of rotating the plane polarized light to the left which is in contrast to glucose which is dextrorotatory.

Fructose has a ketone functional group and the ring closure reaction thus occurs at carbon 2. It therefore forms a five membered ring. The -OH on carbon 5 participates in the formation of the ether linkage to close the ring with carbon 2. This makes a five membered ring with four carbons and one oxygen atom. The anomeric carbon is the center of a hemiketal functional group. It has both ether oxygen and an alcohol group and thus fructose also exists as two anomers. The β-anomer has the C-2 hydroxyl group on the same side of the ring as the C-6 (Figure 1.10). The α-anomer has –OH on the opposite side to the C-6.

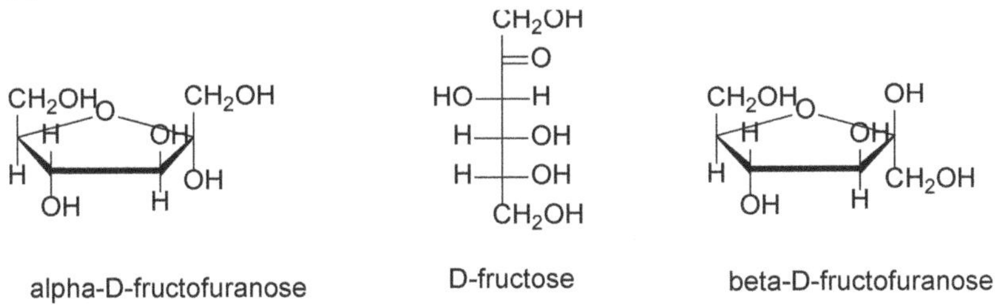

Figure 1.10: Fructose

1.11 Disaccharides

In disaccharides, anomeric carbon of one monosaccharide is linked to the OH of another monosaccharide *via* a condensation reaction resulting in the elimination of water molecule. The bond thus formed is termed as a glycosidic bond. The monosaccharide that bears the anomeric carbon is referred to as the reducing end of the disaccharide.

1.11.1 Maltose

Maltose or malt sugar is a disacchride made up of two molecules of D-glucose connected by α-(1→4)-linkage. Its IUPAC name is 4-O-α-D-glucopyranosyl-β-D-glucopyranose (Figure 1.11). During digestion of starch, enzyme β-amylase cleaves starch to maltose units. Maltose is produced during germination of seeds and thus seedlings contain lots of maltose sugar. Germinating cereals, such as

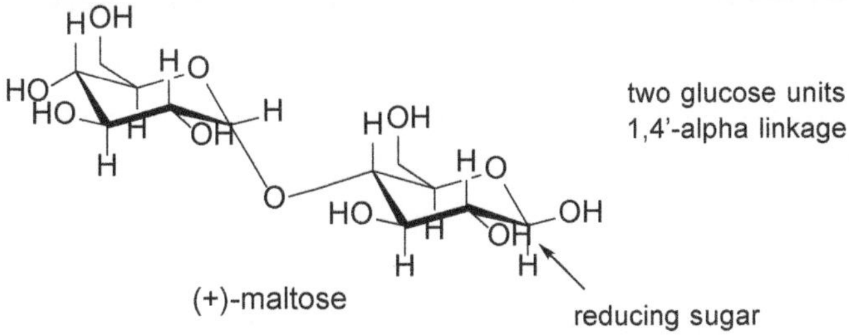

Figure 1.11: Maltose

barley produces maltose during brewing process and therefore beers contain maltose. It is a reducing sugar. The second glucose molecule having free anomeric –OH group is capable of existing in α and β forms.

1.11.2 Cellobiose

Cellobiose bears two molecules of glucose connected by a β-(1→4) glycosidic bond. It's IUPAC name is 4-O-β-D-glucopyranosyl-β-D-glucopyranose. Cellobiose differs from maltose in its configuration at the glycosidic bond (Figure 1.12). Cellobiose cannot be digested by humans and fermented by yeast. Cellobiose is used as an indicator in Crohn's disease and malabsorption syndrome. Enzyme, cellulase or acidic hydrolysis of cellulose and cellulose-rich materials yields cellobiose.

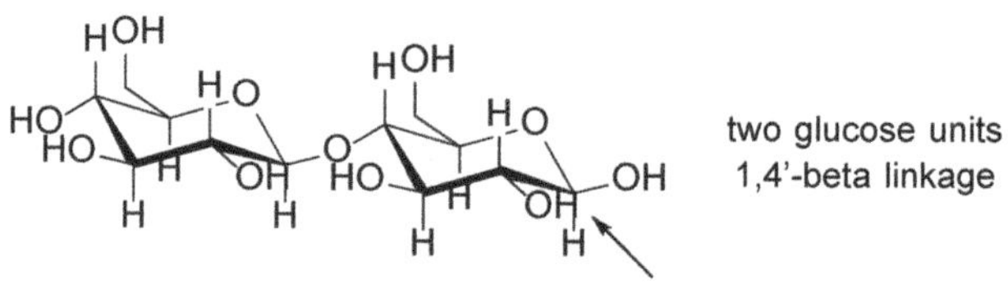

(+)-cellobiose

Figure 1.12: Cellobiose

1.11.3 Lactose

Lactose or milk sugar occurs in the milk of humans, cows and other mammals. Its synthesis occurs in mammary tissue while all other sugars are plant product. It is also a by-product in the manufacturing of cheese.

It is made up of galactose and glucose units joined by β-1,4-glycosidic bond (Figure 1.13). Galactose is at the non-reducing end and glucose at the reducing end. Hydrolysis or catalytic action of enzyme lactase produces the two monosaccharides from lactose. Its IUPAC name is 4-O-β-D-galactopyranosyl-D-glucopyranose.

Lactose intolerance is known as the inability to digest lactose found in milk. This is due to the shortage of the enzyme lactase, which breaks down the lactose into glucose and galactose. This lack of enzyme produces some uncomfortable symptoms because lactose gets fermented by bacteria in the large intestine, producing carbon dioxide gas. The condition can be improved by excluding milk or other sources of lactose from the diet or by consuming them only in small amounts. Galactosemia is another condition with the absence of enzymes needed to convert galactose to glucose. Subsequently, the blood galactose level is significantly elevated, and is also found in the urine. Galactosemia in infants can lead to lack of appetite, weight loss, jaundice, impaired liver function, cataracts, diarrhoea, and mental retardation.

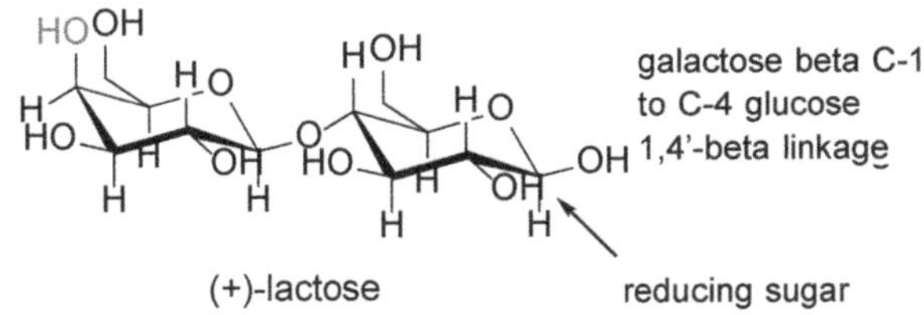

Figure 1.13: Lactose

1.11.4 Sucrose

Sucrose also known as cane sugar, table sugar and *beet sugar,* has α-1,β-2-glycosidic linkage between glucose and fructose. The glycosidic linkage is formed between the OH group of the anomeric carbon of α-D-glucose and the OH group of the anomeric carbon of β-D-fructose. It's IUPAC name is α-D-glucopyranosyl β-D-fructofuranoside (Figure 1.14). Neither of the monosaccharides in sucrose, uncyclizes to form an open-chain structure, therefore it is incapable of demonstrating mutarotation and exists in only one form in the solid state as well as in solution. Sucrose also doesn't display reactions that are typical of aldehydes and ketones. Therefore, it is a nonreducing sugar.

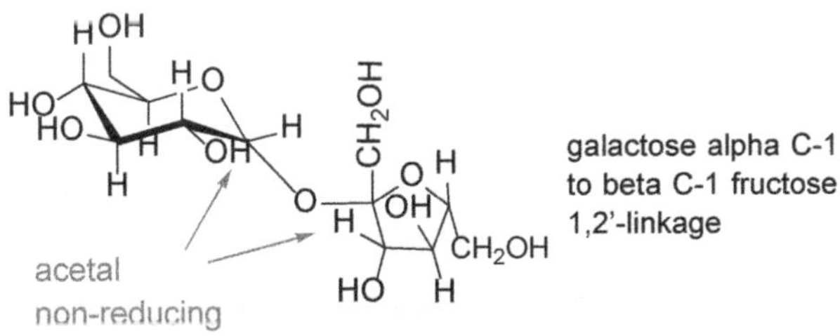

Figure 1.14: Sucrose

The sucrose is hydrolyzed in dilute acid or by enzyme sucrose, also known as invertase and it results in an equimolar mixture of glucose and fructose. This equimolar mixture is known as invert sugar because though sucrose rotates plane polarised light to the right and is dextrorotatory (specific rotation +66.5°), its hydrolysis products, glucose (specific rotation +52.5°) and fructose (specific rotation -92°) rotate the plane polarised light to the left and are laevorotatory.

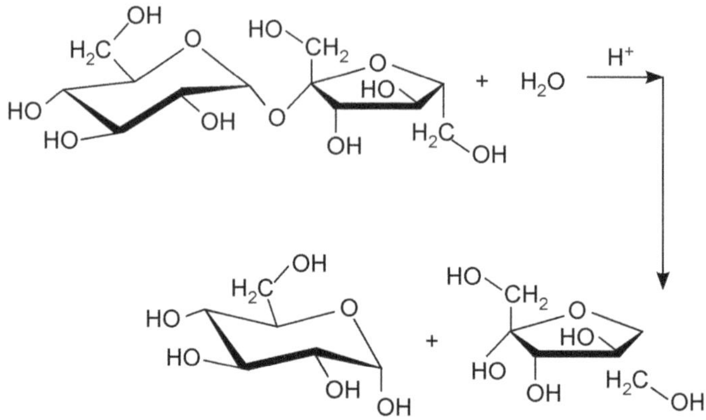

The hydrolysis reaction of sucrose and use of invertase has several commercial applications. It is used commercially for the manufacture of candy, jelly, canning of fruit and bees also make honey by carrying out the hydrolysis reaction using enzyme. In the food industry, fructose is preferred over sucrose as it is comparatively sweeter. Sucrose readily recrystallizes from a solution, but invert sugar does not crystallise. Another enzyme, glucose isomerase can also be used to convert glucose to fructose.

1.12 Polysaccharides

1.12.1 Starch

Starch, a homopolymer of glucose and is present in rice, potatoes, wheat *etc.* The glucopyranose units are bonded by α-linkages. Starch consists of amylose and amylopectin, with amylopectin having a larger share of about 80-90 %. Amylose consists of linear chain of hundreds of glucose molecules with α-linkages between C1 of one glucose and C4 of the next. Amylopectin on the other hand is branched (Figure 1.15). It has α-(1,4) linkage and branching takes place between C6 of one glucose and C1 of the next, at intervals of 20-25 glucose units. Starch can be digested by breaking the alpha-linkages in humans and other animals by the enzyme, amylases.

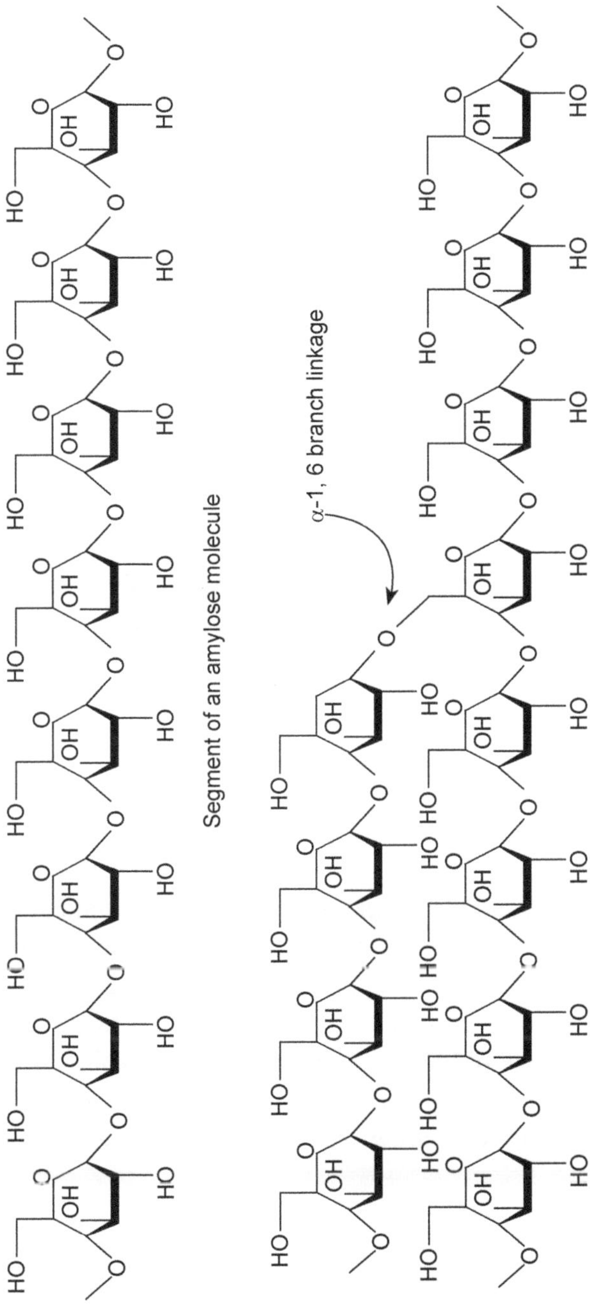

Segment of an amylose molecule

Segment of an amylopectin molecule, showing one α-1, 6 branch linkage

Figure 1.15: Amylose and amylopectin segments in starch, a homopolymer of glucose

1.12.2 Glycogen

Glycogen's structure is similar to amylopectin but with more branching as often as every ten glucose units. It acts as a secondary energy source. Glycogen can be broken down to glucose, when energy is required by the body. It is a glucose polymer with α(1→4) glycosidic bonds and α(1→6) glycosidic bonds at the branches (Figure 1.16). Glycogen is usually found in liver and skeletal muscle.

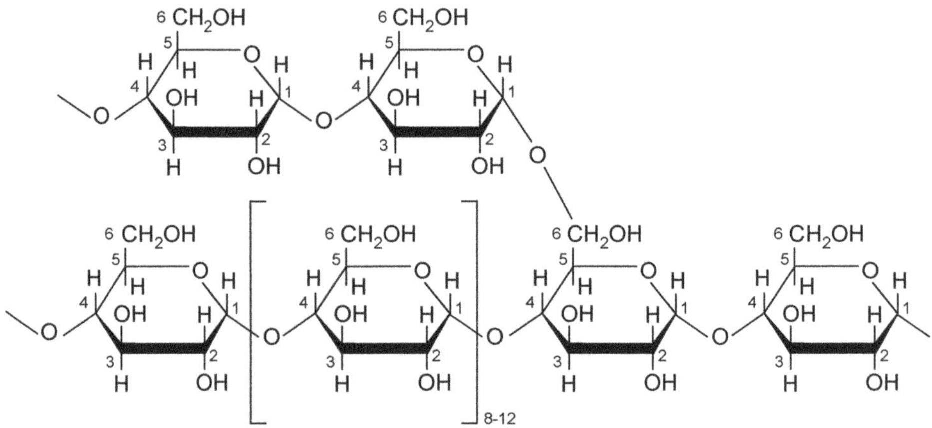

Figure 1.16: Glycogen

1.12.3 Cellulose

Cellulose is the most abundant biomaterial on earth and is the main component synthesized by plants, and also by some bacteria. Paper and cotton are nearly pure cellulose and wood is made up of lignin and cellulose. Like starch and glycogen, cellulose is also a homopolymer of glucose but they are joined by β-1,4 linkages instead of α (Figure 1.17). Cellulose is tough and have high strength and mechanical properties. It fibrous in nature and is water-insoluble owing to the presence of hydrogen bonding between adjacent glucose units. It has a significant role in keeping the structure of cell walls of plant stable and hard.

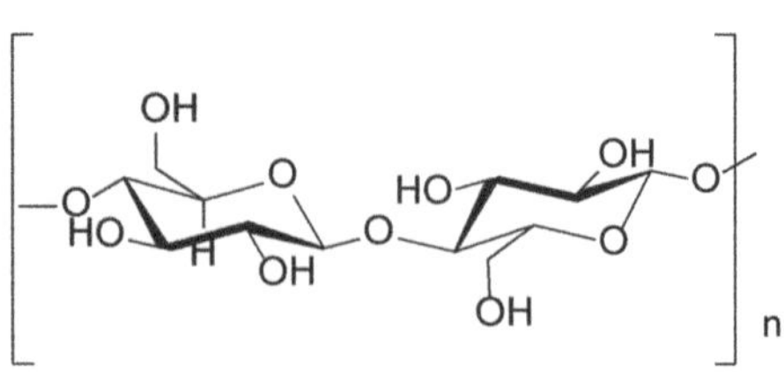

Figure 1.17: Cellulose

Questions

1. Why carbohydrates are known as quick-energy foods?
2. Define carbohydrates.
3. What do the notations D & L refer to in sugar?
4. What is the difference between anomers and epimers?
5. What do you understand by mutarotation?
6. Draw D and L forms of glyceraldehyde, ribose, fructose, glucose and galactose.
7. Explain why D-mannose is optically active but its alcohol derivative, D-mannitol is not?
8. What do you understand by reducing and non-reducing sugars? Give examples.
9. Why is maltose a reducing sugar and sucrose a non-reducing sugar?
10. How are monosaccharide units held together in a disaccharide? Can a di or polysaccharide exist in α and β anomeric forms?
11. Give the structures of sucrose, lactose and maltose. What are the sources of these common disaccharides?
12. How is the classification of polysaccharides related to their composition and function?
13. Why is cellulose insoluble in water and why is it not used as an energy source by human beings?
14. What are glycosides? Explain their formation and importance.
15. Draw structures of lactose and sucrose. Label the anomeric carbons in these and show what type of glycosidic linkages (α or β) they have.
16. Describe the oxidation reactions of monosaccharides.
17. Describe the reduction reactions of monosaccharides.
18. Explain the formation of the furanose ring of fructose.
19. What are the different stereoisomeric relations of glucose to other
20. Define O- and N-glycosidic bonds. How do you classify a glycosidic bond into α o β type?
21. The Draw structure of cellulose and write its biological importance
22. Indicate whether the following pairs of sugars are enantiomers, epimers,

diastereomers or anomers: (i) α-D glucose and β-D glucose (ii) D-glucose and D-talose (iii) D-galactose and D-mannose (iv) L-fructose and D-fructose

23. Explain the products formed on the action of aqueous periodic acid on glucose.
24. Explain why fructose and glucose gives the same osazone.
25. Explain Kiliani-Fischer Synthesis for converting glyceraldehyde to glucose.
26. How can we convert ribose to erythrose.
27. Explain Fischer's proof of the configuration of D-(+)-Glucose
28. Differentiate between starch and glycogen.
29. Draw the structure of fructose, cellobiose, galactose, sucrose, maltose and lactose.
30. What is the difference between cellulose and cellobiose.

❑❑❑

2

Nucleic Acids

2.1 Introduction

Nucleic acids are natural polymers that are made up of many repeating structure units called nucleotides. They are also called polynucleotides. They received their name because they were originally isolated from cell nuclei and have acidic character, hence called nucleic (their origin) acid (nature). They are made up of carbon, hydrogen, oxygen, nitrogen, and phosphorus and are found in all living beings. Nucleic acids have important functions in the cell:

- They store genetic information responsible for the transmission of inherited characteristics from parents to children and from one cell to another.
- They provide necessary information to cell for protein synthesis. They are responsible for the correct sequence of amino acids that is the identity of that family.

The morphological and functional uniqueness of each living being are determined by the information contained in its nucleic acids.

DNA is the genetic material found in all living organism. But some viruses use RNA, not DNA, as their genetic material. Such viruses are intermediates of those living and non-living species as they need host for reproduction and remain in vegetative state outside host.

2.2 Structure Determination of Nucleic Acids

Nucleic acids on hydrolysis give nucleotide. This show that they are made up of nucleotide.

DOI: 10.1201/9781003727453-2

Nucleotide on hydrolysis gives nucleoside and phosphoric acid. This shows that they are made up of nucleoside and phosphoric acid.

Nucleoside on further hydrolysis gives sugar and bases. This shows that they are made up of sugar and bases.

On the basis of above reactions, we can conclude that nucleic acids are made up of sugars, bases and phosphoric acid. This can be summarized as

$$\text{Nucleic acids} \xrightarrow{H^+/H_2O} \text{Nucleotide} \xrightarrow{H^+/H_2O} \text{Nucleoside} + H_3PO_4$$

$$\xrightarrow{H^+/H_2O} \text{Sugar} + \text{Bases}$$

Sugars: Two types of sugars are found in nucleic acid i.e. D-Ribose and 2-Deoxy-D-ribose.

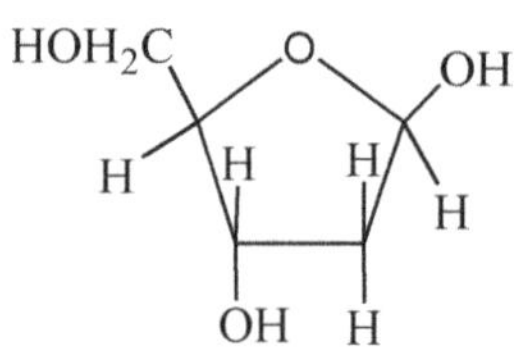

2-Deoxy-D-ribose

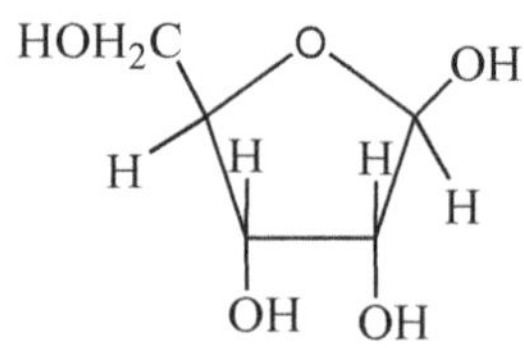

D-Ribose

Bases: There are five types of bases found in the nucleic acids. They are basically derivatives of purine and pyrimidine, hence called purine bases and pyrimidine bases. The basic structures of purine and pyrimidine are as follows.

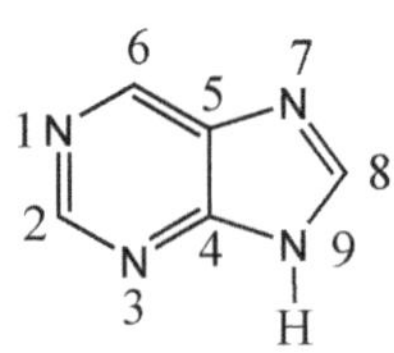

Purine

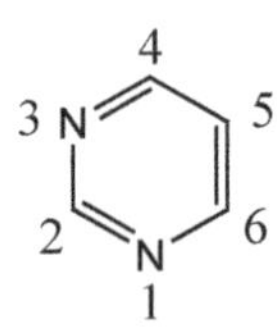

Pyrimidine

Purine bases

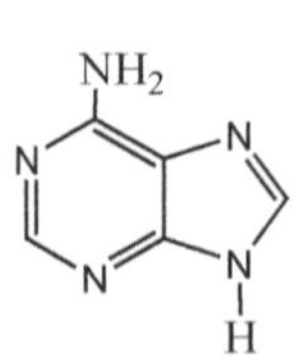

Adenine

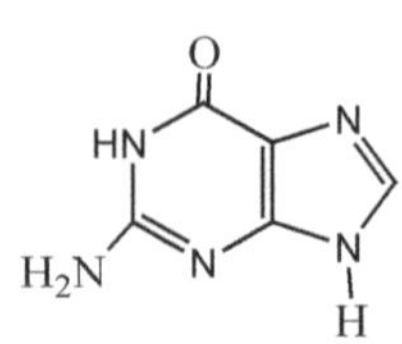

Guanine

Pyrimidine Bases:

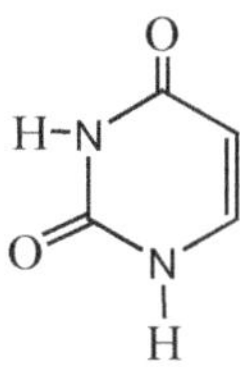

Uracil

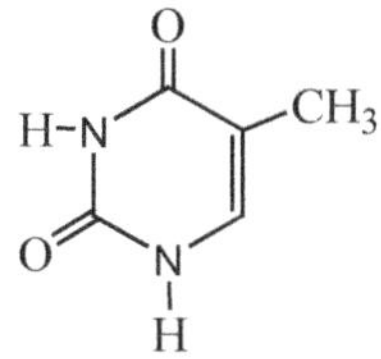

Thymine

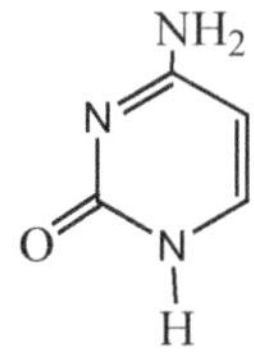

Cytosine

Phosphoric Acids: The structure is given below:

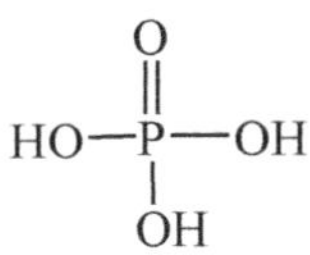

Phosphoric acid

General Structure of Nucleoside Containing:

(a) Pyrimidine base (b) Purine Base

a)

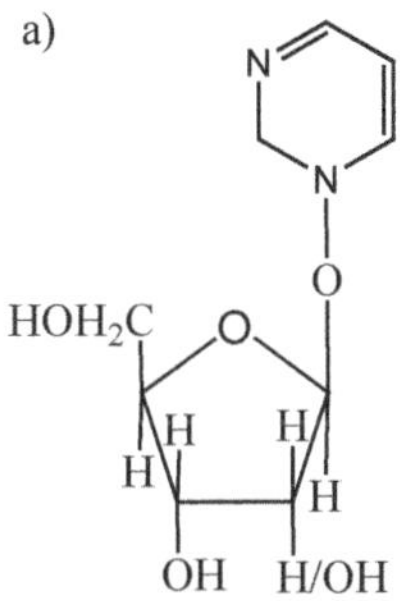

b)

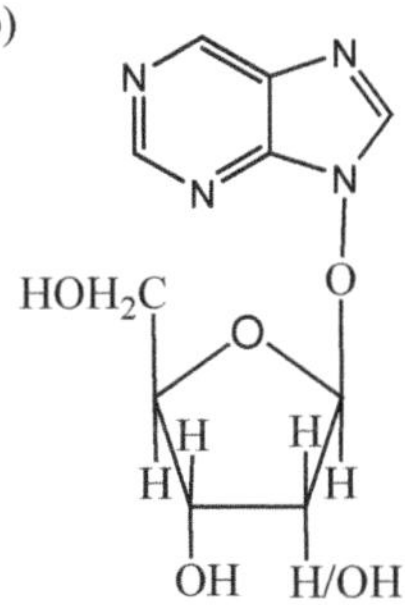

General Structure of Nucleotide Containing:

(c) Pyrimidine Base (d) Purine Base

c)

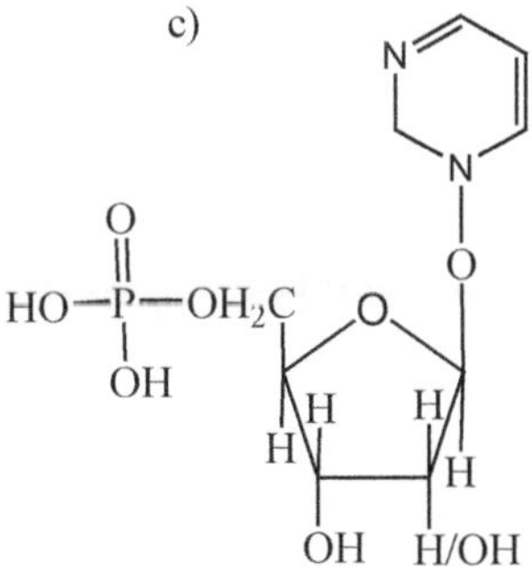

d)

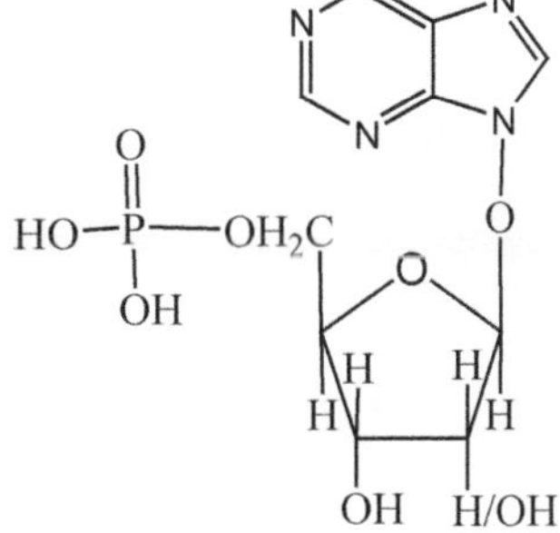

Purine Bases

Adenine: Adenine is a purine derivative and its chemical name is 6-aminopurine. It is one of the four nucleobases present in the DNA. Derivatives of adenine have variety of roles in biological system that includes cellular respiration in the form of both the energy-rich adenosine triphosphate (ATP) and the cofactors nicotinamide adenine dinuclueotide (NAD) and flavin adenine dinucleotide (FAD). It also plays important role in protein synthesis.

2.3 Synthesis of Adenine

Fischer's Method

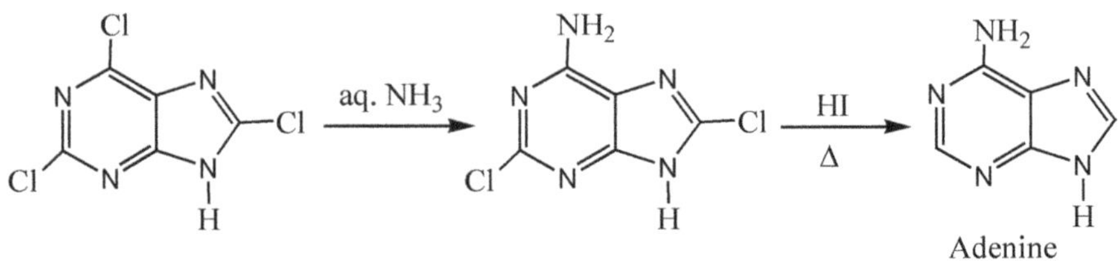

Traube Synthesis

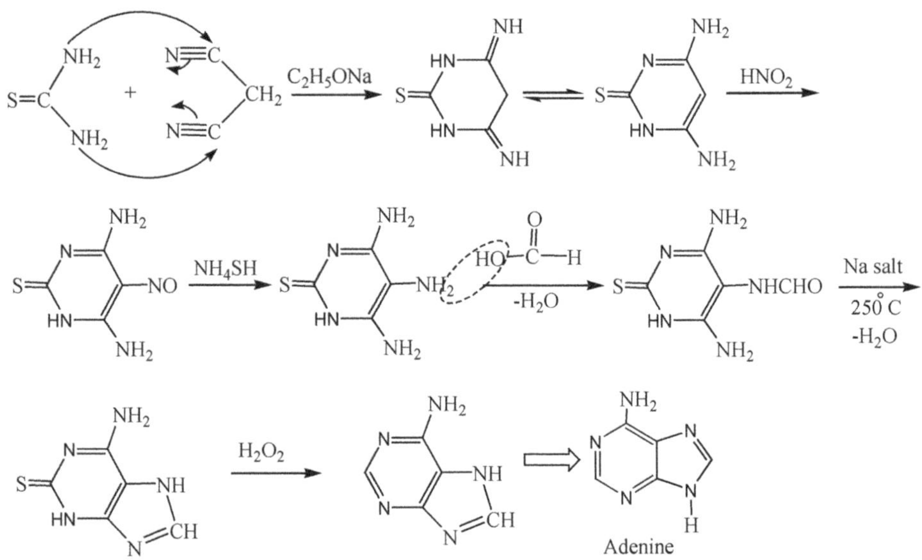

Bredereck *et al*

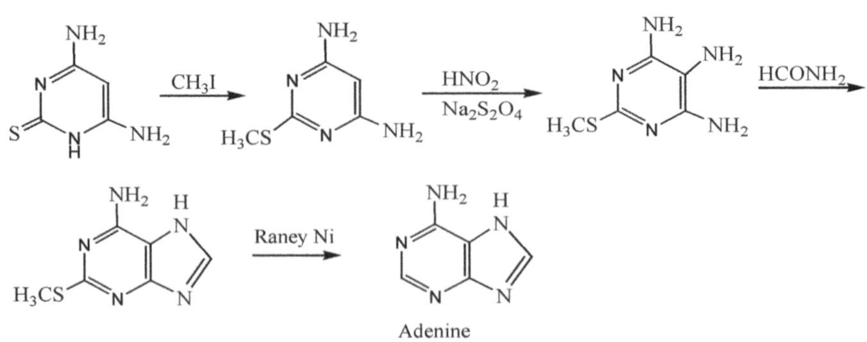

Todd *et al.*

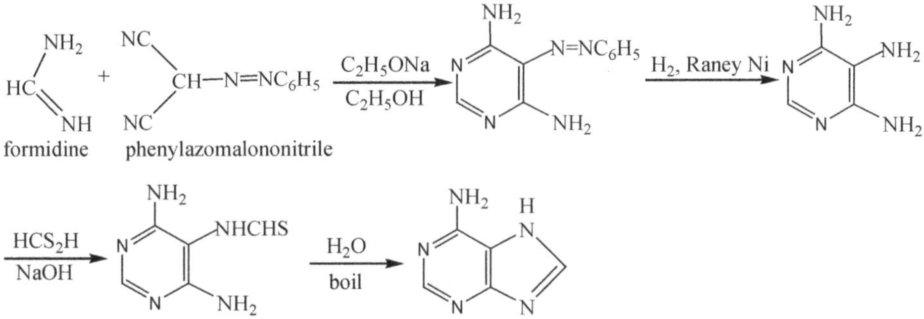

2.4 Guanine

It is 2-amino-6-hydroxypurine or 2-amino-9-H-purin-6-one. It was first isolated by a German chemist Julius Bodo Unger in 1844 from the excreta of sea birds, known as guano. Hence, it was named as guanine. It was used as a source of fertilizer.

It exists in equilibrium between keto and enolic forms. But keto form is more stable so exists in this form.

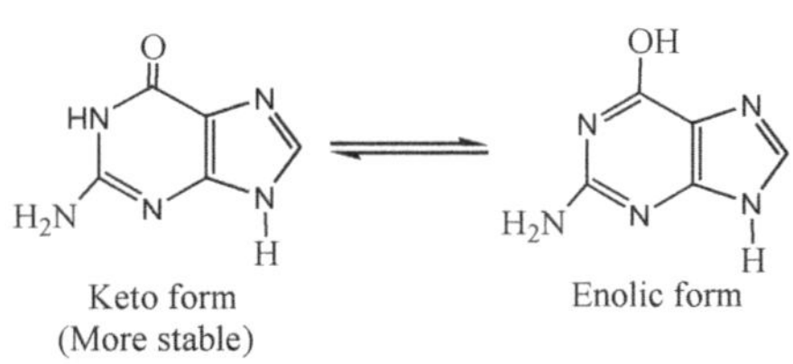

Fischer Synthesis

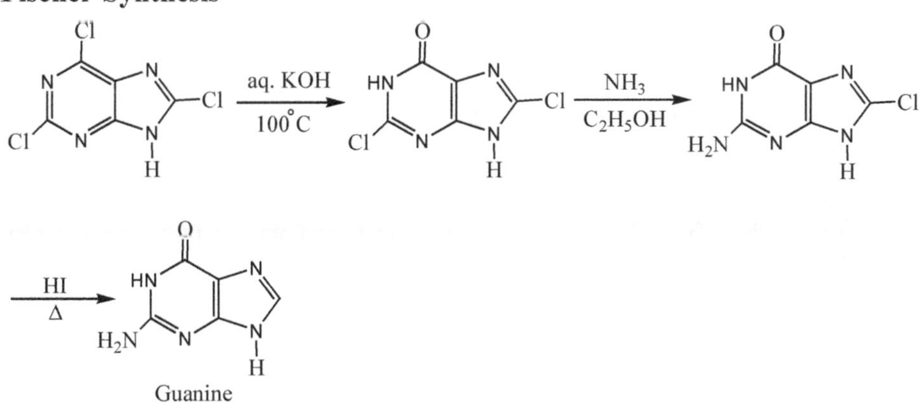

Traube's Synthesis

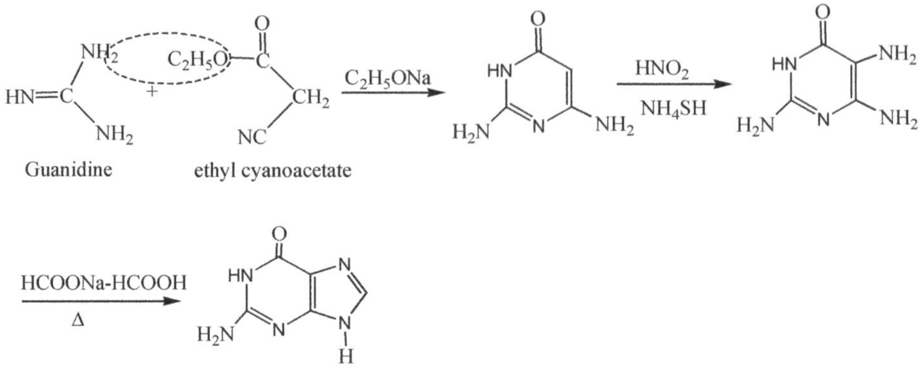

2.5 Properties of Purine and Its Derivatives

Purine is both a very weak acid (pKa 2.39) and an even weaker base (pKa 8.93). If dissolved in pure water, the pH will be halfway between these two pKa values.

Purine is formed by fusing pyrimidine and imidazole nucleus. As we know that pyrimidine being aromatic undergoes electrophilic substitution reaction at 2,4 and 6-positions. Just look at the 2 and 6 positions in purine, they are similar to 2 and 4 positions of the pyrimidine ring. Similarly position 8 may be compared to α-position of the pyrrole. It is also important to note that in purine, there is a continuous delocalization of π electron cloud in both the monocyclic rings and electron density of each ring may actually change as a result of electron donation by the imidazole ring to the pyrimidine.

Hence, purine undergoes electrophilic substitution at 2, 6 and 8-position.

Guanine undergoes hydrolysis with strong acid to give glycine, ammonia, carbon dioxide and carbon monoxide.

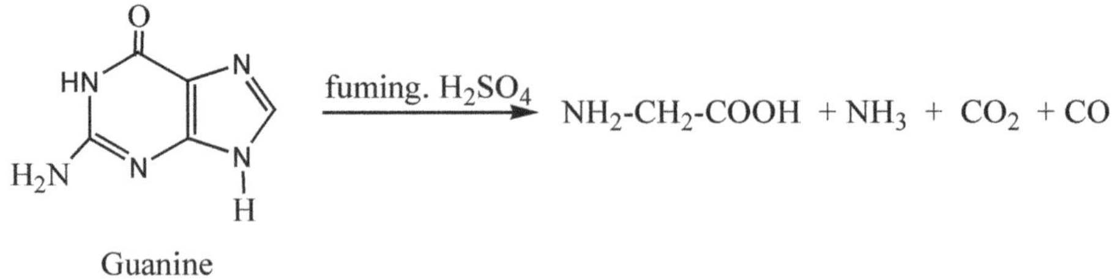

Guanine Easily Deaminated to Form Xanthine

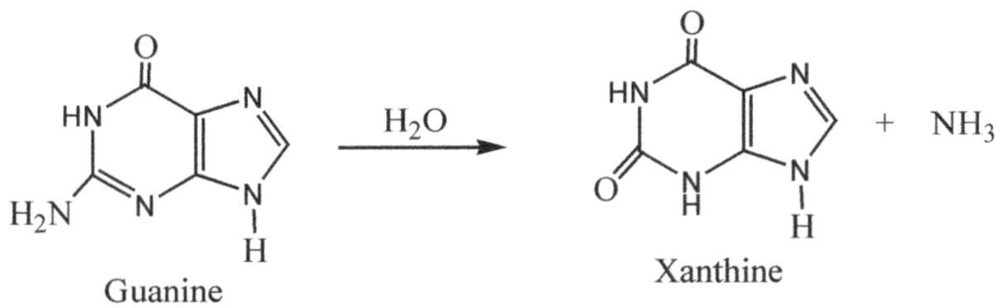

2.6 Pyrimidine Bases

2.6.1 Uracil

Uracil is the naturally occurring pyrimidine derivative. Its chemical name is pyrimidine-2, 4-dione. German chemist Robert Behrend named "Uracil" for this compound in 1885. It was isolated by the hydrolysis of yeast nuclein. It is also found in bovine thymus, spleen and wheat germ. Uracil is a weak acid.

Synthesis

Fischer and Roeder Synthesis

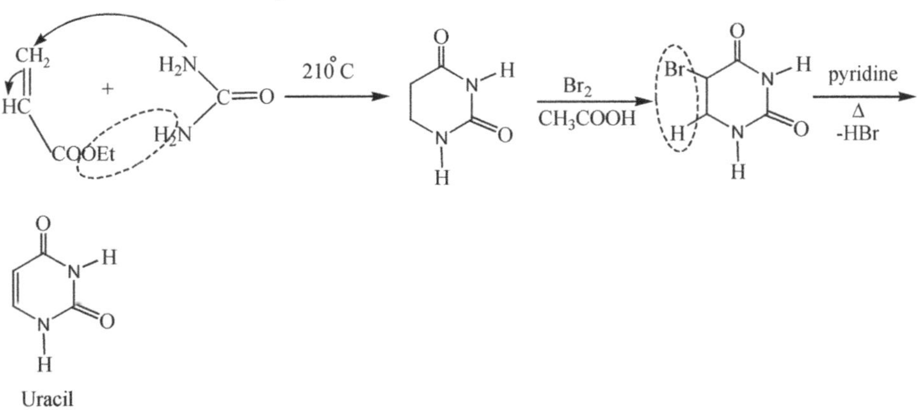

Wheeler and Liddle Synthesis

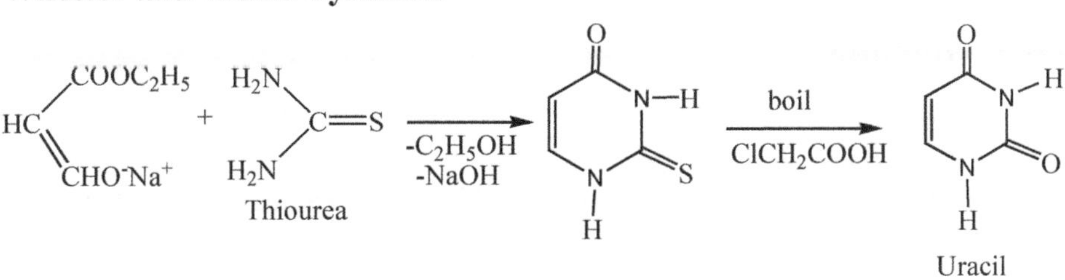

It is also prepared by condensation of maleic acid and urea in presence of dehydrating agent like conc. Sulhuric acid.

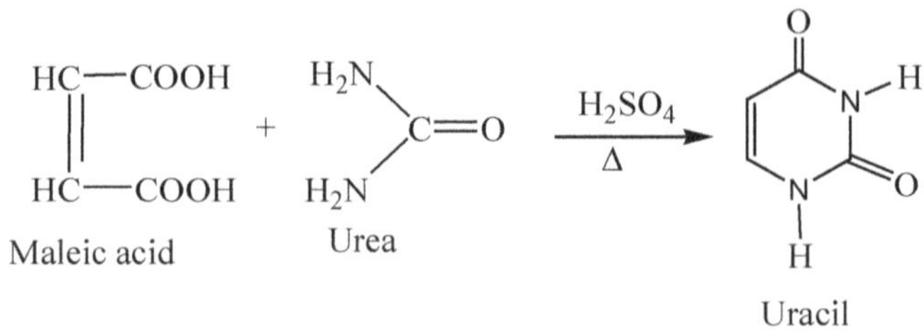

Davidson *et al.*

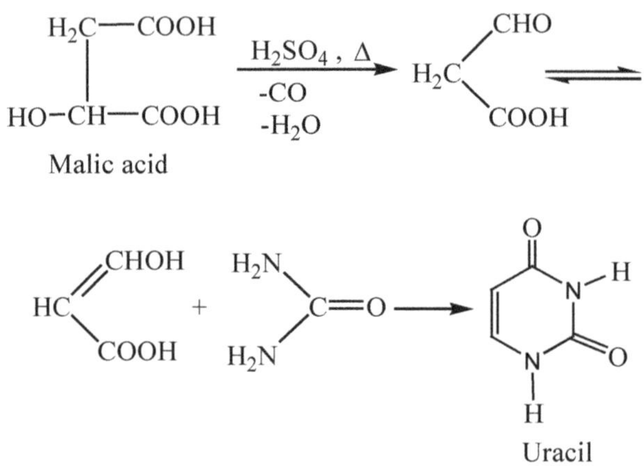

Uracil is synthesized by addition of water to cytosine. It is the simplest method.

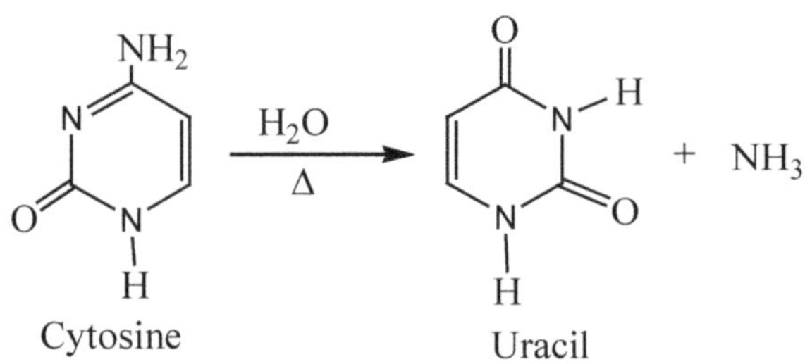

2.6.2 Thymine

Thymine is a pyrimidine nucleobase. It is also known as 5-methyluracil. Its chemical name is 5-Methyl-pyrimidine-2, 4-dione. Thymine was first isolated in 1893 by Albrecht Kossel and Albert Neumann from calf thymus gland, hence, it was named thymine.

Synthesis

Fischer and Roeder Synthesis

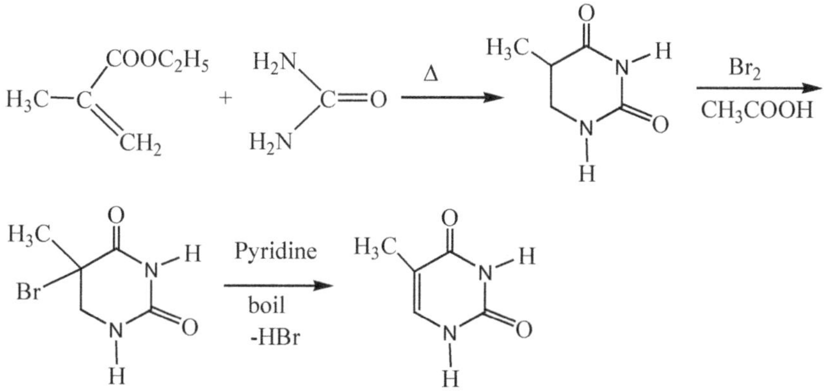

Thymine

Wheeler and Liddle Synthesis

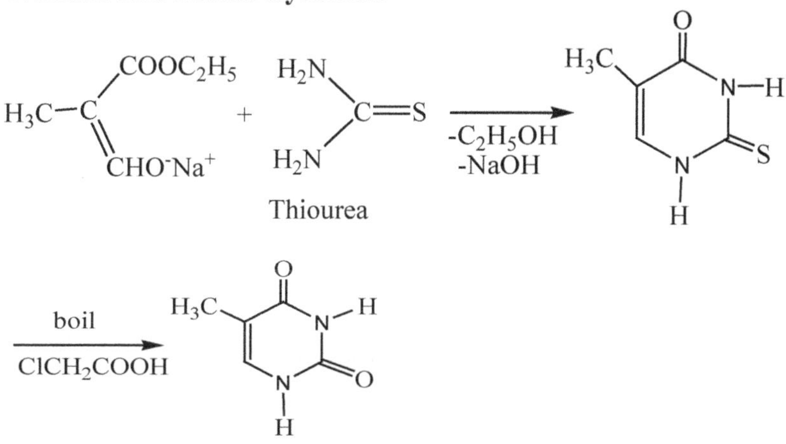

Thymine

Bermann *et al.*

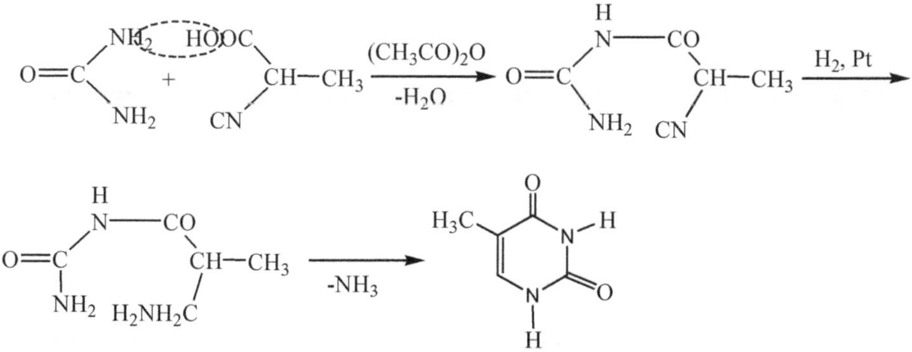

Thymine

2.6.3 Cytosine

Like uracil and thymine, it is also a derivative of pyrimidine. Its chemical name is 4-aminopyrimidin-2-one. Cytosine was also discovered and named by Albrecht Kossel and Albert Neumann in 1894 when it was also obtained by hydrolysis of calf thymus tissues.

Synthesis

Wheeler and Johnson Synthesis

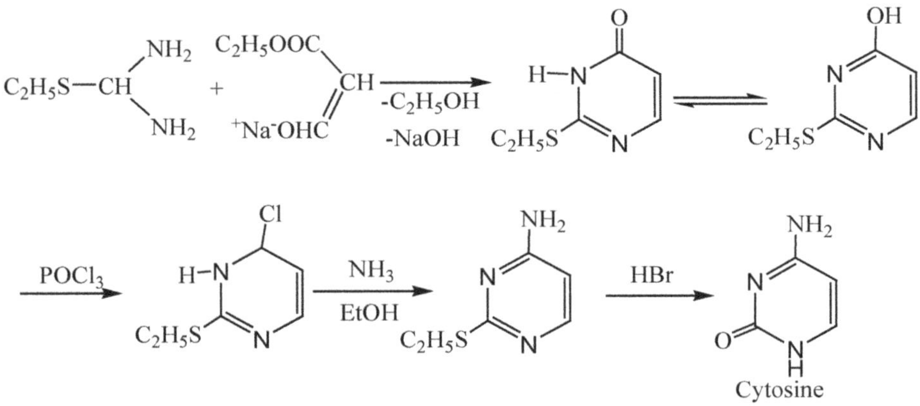

Tarsio *et.al.*

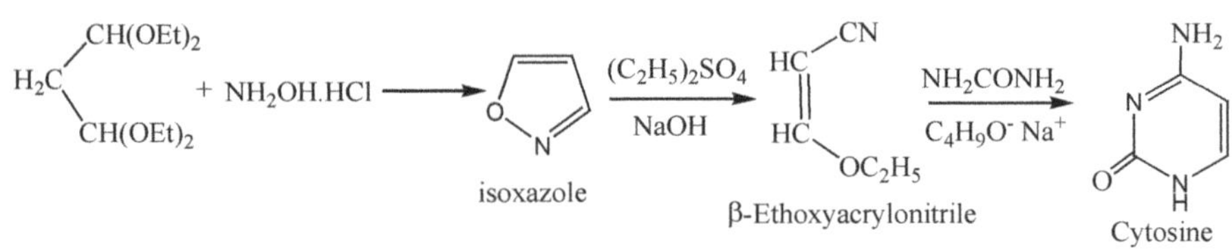

2.7 Properties of Pyrimidine and Its Derivatives

Pyrimidine is an aromatic compound and it exists as a resonance hybrid of following five canonical structures:

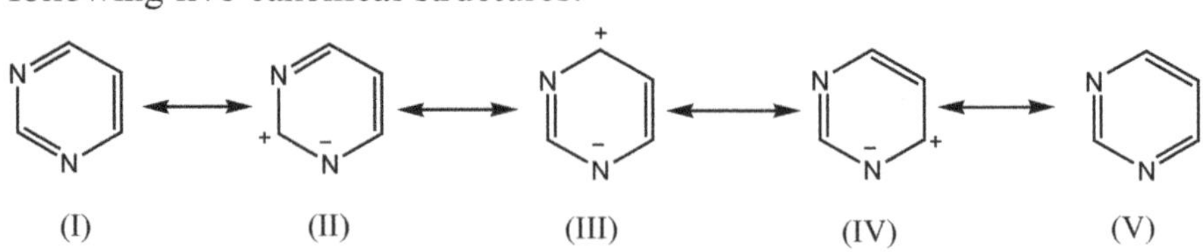

Pyrimidine is highly deactivated ring because of the presence of two electron withdrawing nitrogen atoms. Therefore, it undergoes electrophilic substitution with great difficulty. Look at the charged structures (II, III and IV). In these structures, positive charge is present on the 2, 4 and 6 positions. This shows these positions are electron deficient and therefore nucleophilic substitution can take place on any of these positions. 5th Position has high electron density so it becomes prone towards electrophilic substitution reaction.

Only few reactions have been reported for these bases.

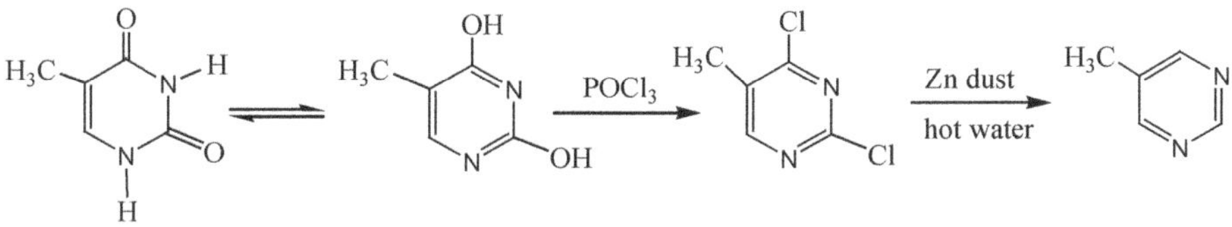

Thymine

Degradation of uracil produces the substrates aspartate, CO_2, and NH_3.

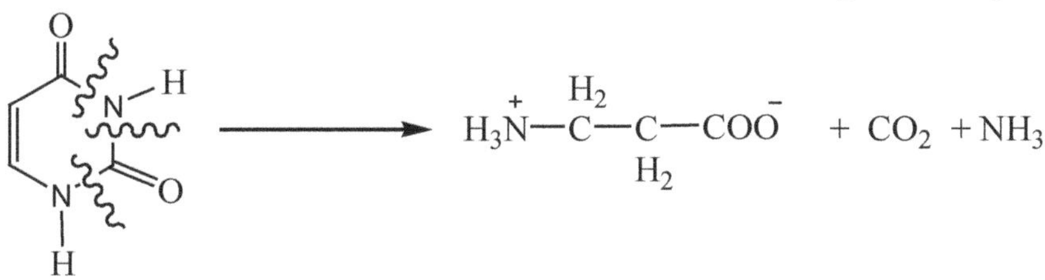

Uracil

Oxidative degradation of uracil in presence of hydrogen peroxide and iron salt produces urea and maleic acid.

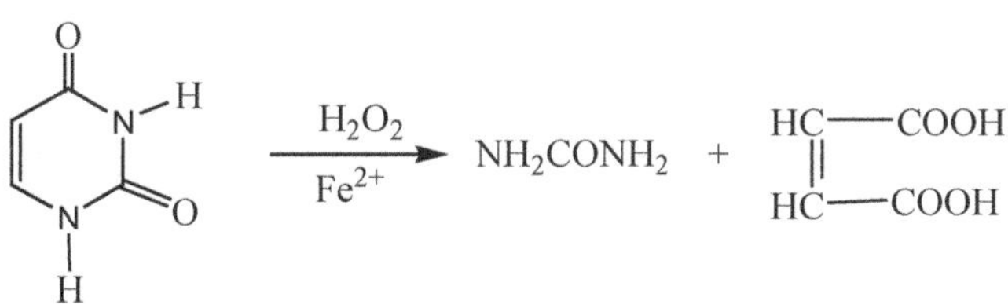

Uracil

Bromination of uracil with bromine in presence of acetate buffer gives 5-bromouracil.

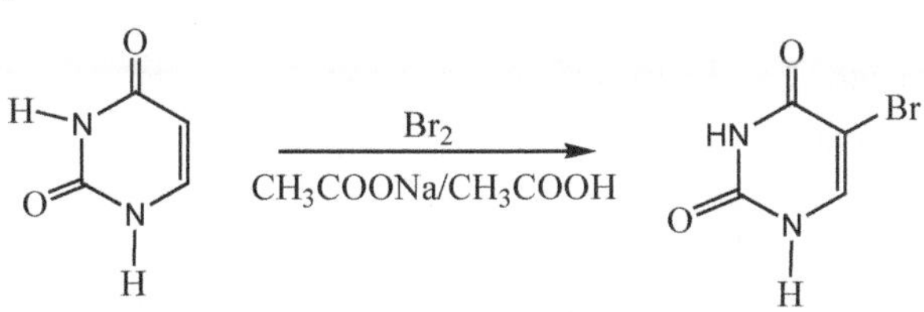

Uracil

5-bromouracil can also be prepared by treating uracil with bromine water. In excess of bromine water it gives 5,5-dibromo-6-hydrohydroxyuracil.

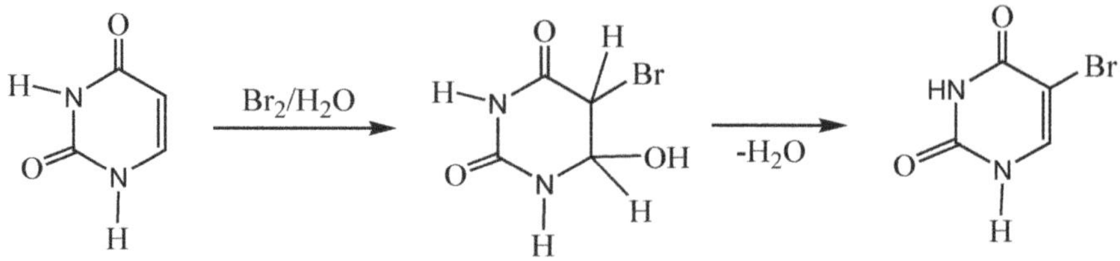

Uracil

Reaction of uracil with phosphoryl chloride gives 2,4-dichloropyrimidine which on reduction with zinc dust gets reduced to pyrimidine.

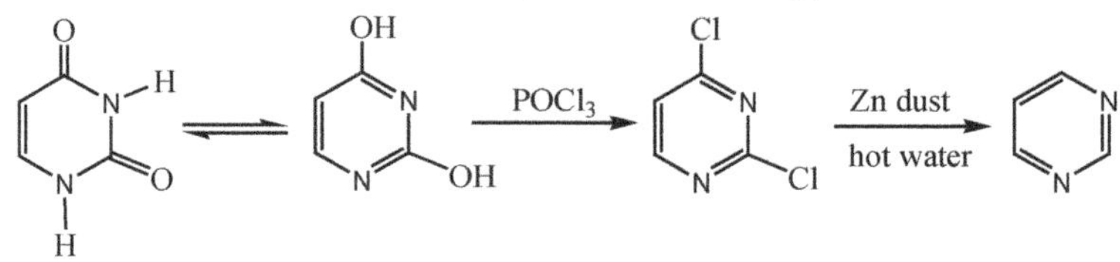

Uracil

Nitration of uracil with nitric acid in presence of acetic anhydride gives 5-nitrouracil.

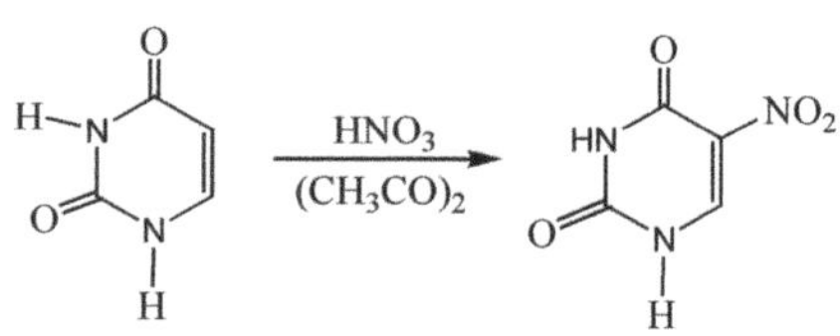

Uracil

Because of the presence of 5-methyl group, thymine is not nitrated or halogenated normally, but with aqueous bromine it gives the dihydropyrimidine derivative.

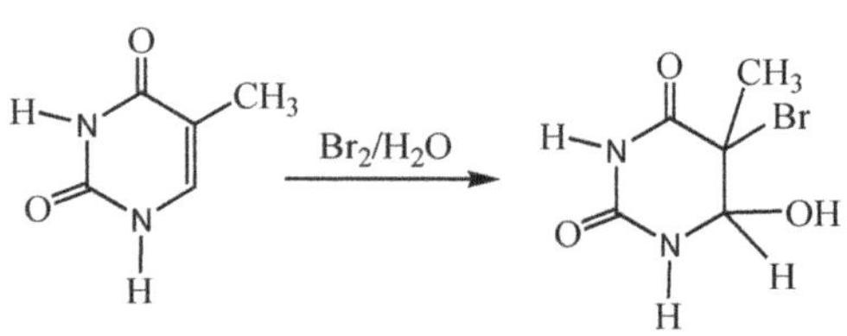

Thymine

Cytosine is prone to deamination reaction i.e. loss of ammonia when hydrolysed with water.

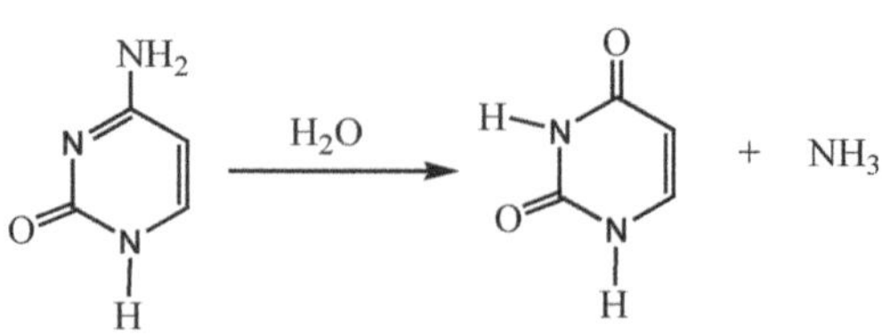

Cytosine

Both purine and pyrimidine bases react with ribose or deoxyribose sugar via glycosidic linkage that is formed between N-H group of base and OH group of sugar.

2.8 Nucleosides

Nucleoside is an organic compound that is made up of a purine or pyrimidine base combined with deoxyribose or ribose by β-N-glycosidic bond. In the formation of this bond, C-1 of sugar is covalently bonded to N-1 of the pyrimidine and N-9 of the purine. For examples, when C-9 of adenine is attached to ribose sugar it forms a nucleoside adenosine (I). Combination of guanine with ribose sugar gives guanidine (II). When C-9 of adenine or guanine is attached to deoxyribose sugar, it forms 2-Deoxyadenosine (III) or 2-Deoxyguanosine (IV). See Table 2.1 for various nucleosides.

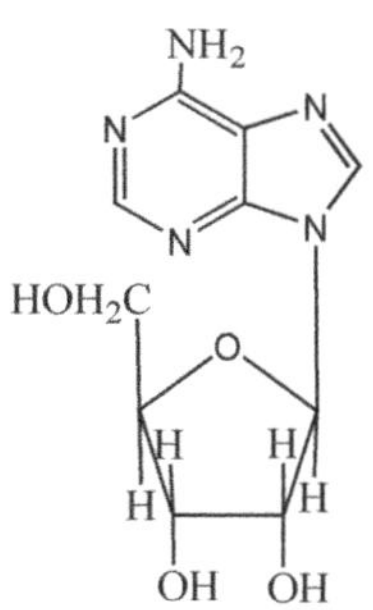

Adenosine (I)

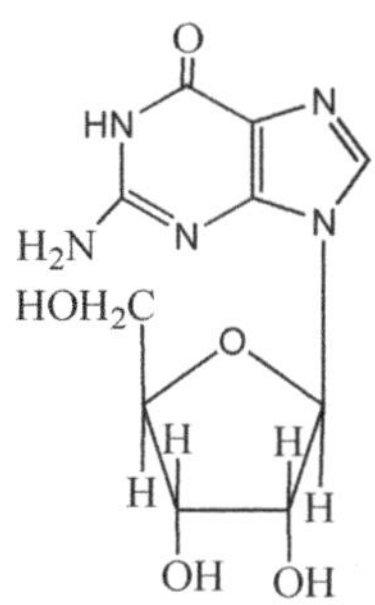

Guanosine (II)

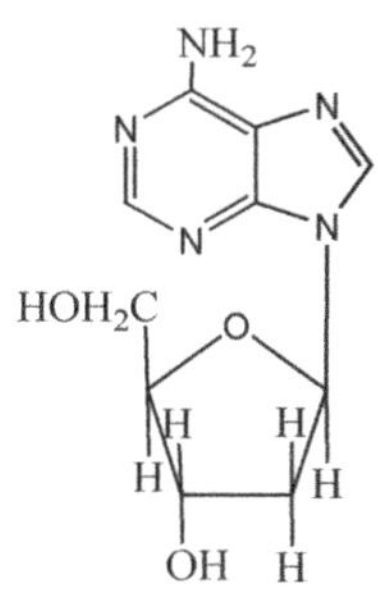

2-Deoxyadenosine (III)

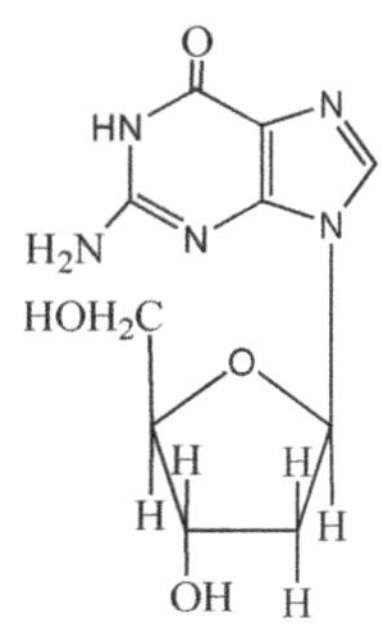

2-Deoxyguanosine (IV)

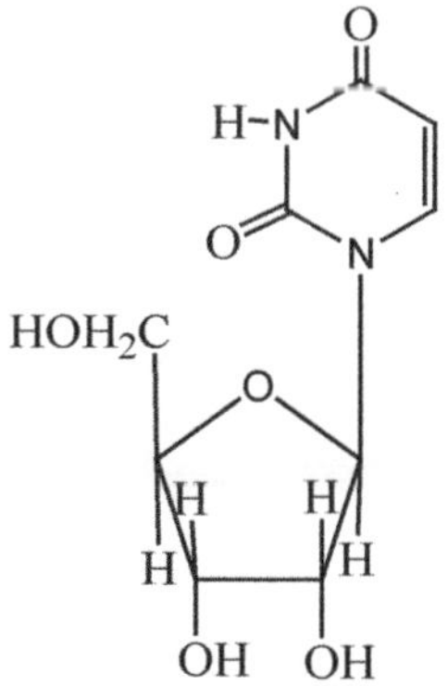

Uridine

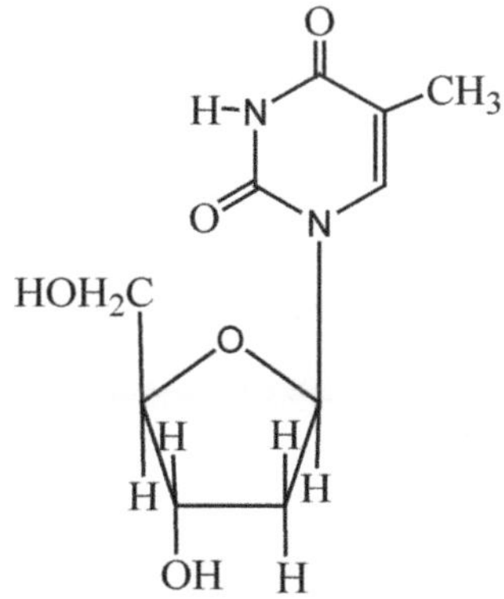

2-deoxythymidine

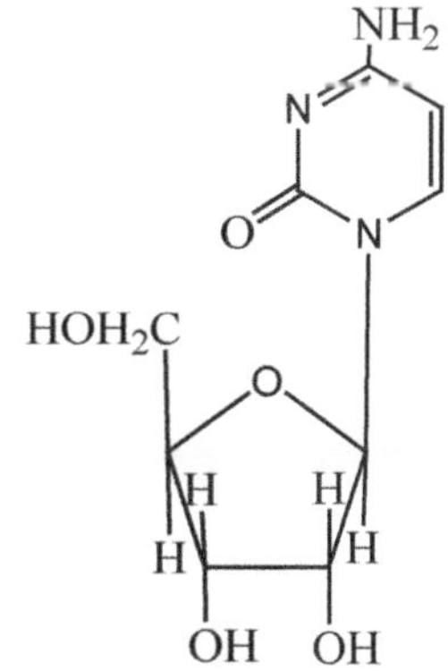

Cytidine

Table 2.1: Nomenclature of various nucleosides

Name of The Nucleoside	Name of The Base	Name of Sugar
Adenosine	Adenine	Ribose
2-Deoxyadenosine	Adenine	2-Deoxyribose
Guanosine	Guanine	Ribose
2-Deoxyguanosine	Guanine	2-Deoxyribose
Uridine	Uracil	Ribose
2-Deoxyuridine	Uracil	2-Deoxyribose
Thymidine	Thymine	Ribose
2-Deoxythymidine	Thymine	2-Deoxyribose
Cytidine	Cytosine	Ribose
2-Deoxycytidine	Cytosine	2-Deoxyribose

2.8.1 Properties of Nucleosides

Several derivatives of nucleoside have been used as antiviral or anticancer agents in medicine. The viral polymerase combines these compounds with non-canonical bases. These bases are then activated in the cell by getting converted into nucleotides. They are administered as nucleosides because charged nucleotides are not able to cross cell membranes.

For example nucleoside analog reverse transcriptase inhibitors (NARTIs or NRTIs) is one class of antiretroviral drugs. This drug inhibits the activity of reverse transcriptase, a viral DNA polymerase enzyme that is required by HIV in order to reproduce. When HIV attacks a cell, reverse transcriptase enzyme copies the single stranded RNA genome of the virus into a double stranded viral DNA molecule. This viral DNA is then integrated into host's chromosomal DNA which allows the host to reproduce the virus. This NRTI blocks the reverse transcriptase enzyme and disrupt the synthesis of the viral DNA and thus prevent HIV from multiplying. Three phosphate groups must be added to activate NRTIs and enable it to be incorporated into the viral DNA. Activated form of NRTI is called NRTI triphosphates.

2.9 Nucleotides

These are the compounds consisting of a nucleoside linked to a phosphate group. They are made up a sugar molecule (either ribose in RNA or deoxyribose in DNA) attached to a phosphate group and a nitrogenous base. The bases used in DNA are adenine (A), cytosine (C), guanine (G), and thymine (T). In RNA, thymine is replaced by uracil (U). These are the building block of nucleic acids. Hence DNA and RNA are biopolymers made up of linear chain of nucleotides. In nutshell we can say that nucleotides are the monomers of nucleic acids.

In nucleotide, nucleoside is attached to phosphoric acid via phosphoester bond. In nucleoside there are three free –OH groups where phosphoric acid can be attached i.e. on C-2, C-3 and C-5 positions (in case of ribose sugar) and two free –OH in case of 2-deoxyribose sugar i.e. C-3 and C-5 positions. These positions are named as 2', 3' and 5'. Both nucleosides and Nucleotides are named after their bases. For example, name of the nucleosides containing adenine and cytosine are adenosine and cytidine respectively. Their corresponding nucleotides are then designated as adenosine 3'-phosphate (if C-3 OH group is involved) or adenosine 5'-phosphate (if C-5 OH group is involved) and cytidine 3'-phosphate (if C-3 OH group is involved) or cytidine 5'-phosphate (if C-5 OH group is involved). Alternatively, they have been named as adenylic acid, guanylic acid, cytidylic acid, thymidylic acid, uridylic acid respectively. The deoxy- form of nucleotides should then be designated 2'-deoxyadenosine 5'-phosphate and so on. The name of the nucleotide also depends upon the number of phosphate group. If one phosphoric acid is esterified then its name is suffixed as monophosphate. Similarly, diphosphate and triphosphate suffixed are used in case of two and three phosphoric acid present. Generally, phosphoric acid is esterified to C-5 position of sugar.

For examples of nucleotides are given below:

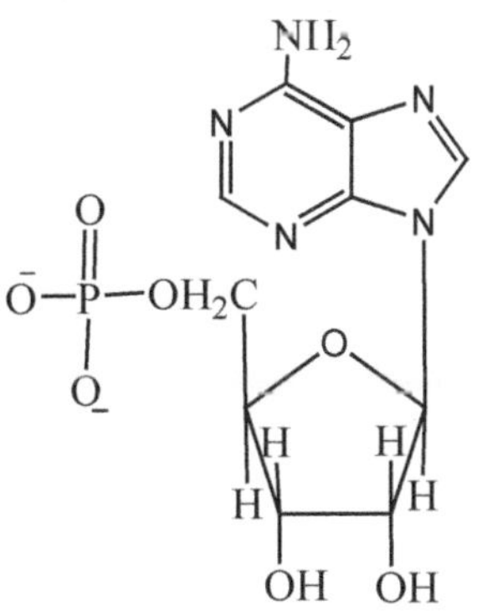

Adenosine 5'-monophosphate
(AMP)

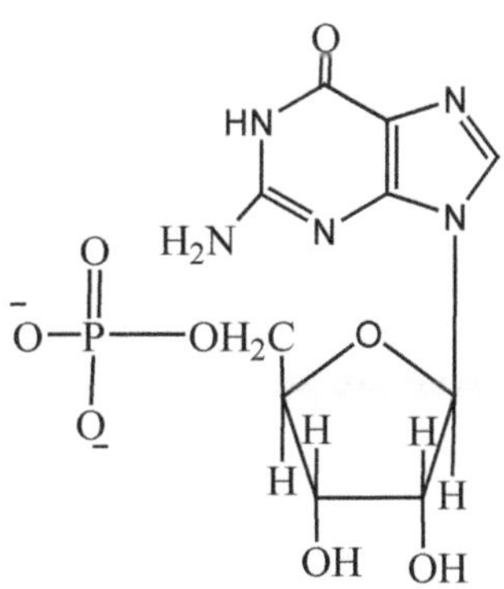

Guanosine 5'-monophosphate
(GMP)

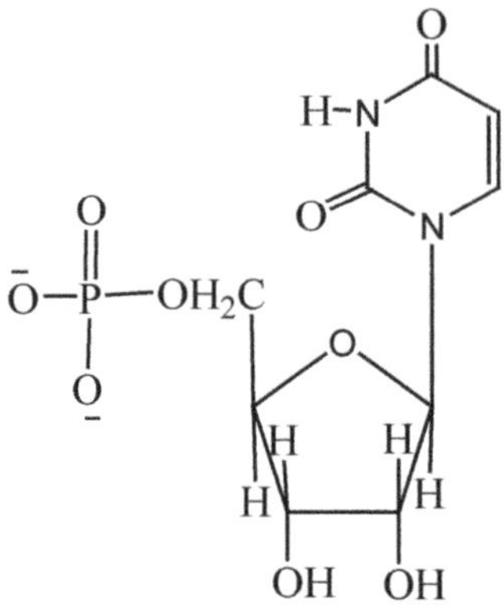

Uridine 5'-monophosphate

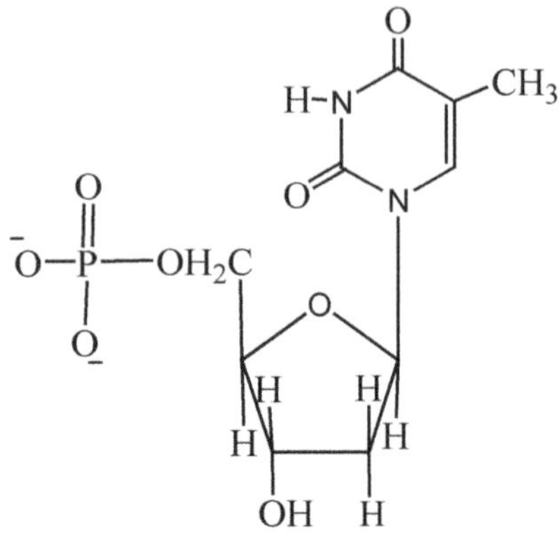

2-deoxythymidine 5′-monophosphate

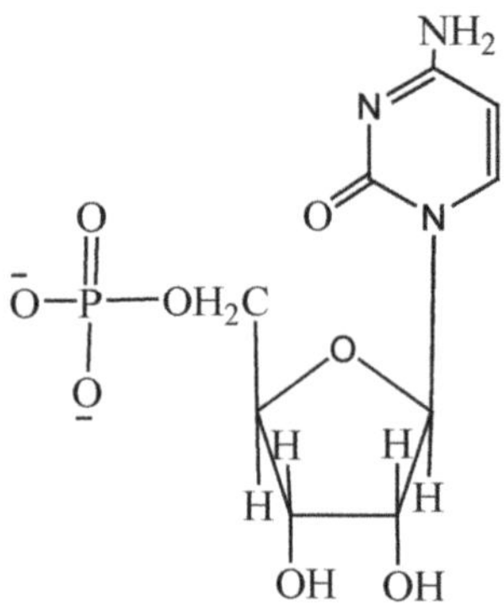

Cytidine 5'-monophosphate

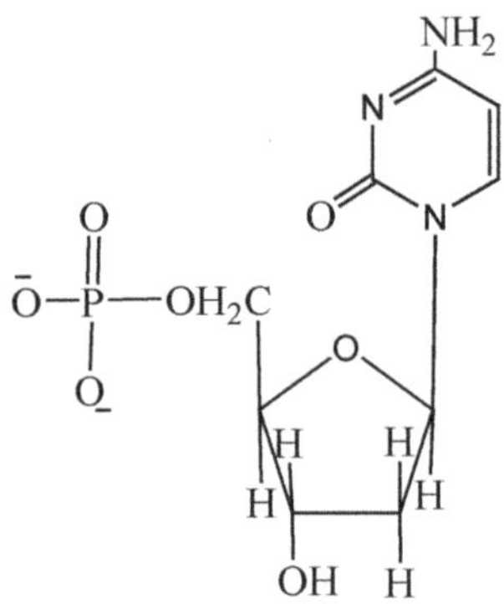

2-Deoxycytidine 5'-monophosphate

Nucleotides play a crucial role in metabolism. They store chemical energy in the form of nucleotide triphosphate. For example Adenosine triphospate (ATP), Guanosine triphosphate (GTP), Cytidine triphosphate (CTP) and Uridine triphosphate (UTP). This energy is available to the many cellular functions that include: synthesis of amino acids, proteins etc. The structure of ADP and ATP are

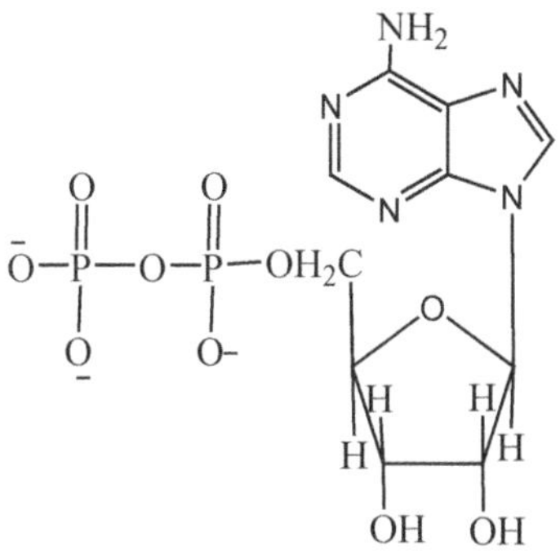

Adenosine 5'-diphosphate
(ADP)

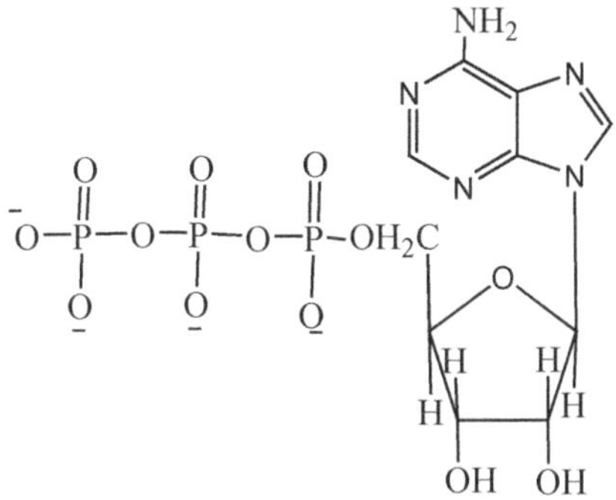

Adenosine 5'-triphosphate
(ATP)

2.10 Adenosine Triphosphate (ATP)

It is an energy-carrying molecule present in the cells of all living things. Mitochondria is also called powerhouse of cell because of the presence of ATP. ATP stores chemical energy obtained from breakdown of food and releases it at the time of need. ATP is made up of nitrogenous base, sugar and a chain of three phosphate groups attached to ribose sugar. Tail of phosphate is the actual source of energy which the cell taps. When cell needs energy, bonds between the phosphates are broken by addition of water molecule. In order to release energy outer phosphate group is removed from ATP and it gets converted into adenosine diphosphate (ADP), the form of the nucleotide having only two phosphates. It also transfers phosphate group to other molecule during phosphorylation process. This process is catalysed by enzymes.

2.11 Polynucleotides or Nucleic Acid

Polynucleotides are obtained by polymerization of nucleotides. These are linear biopolymers in which nucleotides are joined together by phosphodiester bond. **The polynucleotides formed in this way are called nucleic acids.** In polynucleotides, nucleotide residues are numbered in 5'→ 3' direction, i.e.

from left to right. Interaction of the phosphate group of a nucleotide with the 3'-OH-group of another nucleotide leads to the formation of a dinucleotide with a phosphodiester bond. Such a dinucleotide carries a free phosphate group at the 5'-end, and a free OH-group at the 3'-end. Therefore, a new mononucleotide can be linked to the dinucleotide via another phosphodiester bond. Oligonucleotides are the polynucleotides formed in this way and they usually contain less than 30 such units (nucleotides). There are two types of nucleic acids (Table 2.2).

- Deoxyribonucleic acid (DNA)
- Ribonucleic Acid (RNA)

Table 2.2: Difference between DNA and RNA

Deoxyribonucleic Acid (DNA)	Ribonucleic Acid (RNA)
• DNA is a double-stranded molecule. • It contains 2-Deoxyribose sugar. • DNA is stable under alkaline conditions. • DNA stores and transfer genetic information from one generation to another. • It contains four bases such as adenine, guanine, cytosine, thymine.	• RNA is a single-stranded molecule. • It contains ribose sugar. • RNA is not stable under alkaline conditions. • RNA directly codes for amino acids and thus acts as messenger between DNA and ribosomes to make proteins. • It contains four bases such as adenine, guanine, uracil, cytosine.

The backbone of a nucleic acid consists of alternate pentose sugar and phosphate group and bases are attached to the sugar by β-N-glycosidic linkage.

The structure of a fragment of DNA containing four bases adenine, guanine, thymine and cytosine and sugar, 2-Deoxyribose sugar is shown in Figure 2.1. In this structure, linkage is from 5'→3'.

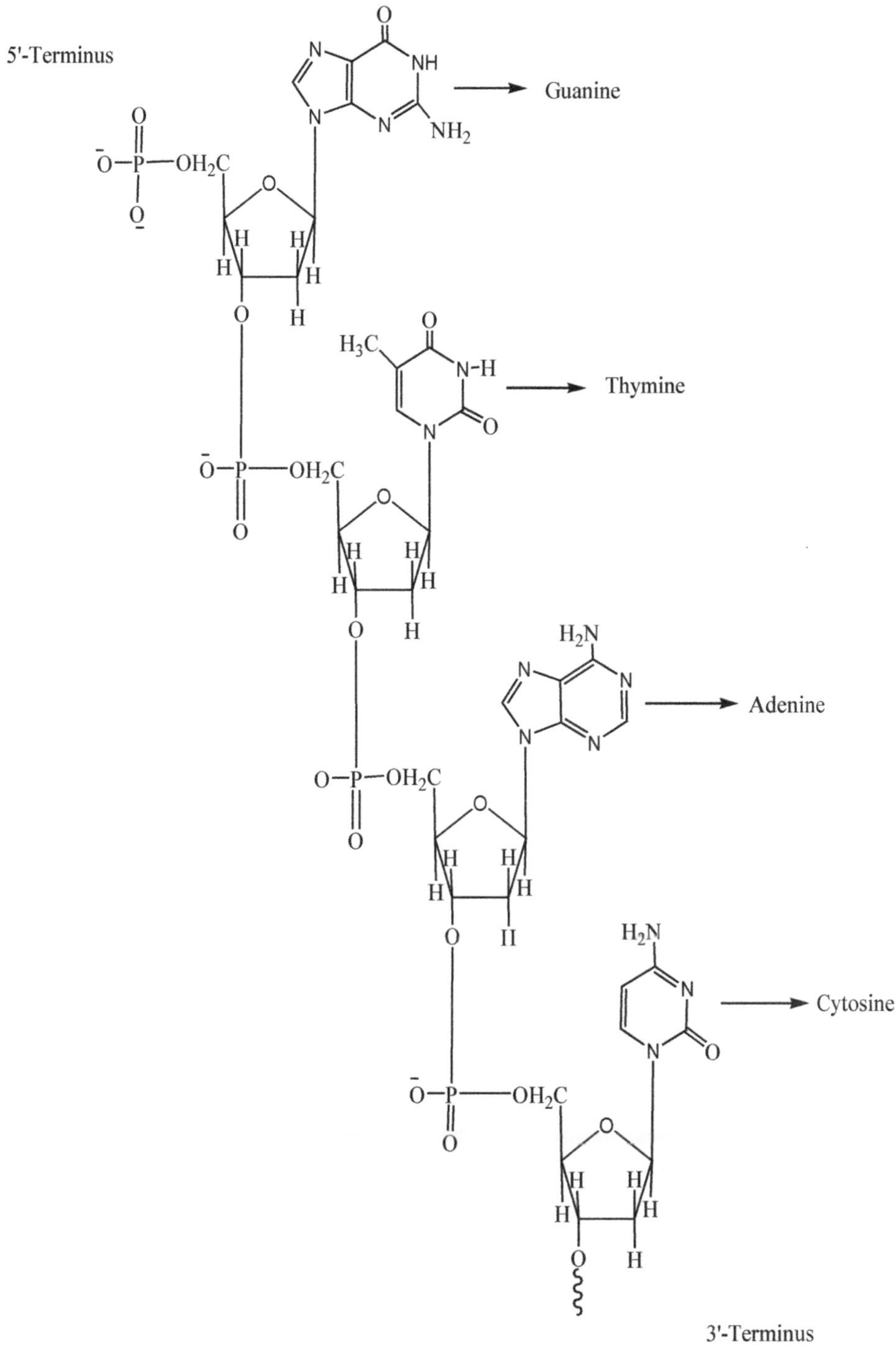

Figure 2.1: Fragment of DNA structure

The structure of a fragment of RNA containing four bases adenine, guanine, uracil and cytosine is shown in figure 2.2. In this structure, linkage is from 5'→3'.

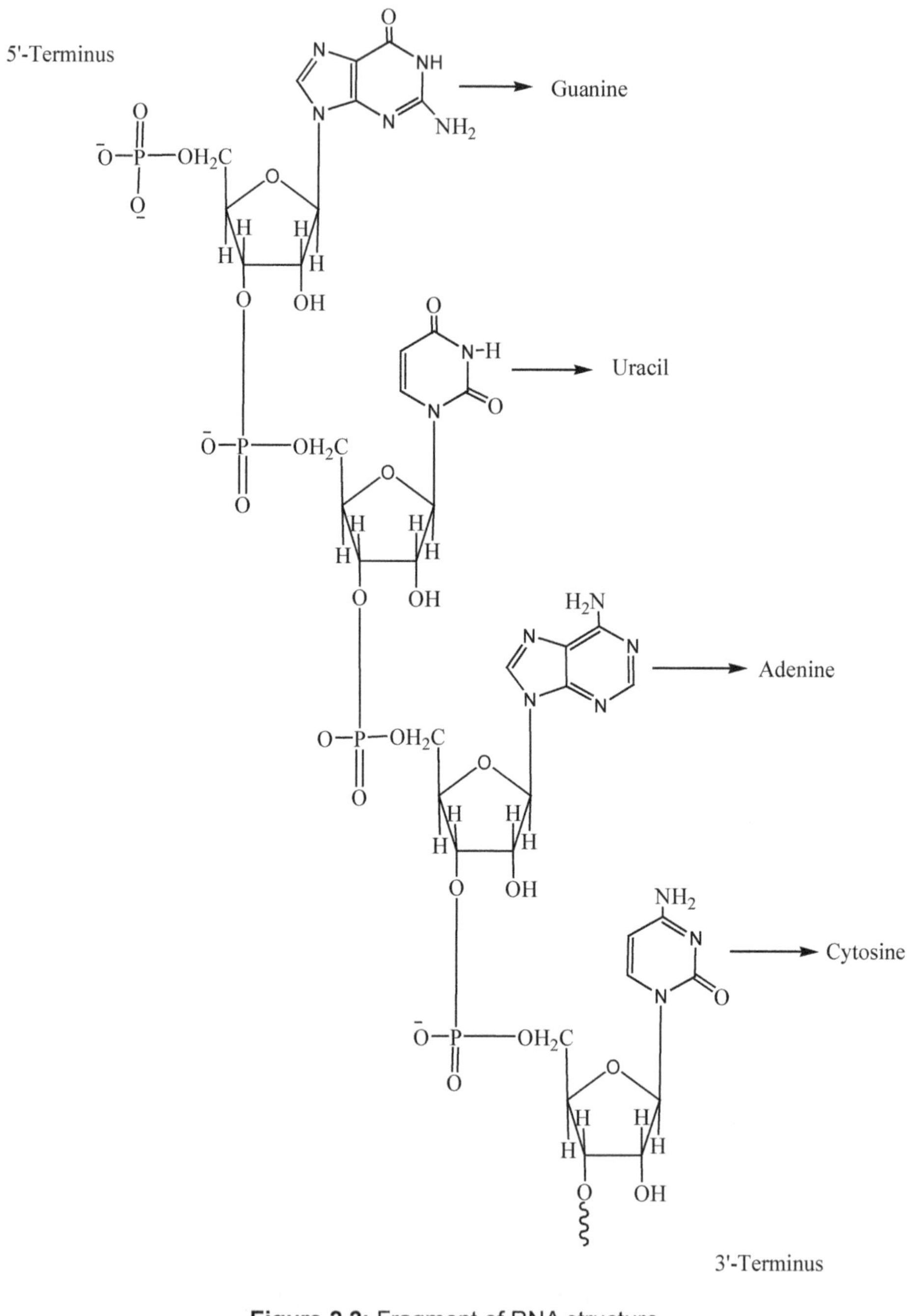

Figure 2.2: Fragment of RNA structure

2.12 Structure of DNA

As we know that DNA is a double helical polynucleotide that is made up of sugar, base and phosphate group. In DNA alternate arrangement of sugar and phosphate group form the backbone of linear polypeptide chain. Nitrogenous bases are stacked together like steps in a spiral staircase by intermolecular hydrogen bond. Many scientists have contributed a lot in determining the structure of DNA.

Structure of DNA was proposed by Watson and Crick in 1953 on the basis of key findings of following scientists.

Watson and Crick did not perform any experiment. They just collected all relevant information, analyzed and proposed the structure of DNA.

1. **Linus Pauling:** He discovered alpha helical structure of proteins, spiraled like a spring coil in 1948.
2. **Erwin Chargaff:** The most important information was provided by this Australian biologist related to the structure of DNA. He analyzed the DNA of different species and determined its composition that DNA is made up of four bases A, T, C, and G. He also made several other key observations which are as follows:
    - A, T, C, and G are not found in equal quantities.
    - The number of adenine units must be equal to thymine units and number of guanine units must be equal to cytosine units.
    - The amounts of the bases varied among species, but not between individuals of the same species.

    These findings are called Chargaff's rules. This information was crucial and helped Watson and Crick to propose the double helical structure of DNA.
3. **Rosalind Franklin and Maurice Wilkins:** The evidence that DNA has double helix was recorded by Rosalind Franklin. Rosalind Franklin was the first who arrived at the double helical structure of DNA on the basis of X-ray study of DNA sample. But she did not announce this finding in order to collect more evidences. Franklin was a meritorious student and belonged to influential family in London. She had earned her Ph.D. in physical chemistry from Cambridge University in 1945 and joined King's College in 1951. There she became extremely skilled in X-ray

crystallography and recorded excellent X-ray images of DNA crystals. On the basis of her study on DNA, she came to conclusion that DNA can exist in two forms depending on the relative humidity in the surrounding air. She recorded x-ray diffractions for "wet" form of DNA (having higher humidity) also and found all the characteristics features of a helix. She presented her research work on certain platform that was attended by James Watson. From there only Watson looked back to continue his study on DNA as he almost decided to leave his research in DNA. Since he was biologist and could not interpret the X-ray images of Rosalind. So, he came across Maurice who was also studying DNA molecule by X-ray crystallography. It is said that Maurice was not happy with the appointment of Rosalind for DNA work as he already started working on it. As a result, the relationship between Rosalind and Maurice was not good. Franklin's excellent X-ray photographs, to which they had gained access without her permission, were critical for DNA structure. In 1962, James Watson, Francis Crick and Maurice Wilkins were jointly awarded the Nobel Prize in Medicine. Unfortunately, by then Franklin had died due to cancer at the age of 37 and Nobel prizes are not awarded posthumously.

2.13 Postulates of Watson and Crick Model of DNA

1. The DNA molecule is made up of two polynucleotide chains also called strands. These strands are twisted around each other and coiled around a common axis to from a right-handed double helix.
2. The two strands are antiparallel i.e. they run in opposite directions in such a way 3'end of one chain face the 5' end of the other.
3. The sugar-phosphate backbones remain on the outside. While core of the helical structure is made up of purine and pyrimidine bases.
4. The two strands in DNA are held together by the formation of hydrogen bonds between purine and pyrimidine bases of the opposite strands.
5. It follows Chargaff's rule i.e. number of adenine is equal to number of thymine (A=T) whereas number of guanine is equal to the number of cytosine (G=C).
6. Adenine (A) always pairs with thymine (T) by two hydrogen bonds while guanine (G) always pairs with cytosine (C) by three hydrogen

bonds. This complimentary base pairing is known as base pairing rule. Therefore, two strands are complementary to one another.

7. The base sequence along a polypeptide chain is variable and is specific for each individual. The sequence of bases carries genetic information.
8. The diameter of DNA is 2 nm or 20 A^0. Adjacent bases are separated by 0.34 nm or 3.4 A^0 along the axis. Complete turn of helix has got length equal to 34 A^0 i.e. each turn contains 10 bases.
9. The DNA helix has two groves, a shallow groove (-1.2 nm) called minor groove and a deep groove (-2.2nm) called major groove (Figure 2.4).

2.14 Base Pairing in DNA

A base pair is two chemical bases bonded to one another forming a "rung of the DNA ladder." The DNA molecule consists of two strands that wind around each other like a twisted ladder. The two strands are held together by hydrogen bonds between the bases, with adenine forming a base pair with thymine and cytosine forming a base pair with guanine (Figure 2.3). There are two intermolecular H-bonds between A and T. whereas G and C are held together by three intermolecular H-bonds.

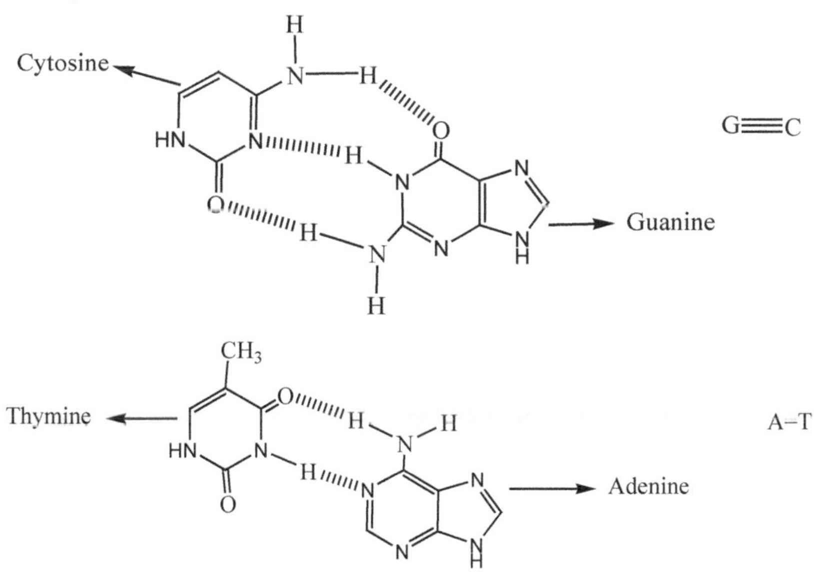

Figure 2.3: Base pairing in DNA

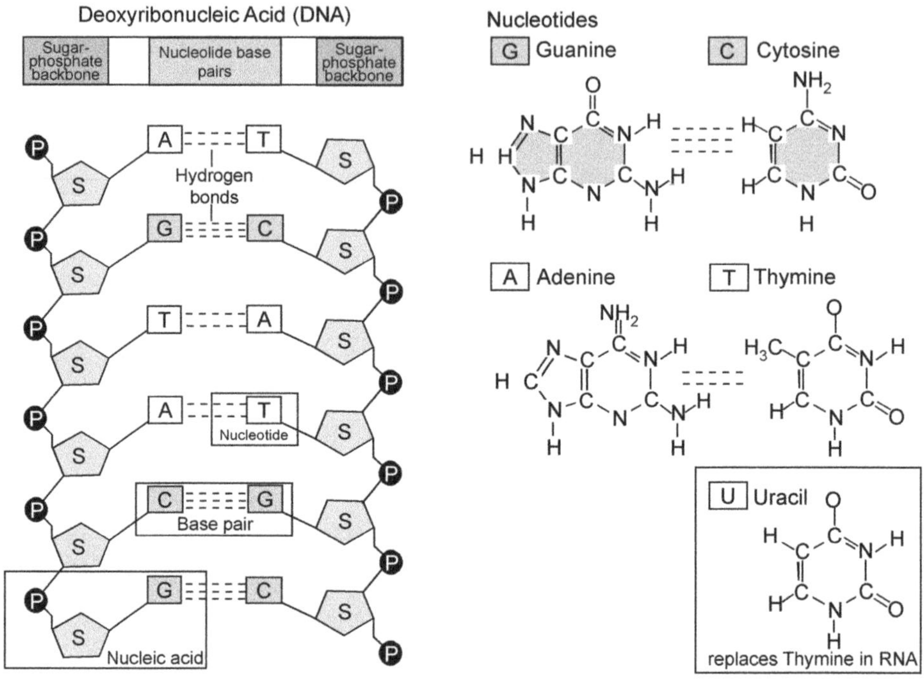

Figure 2.4: Pictorial representation of double helical structure of DNA

2.15 Different Conformations of DNA

DNA can adapt various structural conformations depending upon the following conditions:

1. Hydration level
2. Salt concentration
3. DNA sequence
4. Presence of chemically modified bases
5. Different types of metal ions and their concentrations
6. Presence of polyamine in solution.

Basically, there are three conformations acquired by DNA

B- DNA

It was discovered by Watson and Crick on the basis of X-ray diffraction pattern. It is a right-handed double helical structure. DNA commonly exists in B-form under normal physiological condition i.e. low salt concentration and high moisture level. The two strands run in opposite directions.

The structure comprises of alternate major grooves and minor grooves. The molecule is asymmetrical because the glycosidic bonds of a base pair are not diametrically opposite to each other. There are 10 base pairs in each turn having length of 3.4 nm. The two strands are held together by the formation of hydrogen bonds between nitrogenous bases. The helix width in B-DNA is 2nm. It is narrower than A-DNA.

A-DNA

It is also a right-handed double helix. It is formed when the relative humidity of the environment is less than 75 %. It means that it rarely exists under normal physiological condition. Here also two strands are anti-parallel with each other and not symmetrical. Since it is asymmetrical molecule, therefore major and minor grooves are present in each turn. One turn of the helix contains 11 base pairs having a length of 2.8 nm. Like B-DNA nitrogenous bases form the core of the helix. Its width is 2.3 nm. Overall, A-DNA is wider than B-DNA.

Z-DNA

It was discovered by Rich, Nordheim & Wang in 1984. It is a left-handed double helical structure. Its structure is different when compared with A-DNA and B-DNA. Z-DNA has zigzag backbone. The structure has major and minor grooves. Each turn of Z-DNA has 12 base pairs having a length of 4.56nm. The distance between two adjacent deoxyribonucleotides is 0.37 nm. Its width is 1.8nm which is the narrowest among the three types. It is difficult to observe as it is unstable in nature. Base pairing is same (A=T, G=C). It may take part in expression regulation of some genes or in genetic recombination. It is found in bacteria, eukaryotes and viruses. In some viruses, they require Z-DNA binding proteins for pathogenesis.

2.16 Replication

DNA replication is the biological process that produces two identical replicas of DNA from one original DNA molecule. It occurs in all living organism that forms the basis for biological inheritance. In this process, DNA makes a copy of itself during cell division. As we know that DNA has double helical structure and made up of two complementary strands. During replication process, two strands get separated from each other. Each strand acts as a template for the synthesis of other i.e. its counterpart. This process is called semiconservative replication. Because of the semiconservative replication, new helix will be made up of an original DNA and a newly synthesized strand.

We Can Summarize The Process of DNA Replication as Follows:

1. In DNA replication first of all, unzipping of double helical structure takes place. This is done by Helicase enzyme which breaks down the hydrogen bonds between the complementary bases that holds two strands together.
2. The separation of two strands of DNA forms a 'Y' shape which is called replication fork. These separated strands then act as template for making new strands.
3. One of the strands is oriented in the 3' to 5' direction i.e. towards the replication fork. This is called leading strand. The other strand is oriented in the 5' to 3' direction i.e. away from the replication fork. This is called lagging strand. These two strands are different hence replicate differently.
4. Meanwhile Helicase enzyme separates two strands, another enzyme called primase briefly attaches to each strand and assembles a foundation so that replication can start. This foundation is a short chain of nucleotides called primer.
5. After the primer is in place on a single polynucleotide chain, DNA polymerase enzyme wraps itself around the strand and start attaching new nucleotides to the exposed nitrogenous bases. In this way, polymerase enzyme assembles a new DNA strand on the top of the existing one.
6. The formation of new strand relies upon availability of nucleotides surrounding the existing strand as DNA polymerase moves down the unwounded DNA strand. The nucleotides that form the new strand always paired with its complementary nucleotides in the template strand like A pairs with T nucleotides and G pairs with C. This phenomenon is called base pairing and it is significant in the formation of two complementary strands of DNA.
7. Base pairing ensures that the sequence of nucleotide in the new strand is complementary to the sequence of nucleotide in the existing template strand. This is known as anti-sequence of the template strand. Because of this complementary base pairing, this replication process goes on as a series of sequence and anti-sequence copying which becomes the basis that preserves the coding of the original DNA.

2.17 Structure of RNA

RNA is a single stranded molecule. It is made of ribonucleotides that are linked together by phosphodiester bonds. Ribonucleotide in RNA contains ribose sugar, one of the four nitrogenous bases (A, U, G and C) and a phosphate group. The structural difference between the sugars gives DNA added stability and make it more suitable for storing of genetic information. As compared to DNA, RNA is less stable and thus it is more suitable for short term functions. In RNA, uracil forms a complementary base pair with adenine instead of thymine in DNA. Complementary base pairing in RNA is given Figure 2.5.

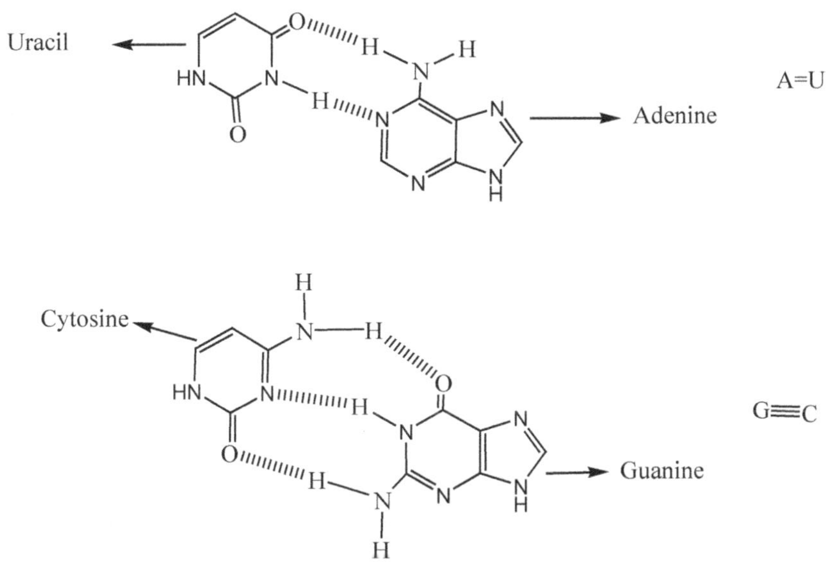

Figure 2.5: Base pairing in RNA

Single-stranded RNA can also exist in many secondary structures in which a single RNA molecule folds over and forms hairpin loops, stabilized by intramolecular hydrogen bonds formed between complementary bases. Such base-pairing of RNA is critical for many RNA functions, such as the ability of t-RNA to bind to the correct sequence of m-RNA during translation process.

2.17.1 Types of RNA Molecules

Cell uses the information stored in DNA by creating RNA that direct the synthesis of proteins by the process of translation. Proteins in the cell serve many functions like building of cellular structures, acts as enzyme for catalyzing many biological reactions. There are mainly three types of RNA directly involved in protein synthesis:

- Messenger RNA (m-RNA)
- Ribosomal RNA (r-RNA)
- Transfer RNA (t-RNA)

Two French scientists François Jacob and Jacques Monod in 1961 suggested the presence of an intermediatory between DNA and protein synthesis which they named messenger RNA. Soon they gathered evidence in support of it. They said that m-RNA helps in transmitting of information stored in DNA for protein synthesis. If DNA is a complete library of cellular information then m-RNA is a photocopy of specific information that is needed for protein synthesis at the time of requirement. Functions of three types of RNA as follows:

- The m-RNA plays important role in protein synthesis. It is a single stranded molecule that copies information from DNA during the process of transcription. Then it moves from nucleus of cell to the cytoplasm which contain ribosomes. Here ribosomes read the m-RNA and make protein by following the instructions. m-RNA is less stable and short lived in the cells especially in prokaryotic cells.
- Transfer RNA (t-RNA) has got very specific function i.e. bring amino acids to the ribosomes where proteins are synthesized. It acts as a key in interpreting the code words in m-RNA. It reads the m-RNA in 3 letter nucleotide sequence called codons. Each amino acid has its own t-RNA which binds it and carries to the growing polypeptide chain if m-RNA calls for it. It is a stable type of RNA and one of the smallest which is made up of 70-90 nucleotides. Because of base pairing between t-RNA and m-RNA, correct amino acid is inserted in the long polypeptide chain. Any change (mutation) in the t-RNA and r-RNA affects the whole activity of the cell because both are essential for proper protein synthesis.
- Ribosomal RNA (r-RNA) is the main component of ribosomes where the synthesis of proteins takes place. It is important for all functions of ribosomes like from binding to m-RNA and assigning t-RNA to catalyze the formation of peptide bond between amino acids. They are rod shaped organelles often called power generator of the cell. During aerobic respiration, chemical energy stored in ATP is given out to perform various metabolic activities. The existence of higher animals would have

been difficult without r-RNA under anaerobic respiration. It is because of r-RNA cell produces 15 times more ATP as compared to under normal condition. More than 60 percentage weight of ribosomes is composed of r-RNA and even the shape of ribosomes is determined by it. r-RNA has many binding sites for a molecule to get attached. Even the procaryotic r-RNA is the target site for many antibiotics such as tetracyclines and streptomycin.

Important Point to Remember

It is reported that RNA does not act as hereditary material in most of the cells. But it does so for many viruses that do not contain DNA. It clearly shows that RNA has additional capacity to carry genetic information. Though RNA is a single stranded molecule but there is significant diversity in viruses. For example Rhinoviruses, which causes common cold, influenza viruses and the Ebola virus are single-stranded RNA viruses. Rotaviruses is another one responsible for severe gastroenteritis in children and other immune compromised individuals are examples of double stranded RNA viruses. Double stranded RNA is uncommon in eukaryotic cells and its presence in cell is a clear indication of viral infection.

2.18 Transcription

Transcription is the first step of gene expression. In this process, information stored in a strand of DNA is copied into a messenger RNA (m-RNA). DNA safely and stably stores genetic material in the nuclei of cells. Meanwhile, m-RNA is just like a photocopy of reference book (DNA) because it carries the same information that is stored in DNA. But it is not used for long term storage and can freely exit the nucleus. Although the m-RNA contains the same information, it is not an identical copy of the DNA segment, because its sequence is complementary to the DNA template. RNA polymerase is the main enzyme involved in transcription process and it uses a single-stranded DNA template to synthesize a complementary strand of RNA. Specifically, RNA polymerase builds an RNA strand in the 5' to 3' direction, adding each new nucleotide to the 3' end of the strand. Transcription takes place in the following three steps:

Initiation: RNA polymerase binds to a sequence of DNA called the **promoter**, found near the beginning of a gene. Each gene has its own promoter. Once bound, RNA polymerase separates the DNA strands, providing the single-stranded template needed for transcription.

Elongation: One strand of DNA acts as a template for RNA polymerase. As it "reads" this template one base at a time, the polymerase builds an RNA molecule out of complementary nucleotides, making a chain that grows from 5' to 3'. The RNA transcript is nearly identical to the non-template (**coding**) strand of DNA, but the only difference is that instead of thymine base it contains uracil.

Termination: RNA polymerase enzyme keeps transcribing until it gets a signal to stop. So, this process of ending transcription is called termination. When its job of transcribing is over, they cause the transcript to be released from the RNA polymerase.

2.19 Translation

The process of synthesis of protein from the information present in the molecule of messenger RNA (m-RNA) is called **translation.** The translation process involves reading of genetic code in m-RNA to make protein. Genetic code is a set of rules that defines how an m-RNA sequence is to be translated into the 20-letter code of amino acids, as we know amino acid are the building blocks of proteins. The genetic code is a set of 3-letter combination of nucleotides. This combination is called codon. Each codon contains specific amino acid or stop signal. Translation process occurs in ribosome which is the manufacturing unit of proteins. The ribosome comprises of many small and large units. It is a complex molecule made up of several ribosomal RNAs and number of proteins. Translation process takes place in three stages: initiation, elongation, and termination.

During initiation, the small ribosomal subunit binds to the start of the m-RNA sequence. Then a transfer RNA (t-RNA) molecule carrying the amino acid methionine binds to what is called the start codon of the m-RNA sequence.

The start codon in all m-RNA molecules has the sequence AUG and codes for methionine. Then large ribosomal subunit binds to it from initiation complex. During elongation stage ribosome continues to translate each codon one by one. Charged t-RNA enters the ribosomal complex and gets inserted. As a result, polypeptide chain becomes one amino acid longer while uncharged t-RNA gets departed. This elongation steps goes on until all of the codon are read. In the last termination happens when the ribosome reaches a stop codon (UAA, UAG, and UGA). t-RNA molecules don't recognize these codons that is the indication to the ribosome that translation is complete. The new protein is then released, and the translation complex comes apart.

Questions

1. What are nucleosides? Give one example.
2. Write down the composition of nucleic acid.
3. Write down the name and structure of the sugars present in nucleic acid.
4. Write down the name and structure of all the bases present in nucleic acid.
5. What are nucleotides?
6. Write down the structure and name of nucleoside containing D-ribose and uracil.
7. Write down the structure and name of nucleotide which on hydrolysis gives 2-Deoxyribose, phosphoric acid and adenine.
8. Give difference between nucleotide and nucleoside.
9. What are AMP, GMP, CMP stand for?
10. Give full form of ATP with its structure. Why is it called energy currency of cell?
11. What are the different roles of ATP?
12. Give difference between DNA and RNA.
13. Write down the structure of ADP, TMP, UMP.
14. Write down the postulates of Watson and Crick model of DNA.
15. Name the different bases present in DNA.
16. Name the different bases present in RNA.
17. What are Chargaff's rules?
18. Explain transcription.
19. Explain the concept of DNA replication.
20. What is translation?
21. What are the different types of RNA found in cell? Write their function.
22. Write the structure of complementary base pairing in DNA.
23. How would you synthesize adenine?
24. Write down the wheeler and Johnson synthesis of cytosine.

25. Write down the Fischer synthesis of adenine.
26. Write down Fischer and Rodder's synthesis of uracil.
27. Write down Traube's synthesis of guanine.
28. Write the complementary bases of DNA strand 5'-ATCCCGTAGAGG-3'.
29. Write down the fragment of DNA containing four bases.

❑❑❑

3

Amino Acids, Peptides and Proteins

3.0 Amino Acids

Amino acids are organic compounds that are made up of elements like carbon, hydrogen, oxygen and nitrogen along with a variable side chain group. They play many critical roles in our body. They are required for various life processes such as building of proteins, hence, referred as building blocks of proteins, hormones and neurotransmitter synthesis etc.. Some of them may be taken as food supplement to enhance the performance of athlete or mood improvement. In nutshell,

"Amino acids are organic acids that contain amino group attached to the chain carrying carboxylic acid".

3.1 Types of Amino Acids

There are 3 Types of Amino Acids:

1. **α-amino Acid:** The amino acids having both the amine and the carboxyl groups are attached to the first carbon (α) atom are known as **α-amino acids.** They are generally represented by the following structure.

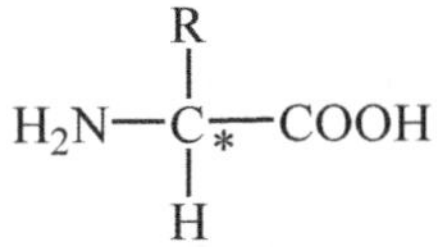

There are 20 amino acids which are called proteinogenic ("protein-building"). These amino acids combine together by peptide bond to

DOI: 10.1201/9781003727453-3

form polypeptides that are building-blocks of a vast array of proteins. These are all L-stereoisomers (left handed isomers). Although few are D-amino acids ("right-handed") found in bacterial envelopes as neuromodulator (D-serine) and in some antibiotics such as penicillin.

2. **β-amino Acids:** The amino acids that contain amine group attached to a β- carbon atom of a chain carrying a carboxyl group are known as **β-amino acids.** They are generally represented by the following structure.

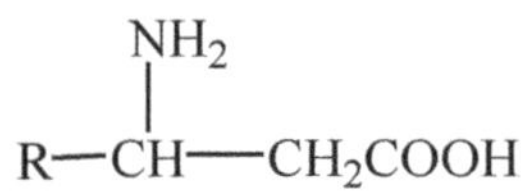

e.g. β- alanine is the one which is found in nature but not in protein. They are crucial in medicinal chemistry.

3. **γ-amino Acids:** The amino acids that contain amine group attached to a γ- carbon atom of a chain carrying a carboxylic acid group are known as **γ-amino acids.** They are non-proteinogenic amino acids. They are generally represented by the following structure.

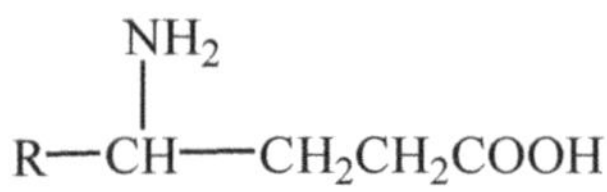

e.g. γ-amino butyric acid (GABA) is a non-standard and non-proteinogenic amino acid that has biological function. It acts as neurotransmitter in brain.

3.2 Classification of Amino Acids

There are Two Ways to Classify Amino Acids:

I. On the Basis of Nature

They are of three types

1. **Neutral Amino Acids:** Those amino acids which contain one amino group and one carboxylic group are called neutral amino acids. They are generally represented by the following structure-

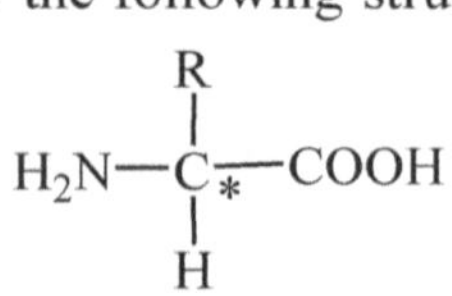

Examples are glycine, valine, alanine, leucine etc.

2. **Acidic Amino Acids:** Those amino acids which contain one amino group and two carboxyl groups are called acidic amino acids. Second carboxylic group is present in the side chain. They are generally represented by the following structure-

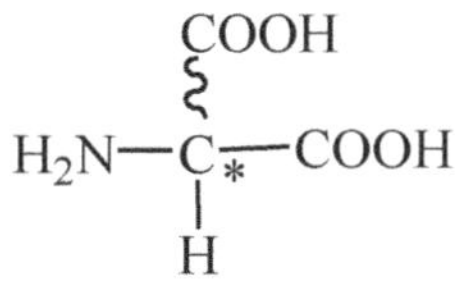

Examples are aspartic acid, glutamic acid

3. **Basic Amino Acids:** Those amino acids which contain two amino groups and one carboxyl group are called basic amino acids. Second amino group is present in the side chain. They are generally represented by the following structure-

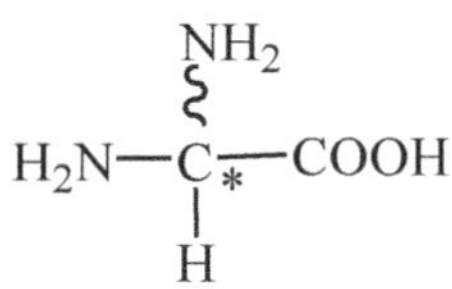

Examples are lysine, arginine, histdine .

II. On the Basis of Need/Requirement

There are 20 amino acids which are required by our body for proper growth and functions. They are further classified as essential, non-essential and conditionally essential amino acids depending upon several other factors.

1. **Essential Amino Acids:** The amino acids that cannot be synthesized from scratch by the organism in adequate amount as required by the body and therefore, must be supplied in the diet are called essential amino acids. They are also called **indispensable amino acids**. Although 20 amino acids are important for our good health, but only 9 of them are classified as essential amino acids. These are valine, leucine, isoleucine, histidine, lysine, methionine, phenylalanine, tryptophan and threonine.
2. **Non-Essential Amino Acids:** The amino acids that can be synthesized by the body of an organism in adequate amount are called non-essential amino acids. They are also called **dispensable amino acids.** These are alanine, serine, cystine, aspartic acid etc.
3. **Conditionally Essential Amino Acids:** There are six amino acids that are called conditionally essential in the human diet, meaning their synthesis can be limited under special pathophysiological conditions,

such as prematurity in the infant or person in severe catabolic problem. These are arginine, cysteine, glycine, proline, glutamine and tyrosine.

3.3 Structure of Amino Acids

1. Neutral Amino Acids

Name of Amino Acid	Structure	3-Letter Abbreviation
Glycine	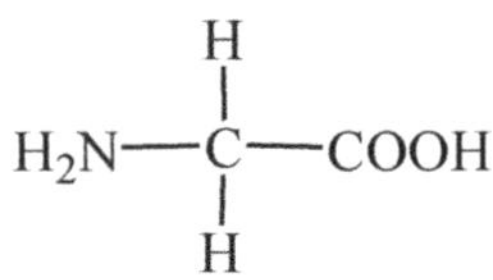	Gly
Alanine	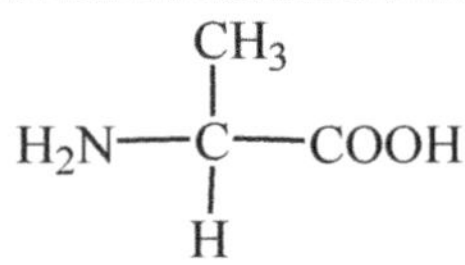	Ala
Valine	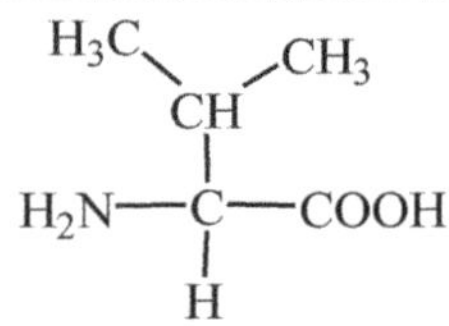	Val
Leucine	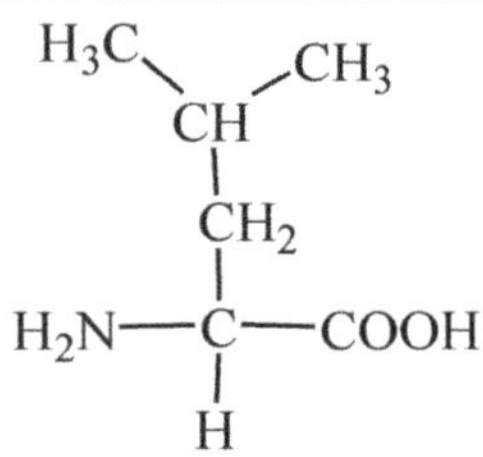	Leu
Isoleucine	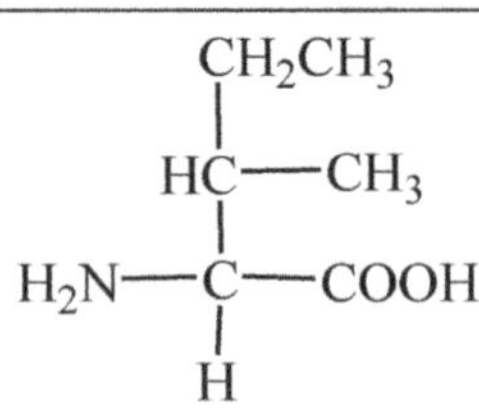	Ile
Phenylalanine	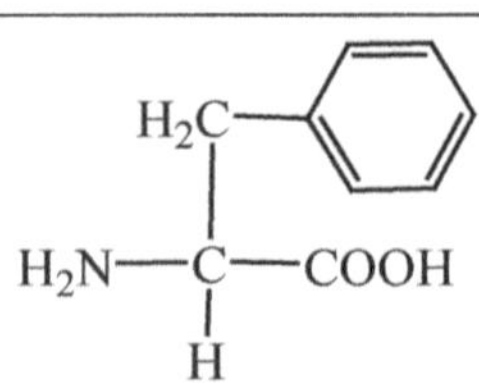	Phe

Name of Amino Acid	Structure	3-Letter Abbreviation
Tyrosine	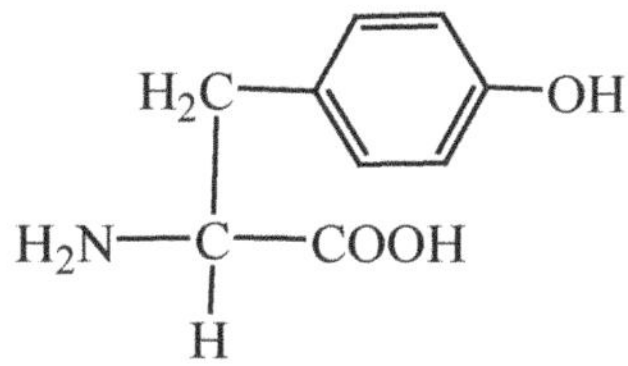	Tyr
Tryptophan	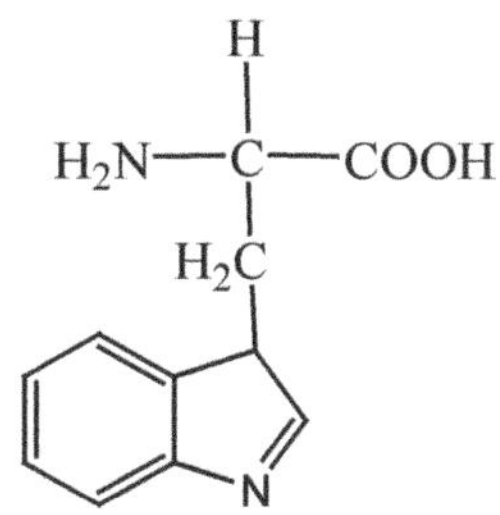	Try
Serine	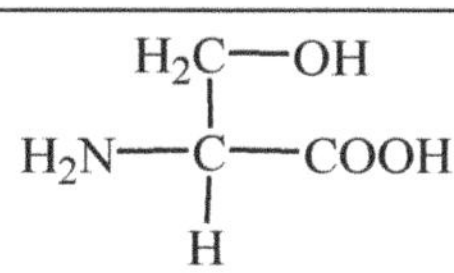	Ser
Threonine	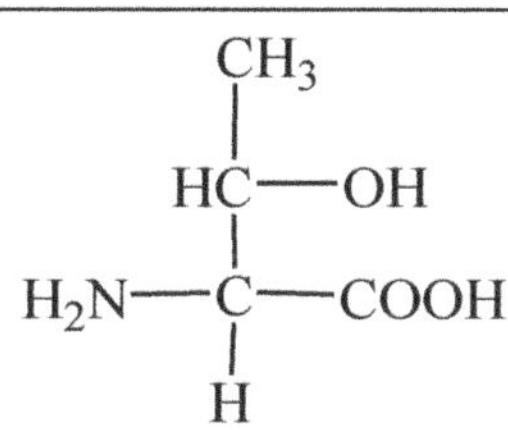	Thr
Proline	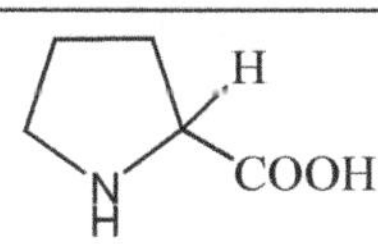	Pro
Hydroxyproline	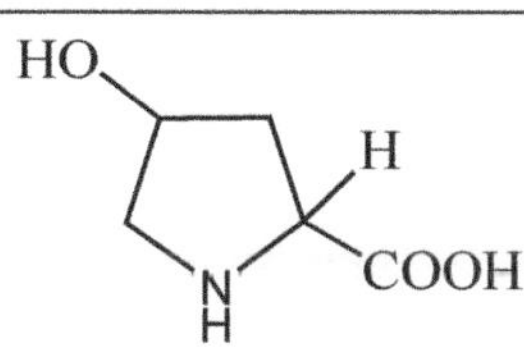	Hyp
Cysteine	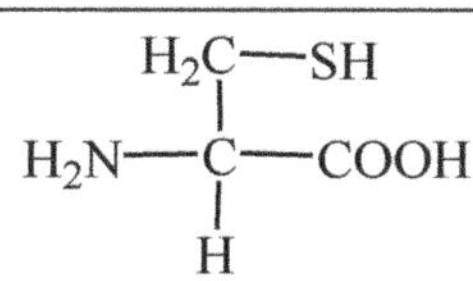	Cys

Name of Amino Acid	Structure	3-Letter Abbreviation
Cystine	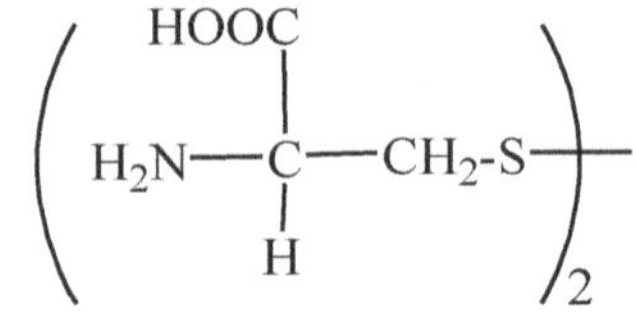	Cys or Scy
Methionine	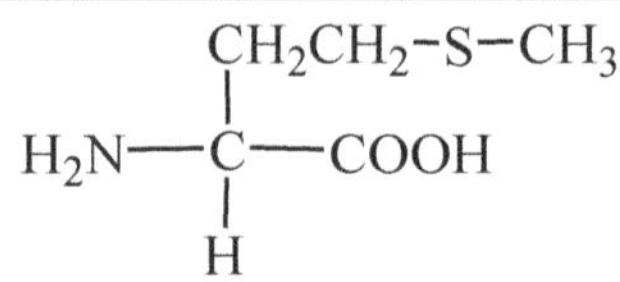	Met

2. Acidic Amino Acids

Name of Amino Acid	Structure	3-Letter Abbreviation
Aspartic acid	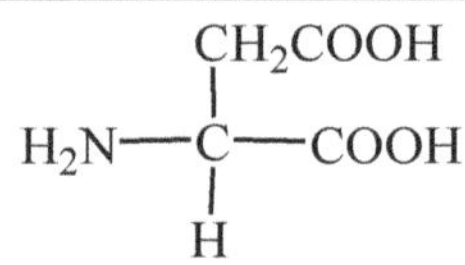	Asp
Glutamic acid	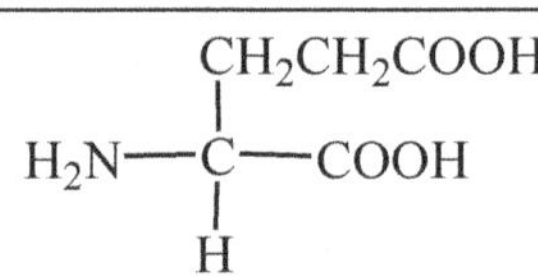	Glu

3. Basic Amino Acids

Name of Amino Acid	Structure	3-Letter Abbreviation
Lysine	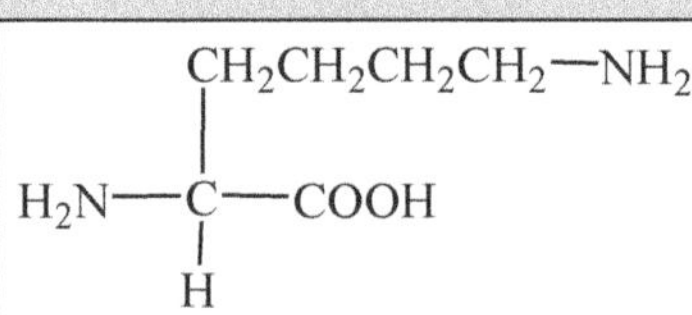	Lys
Arginine	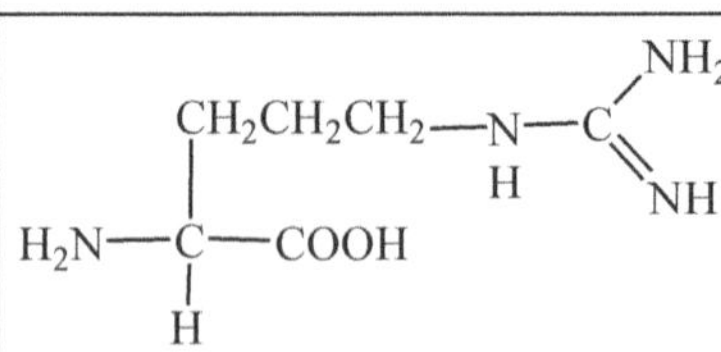	Arg

Name of Amino Acid	Structure	3-Letter Abbreviation
Histidine	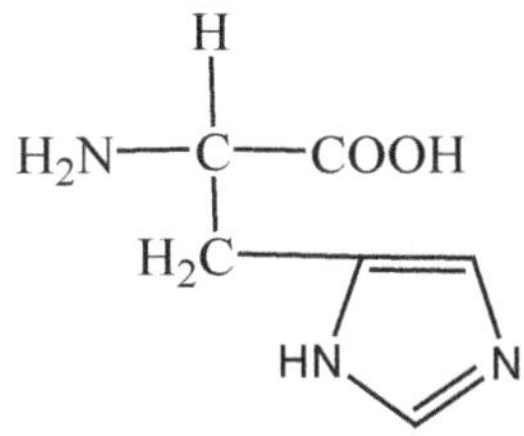	His

3.4 Optical Activity of Amino Acids

All amino acids except glycine contain chiral center, so they exist in two stereoisomers i.e. D-amino acid and L-amino acid (a pair of enantiomers).

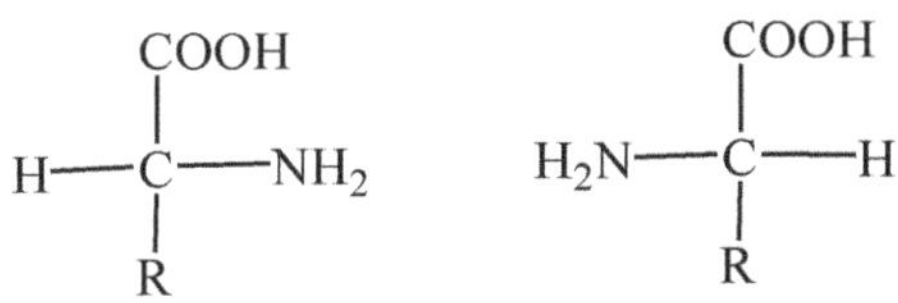

D-amino acid L-amino acid

All naturally occurring amino acids have L- configuration (S) except cysteine which has R configuration.

e.g. alanine

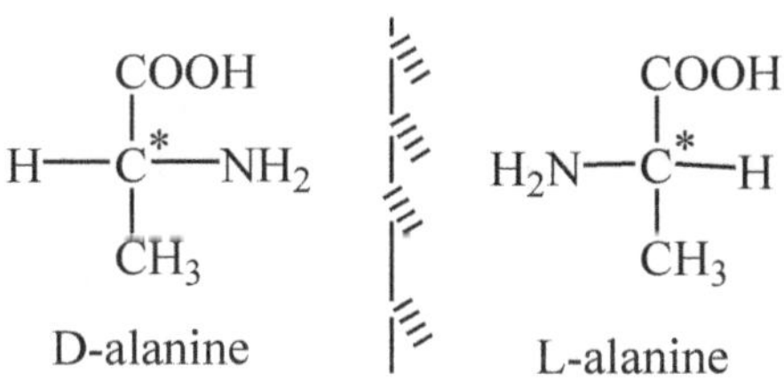

D-alanine L-alanine

Threonine contains two chiral centers so exists as 4 stereoisomers

No of stereoisomers /optical isomers = $2^n = 2^2 = 4$

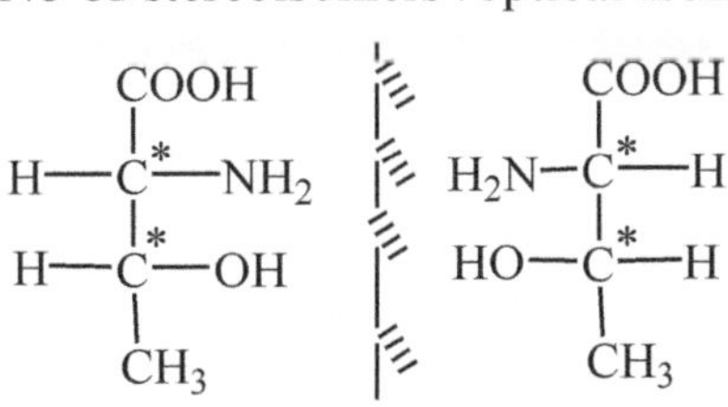

D-allothreonine L-allothreonine

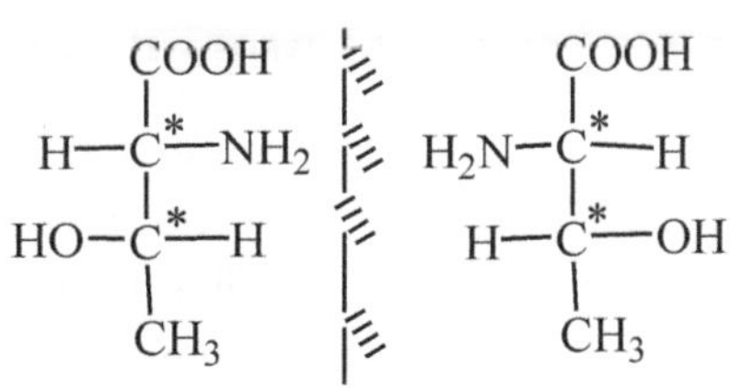

D-threonine L-threonine

3.5 Synthesis of Amino Acids

There are various methods to synthesize amino acids. Few are listed below

3.5.1 Hydrolysis of Proteins

Proteins on heating with dilute hydrochloric acid are hydrolyzed to give a mixture of amino acids which can be separated by chromatography (Column chromatography)

$$\text{Proteins} \xrightarrow[\Delta]{\text{dil. HCl}} \text{mix. of amino acids}$$

3.5.2 Amination of α-haloacids

When α-haloacids are heated with excess of ammonia, they undergo amination to form amino acids. Reaction takes place as follows:

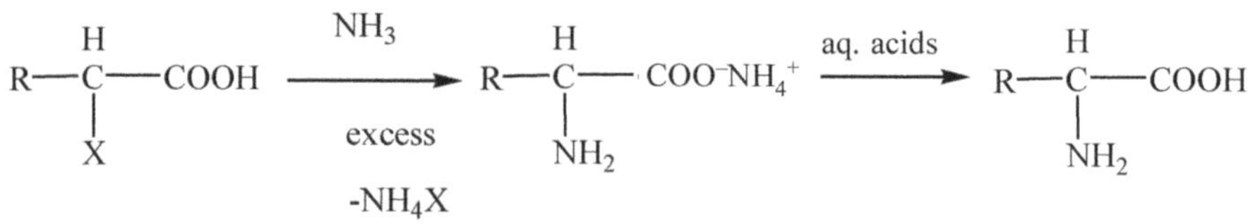

R = side chain of amino acid.

Preparation of α-haloacids

α-Haloacids are prepared by Hell Volhard Zelinsky (HVZ) reaction. In this, carboxylic acid containing α-hydrogen is treated with chlorine or bromine in presence of red phosphorus and It gets converted into α-substituted haloacid.

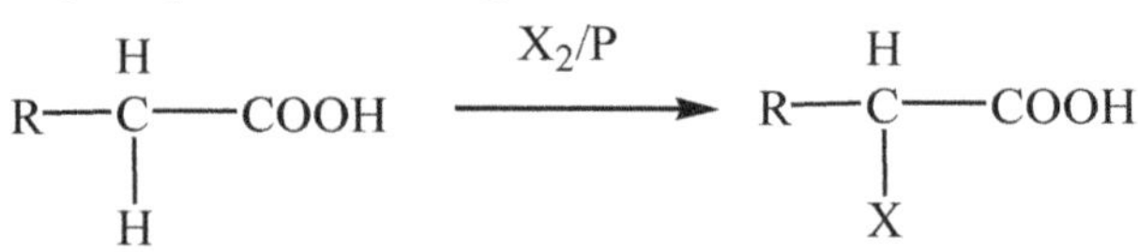

$X_2 = Cl_2, Br_2$

3.5.3 Gabriels Phthalimide Synthesis

In this method, potassium salt of phthalimide (prepared by fusing phthalimide with potassium hydroxide) reacts with α-haloester. Intermediate formed is then subjected to alkaline hydrolysis i.e. heated with alkali like sodium hydroxide followed by acidification with dilute hydrochloric acid to liberate free amino acid. The reaction takes place as follows:

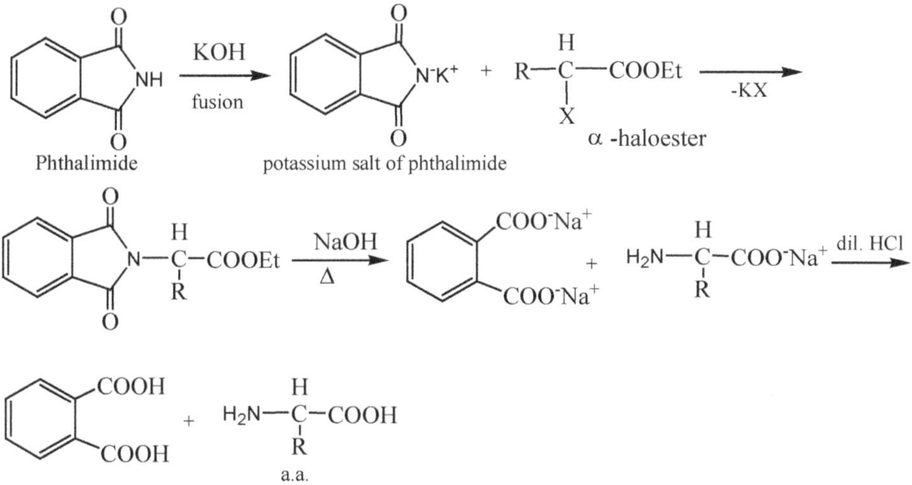

3.5.4 Strecker's Method

In this method, aldehyde is heated with a mixture of potassium cyanide and ammonium chloride to get amino acid. The reaction takes place as follows:

$$NH_4Cl + KCN \longrightarrow KCl + NH_3 + HCN$$

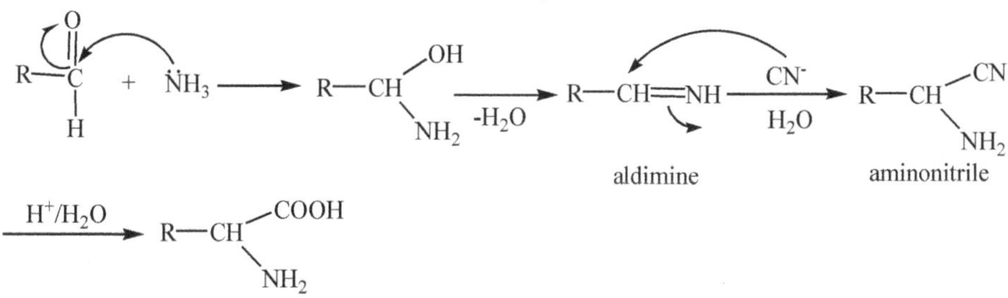

3.5.5 Malonic Ester Synthesis

Amino acids can also be synthesized from malonic ester (DEM). In this method, malonic ester reacts with alkyl halide in presence of base to form alkyl malonic ester which on hydrolysis gives dicarboxylic acid. This on reaction with bromine followed by addition of ammonia gives amino acid. The reaction takes place as follows:

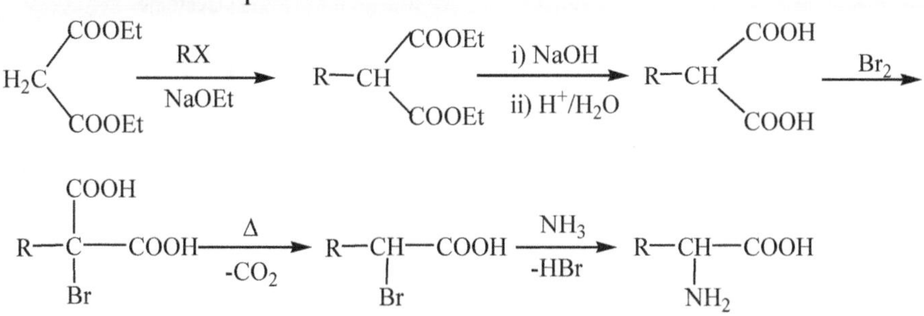

3.5.6 Phthalimide-malonic Ester Method

It is also called combined method as in this method malonic ester method is combined with Gabriel's phthalimide method. This is a good method to synthesize amino acids with better yield.

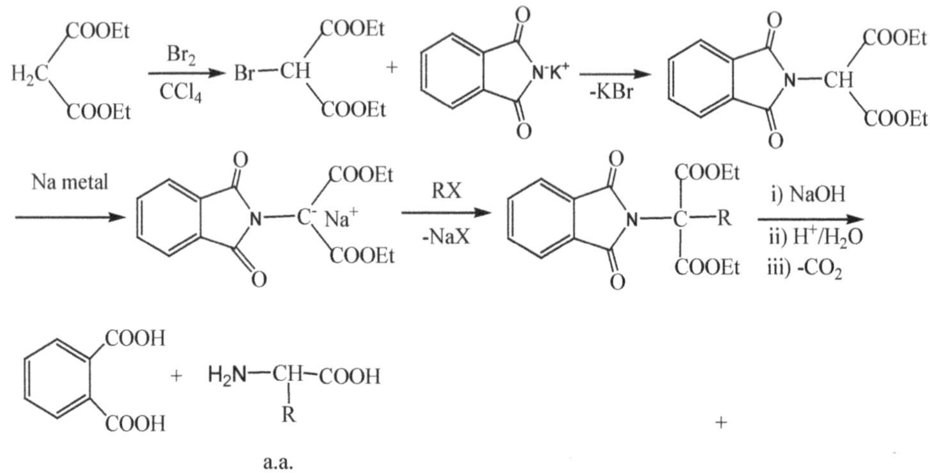

Examples

(a) Cysteine and Cystine

Synthesis of side chain

$C_6H_5CH_2SH + HCHO + HCl \longrightarrow C_6H_5CH_2SCH_2Cl$

benzylthiol — benzylthiomethylchloride

Cysteine/Cystine

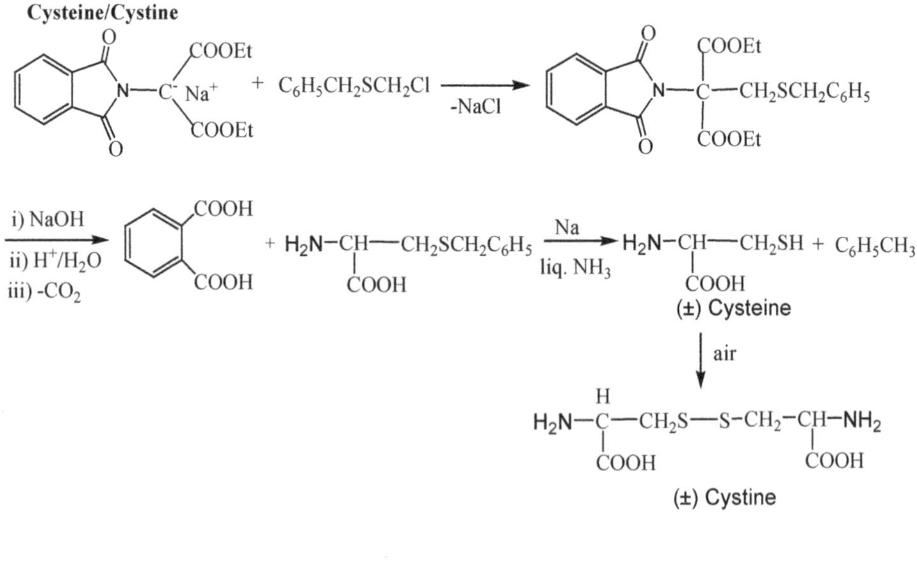

Proline

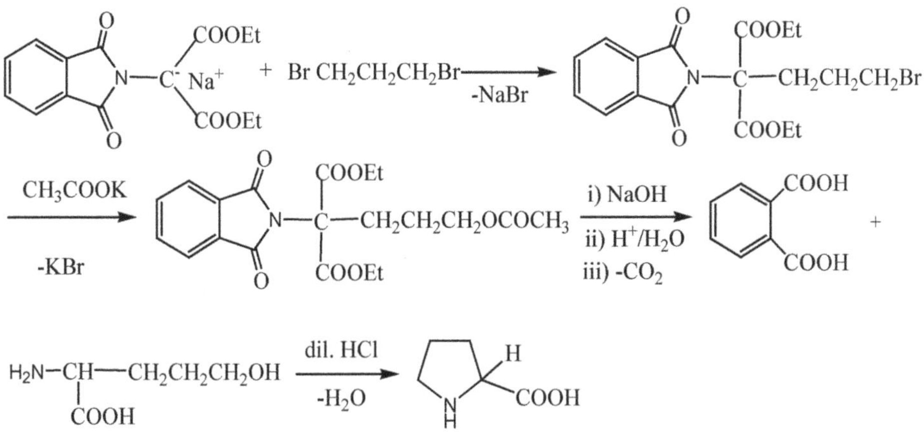

Proline

Aspartic acid

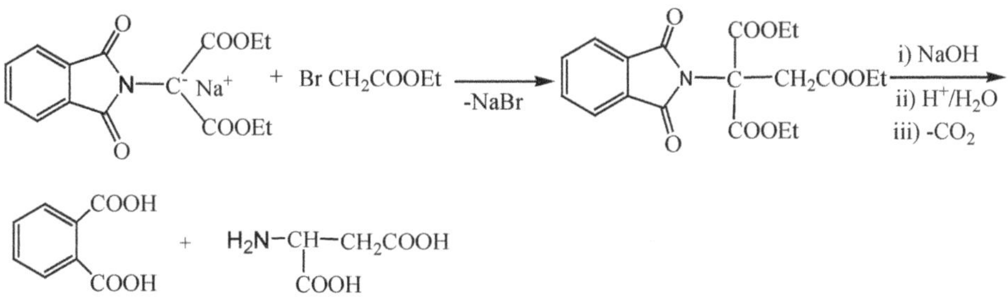

Glutamic acid

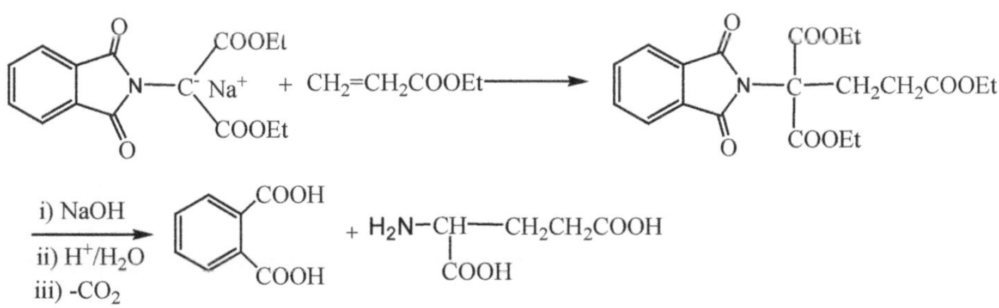

3.5.7 Acyl Amido Malonic Ester or Benzoyl Amido Malonic Ester Synthesis

In this method, malonic ester reacts with nitrous acid followed by catalytic reduction to give amino malonic ester. This amino malonic ester is short of alkyl group (side chain) and this can be introduced by alkylation reaction.

Before introduction of side chain, amino group needs to be protected which can be done by acylation or benzoylation. The reaction takes place as follows.

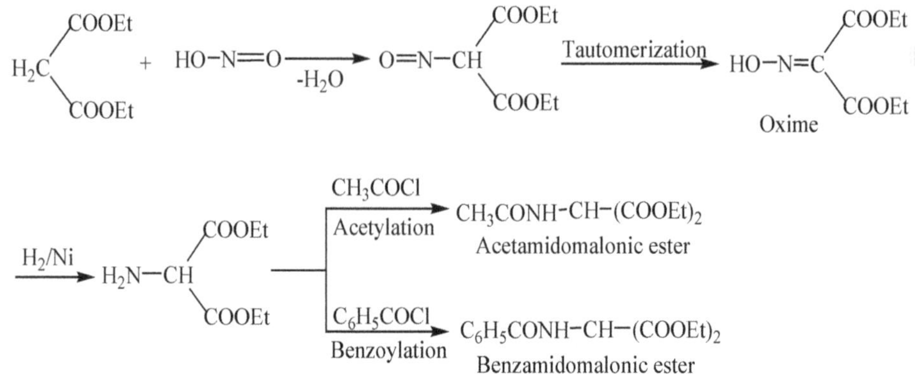

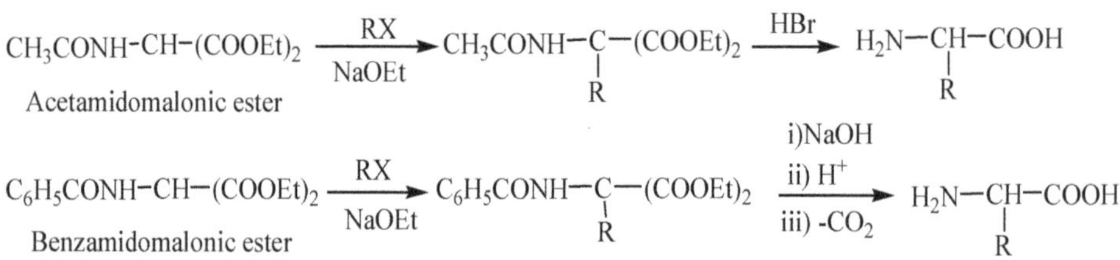

Examples are

Alanine

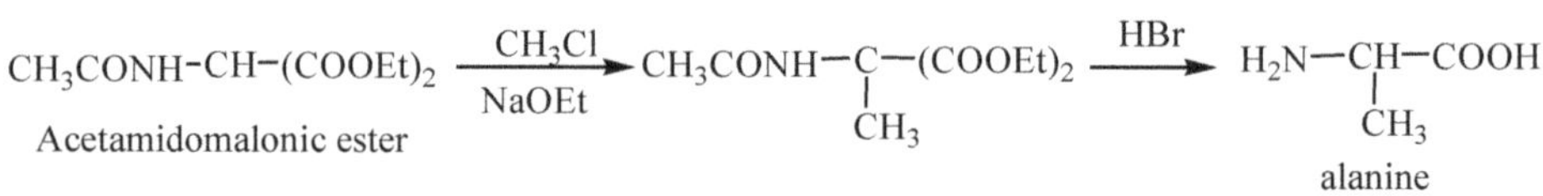

Leucine

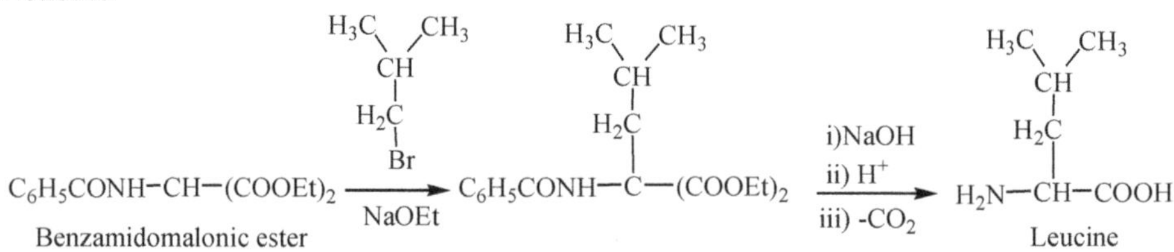

Valine

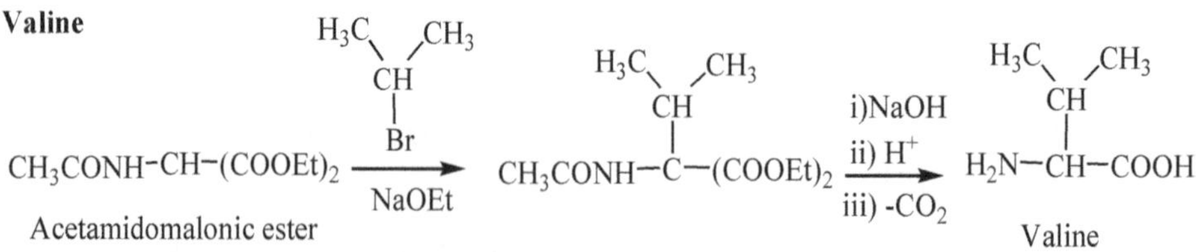

3.5.8 Curtius Azide Method

In this method, malonic ester is converted into acid azide by Curtius reaction.

The reaction takes place as follows:

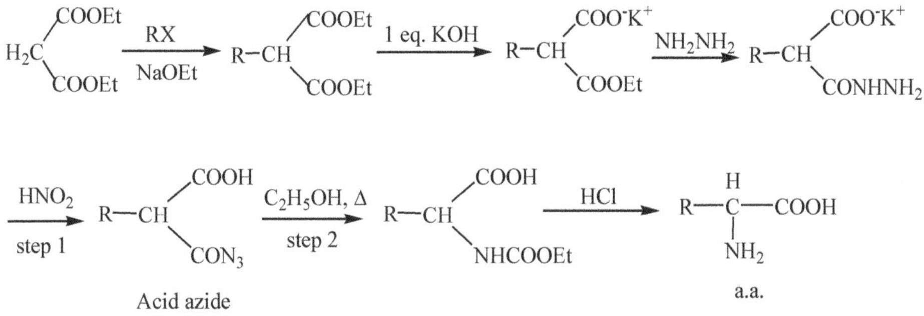

Mechanism of Step 1:

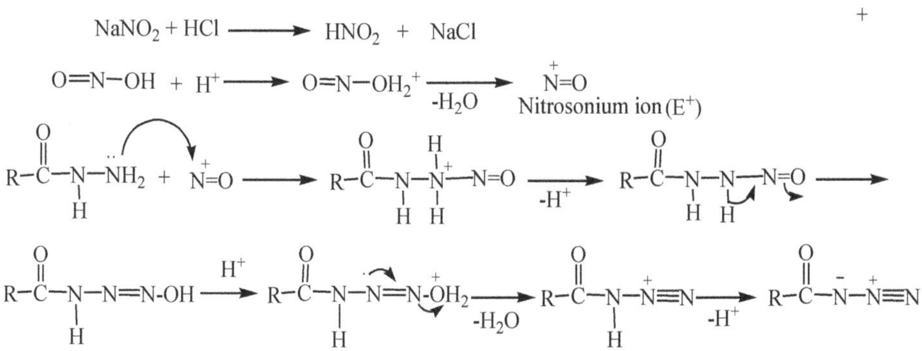

Mechanism of Step 2: This step is called Curtius rearrangement reaction. It proceeds via the formation of isocyanate.

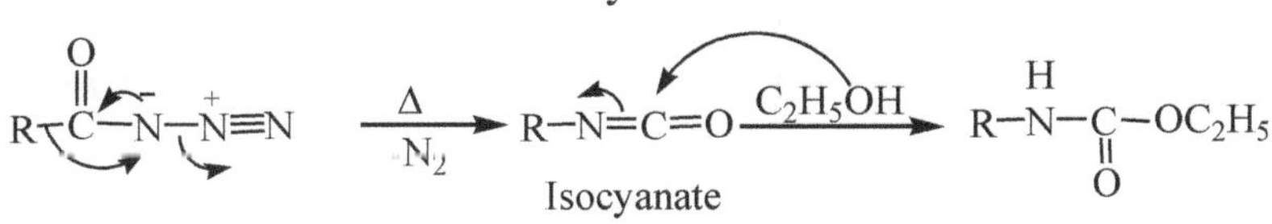

3.5.9 From Ethyl Cyanoacetate

Instead of malonic ester, we can also synthesize amino acids from ethyl cyanoacetate.

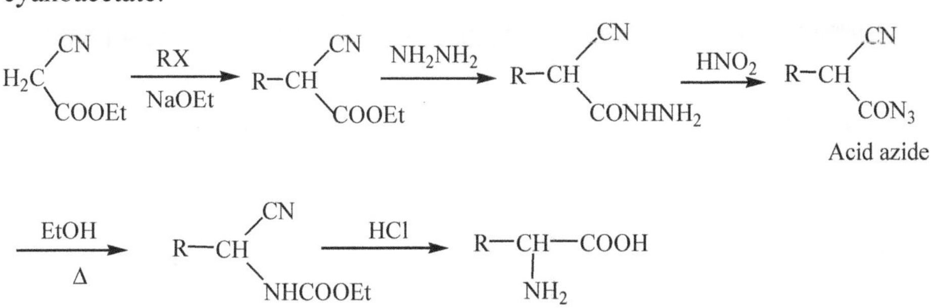

3.6.0 Darapsky Synthesis

It is similar to Curtius method. Condensation of aldehyde with ethyl cyanoacetate followed by catalytic reduction gives an intermediate. After that reaction proceeds similar to Curtius method.

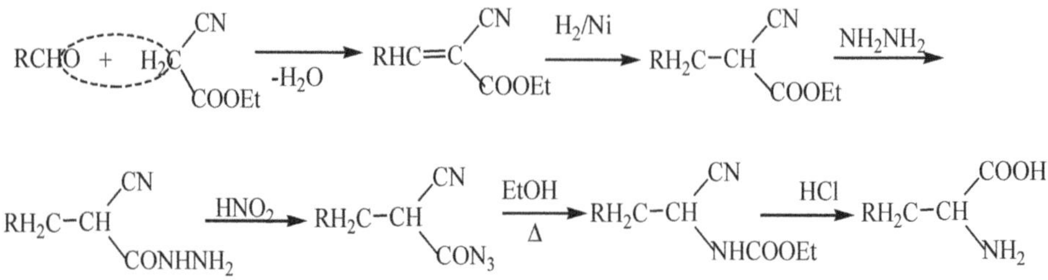

3.6.1 Erlenmeyer Azlactone Synthesis

This method was introduced by Erlenmeyer. It is used to synthesize aromatic amino acids by condensing aromatic aldehydes with Hippuric acid (benzoyl derivative of glycine) or aceturic acid (acetyl derivative of glycine). Azlactone formed is then warmed with dilute NaOH to open the ring. Reduction of intermediate with sodium amalgam (Na-Hg) followed by hydrolysis with hydrochloric acid gives amino acid. We can also synthesize aliphatic amino acid by this method.

Phenylalanine

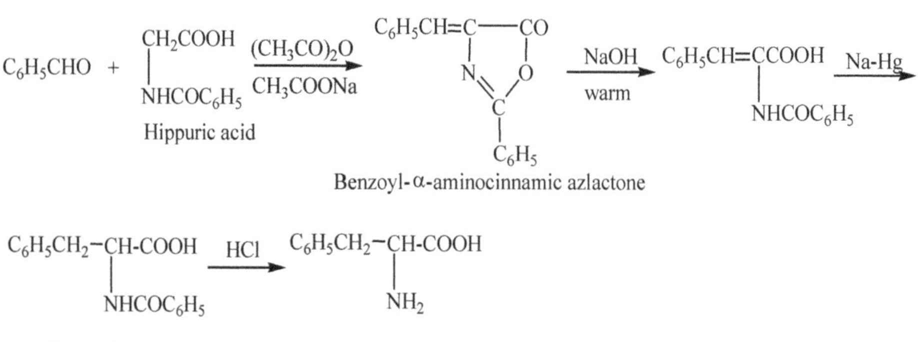

Tryptophan

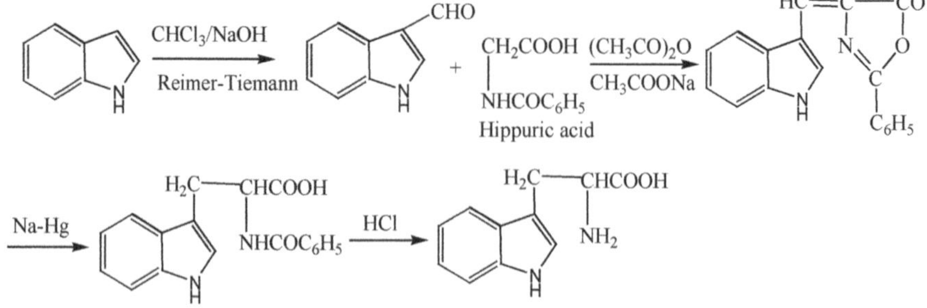

Tyrosine

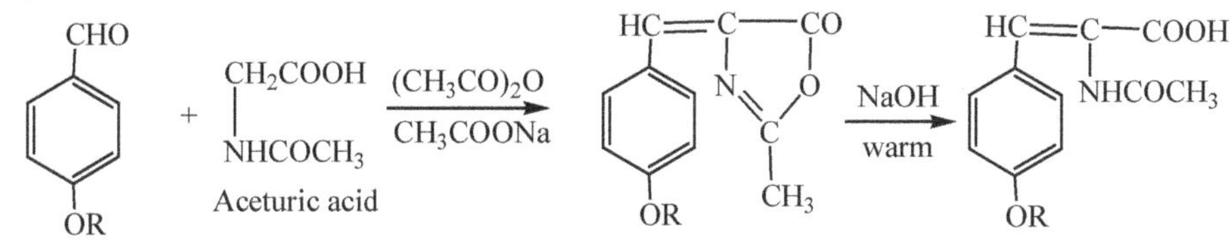

where R= protecting group

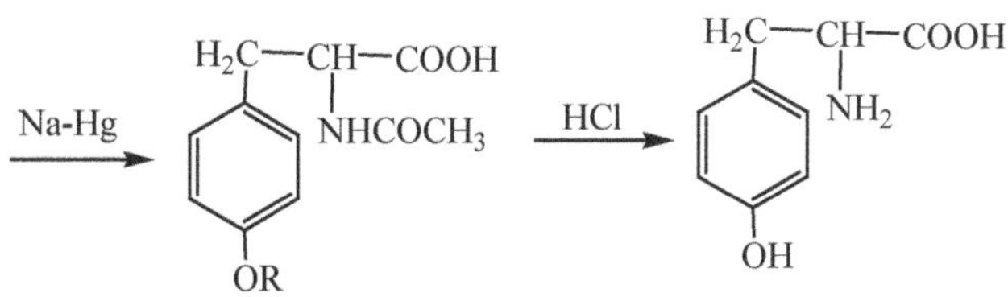

3.6.2 Hydantoin Synthesis

It is similar to Erlenmeyer azlactone method. But in this method, aromatic aldehyde is heated with hydantoin instead of hippuric or aceturic acid.

Tryptophan

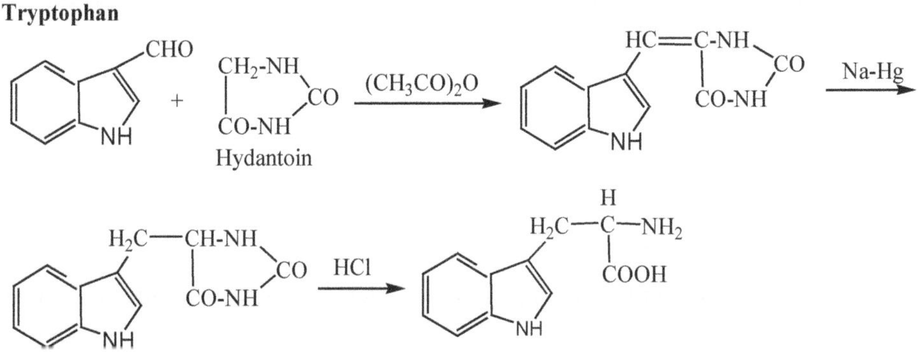

This method is improved by using acetylthiohydantoin instead of hydantoin to get better yield of amino acids.

3.6.3 Bucherer Hydantoin Synthesis

It is a modified hydantoin method. In this, carbonyl compound is converted into substituted hydantoin using sodium cyanide and ammonium carbonate then on hydrolysis to get amino acid.

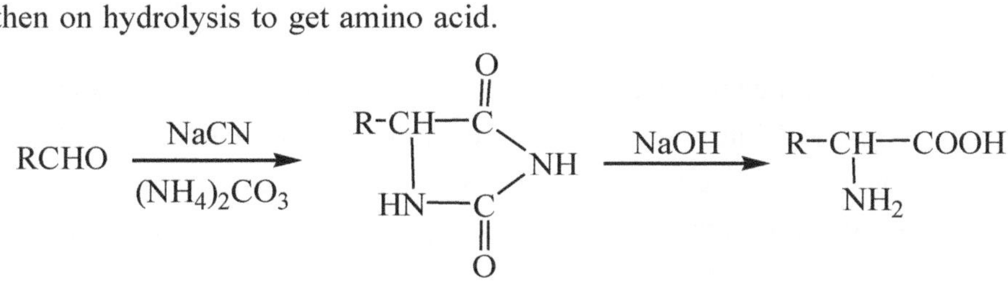

Example is

Methionine

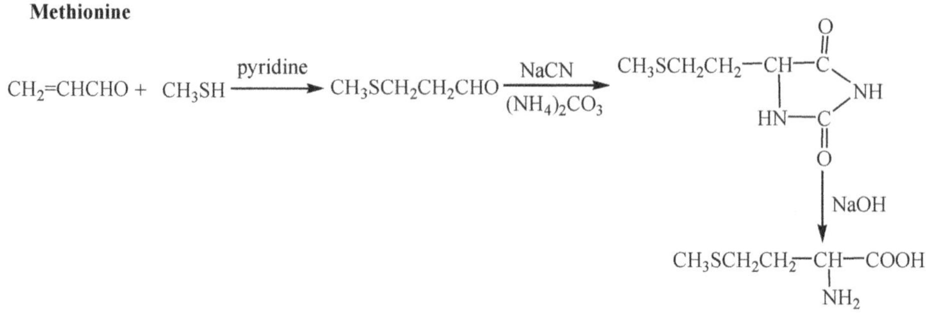

3.6 Physical Properties of Amino Acids

1. They are colorless, crystalline solids having high melting point.
2. They are polar thus soluble in water but insoluble in organic solvent.
3. They do not possess properties that are characteristics of simple amines or simple carboxylic acids but more like those of salts. The reason for this anomalous behavior is that they contain acidic carboxylic group and basic amino group within the same molecule. In aqueous solution, acidic carboxyl group loses proton which is taken up by basic amino group to form a dipolar ion known as zwitterion or inner salt or amphotate. Because of the formation of this inner salt they are freely soluble in water but insoluble in organic solvent.

This Zwitterion is amphoteric in nature, hence reacts with acid as well as with base. In acidic medium it exists as cation whereas in basic medium it exists as anion.

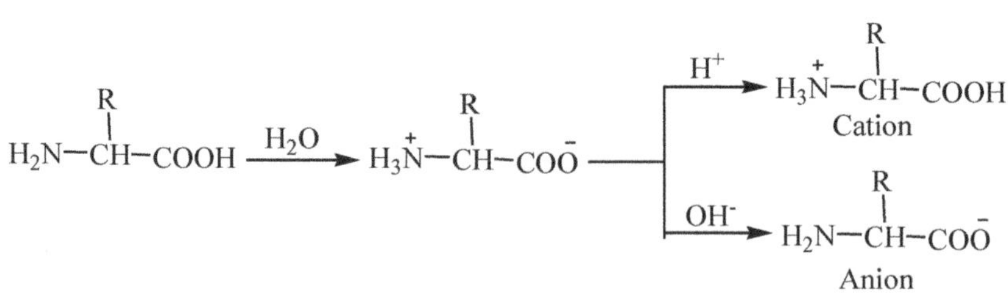

These ions on passing the electric current, move towards respective electrodes. Positively charged ion (cation) moves towards cathode whereas negatively charged ion (anion) moves towards anode. This process of separation of amino acid on passing the electric current is known as electrophoresis.

3.7 Electrophoresis

The term Electrophoresis means motion of charged particles under the influence of uniform electric field. It is based on isoelectric point of amino acid, size and their binding affinity. The technique is mainly applied to separate and analyze biomolecules such as DNA, RNA, proteins, etc.

3.7.1 Types of Electrophoresis

Electrophoresis is of two types depending upon the presence or absence of supporting media. i.e. free flow electrophoresis and zone electrophoresis.

1. **Free Flow Electrophoresis:** It is carried out without solid support material like gel. Free electrolyte is used in this technique. Nowadays, it has become out- date technique. It is mostly used in non-biological experiments. There are two types of free electrophoresis.
   - **Micro-electrophoresis:** It is a continuous separation technique in which sample is continuously flowed through a perpendicularly applied electric field. Zeta potential can be calculated by this method. Zeta potential is a colloidal property in a liquid medium.
   - **Moving Boundary Electrophoresis:** This is one of the oldest methods developed by Arne Tiselius. It uses a U-shaped cell filled with buffer solution containing electrodes at its ends. It has been used for quantitative analysis of complex mixtures of macromolecules, especially proteins.
2. **Zone Electrophoresis:** It is the widely used technique. In this method, separation is carried out on a stabilizing media. There are 4 types of zone electrophoresis.
   (a) Paper electrophoresis
   (b) Cellulose acetate electrophoresis
   (c) Capillary electrophoresis
   (d) Gel electrophoresis: Examples are SDS page, agarose gel electrophoresis etc.

3.7.2 Paper Electrophoresis

Chromatography paper generally called Whatman paper no. 1 having adsorption capacity is used as supporting medium in this process. In this, vertical or circular paper strip is taken and mixture of charged molecules or amino acid is placed at the centre of the paper at a particular pH. pH is maintained by saturating a strip with buffer. Amino acids or charged molecules migrate towards respective electrode on applying high voltage. Charged molecules appear as spots or concentric rings. After electrophoresis separated molecules can be identified by various techniques like staining techniques

Applications:

1. Analysis of serum for diagnostic purpose.
2. Analysis of muscle protein (myosin), egg albumin, casein of milk, insect and snake venom.

Disadvantage:

It takes very long time for completion generally needs 14-16 hours.

3.7.3 Cellulose Acetate Electrophoresis

Kohn developed this technique in 1958. In this technique, bacteriological acetate membrane filters are used in place of chromatography paper.

Advantage of This Method Over Paper Electrophoresis:

- Cellulose acetate (CA) strips used for this method are free from lignin and hemicellulose which act as barrier in free movement of bulkier molecules.
- It also has low glucose content thus best for electrophoresis of polysaccharides.
- CA strips are not hydrophilic in nature which is good for better resolution.

Applications:

It mainly finds application for clinical purposes such as separation of glycoproteins, hemoglobin from blood etc.

3.7.4 Capillary Electrophoresis

A narrow hollow tube called capillary tube is used for separation in this method. It is based on sample size : charge ratio. It is a new technique as compared to agarose gel electrophoresis or SDS -PAGE. We can separate charged as well as non-charged molecules by this technique, hence more advantageous over other techniques. In capillary method, mixture of ions is carried out in electrolytic solution present in a narrow fused silica capillary. Both ends of the capillary are then put into vials (inlet and outlet) filled with electrolytic solution containing electrodes attached to high voltage supply. Analyte is then introduced in the capillary by hydrodynamic injection or electro kinetic injection.

When voltage of 5-30 kV is applied across the capillary, zones of components are formed due to their different mobility. These zones then migrate towards outlet side of the capillary. We can detect separated zones directly through the capillary wall.

Advantages

- It has high separation efficiency
- Analysis time is short
- Low amount of sample and electrolyte is required
- Generation of waste is minimal
- It is much easy technique

Disadvantages

It has poor resolution because of the small diameter of capillary tube as heat is dissipated and causes diffusion.

Applications

It finds application in food analysis & analysis of pharmaceutical products and environmental pollutants.

3.7.5 Gel Electrophoresis

It is the most commonly used laboratory technique to separate charged molecules like DNA according to size. Gel is used as supporting media in this method. On applying an electric current across gel, one end of it becomes positively charged and other end acquires negative charge. Sometimes, it is used for purifying the molecules prior to other methods such as mass spectrometry, PCR, DNA sequencing, cloning etc.

Principle

The electromotive force i.e. EMF generated across the electrodes pushes the molecules to move through gel which is a permeable matrix just like a sieve.

Positively charged molecules moves towards cathode and negatively charged molecules moves towards anode.

Types of Gel Electrophoresis: There are two types of gel electrophoresis

(a) Agarose Gel Electrophoresis

This method is used to separate a mixture of macromolecules such as DNA, RNA and proteins. Agarose gel one of the two main components of agar is used in this technique.

Principle

On applying the electric current across the electrodes of an electrophoretic tank containing agarose gel and sample loaded on gel, biomolecules get separated on the basis of their molecular size and start moving towards respective electrodes. In this, agarose gel acts as a sieve. Bigger molecules stay above sieve and smaller molecules move faster towards respective electrodes.

Requirements

Supporting Media: Agarose is used as supporting media. It is a polysaccharide isolated from sea weed. It is a linear polymer made of alternate units of D- and L-galactose. It melts at 90°C and forms gel approximately at 40°C (called gelation). During gelation mesh of channels are formed varying from 50-200 nm in diameter. The diameter of these channels determines the porosity of the gel, thus acts as sieve. DNA of different sizes can be resolved using different concentration of gel. Higher concentration of gel enables separation of small DNA molecules while lower concentration is used for separation of bigger DNA molecules.

Buffer: Buffer is used to maintain the pH during separation. Hence, gel is dipped into buffer and it is necessary to convert molecules into charged species.

For example,

(i) **Tris Acetate EDTA Commonly Called TAE Buffer** is generally used in agarose electrophoresis.

Advantage

- It is best for resolution of bigger DNA molecules.
- It requires low voltage

Disadvantage

- It takes more time for completion.

(ii) Tris Borate EDTA Commonly Called TBE: It used for smaller DNA molecules.

(iii) Sodium Borate Commonly Called SB

Advantage

- low conductivity thus allows higher voltage
- requires less time

Disadvantage

- Poor resolution

Electrophoresis Buffer: It is used to bathe gel. Examples are Tris acetate EDTA, Tris borate EDTA

Loading Buffer: Dense medium is required to allow the sample to fall into sample wells. So, glycerol is used for this purpose along with one or two dyes are added to allow monitoring of the extent of electrophoresis visually.

Dye: Dyes are used as a tracking or monitoring agent. It helps in visual monitoring of separated molecules after the process of electrophoresis is over. For example, Ethidium bromide is added as it is a fluorescent dye that intercalates between the stacked bases of DNA. On irradiation with UV light (302 nm wavelength), it emits orange light of wavelength 590 nm.

Procedure

1. At first agarose powder is mixed with gelatin (electrophoretic buffer) to the desired concentration. It is then heated in a microwave till it melts completely. Ethidium bromide dye is also added to the gel to enable visualization of DNA fragment after electrophoresis.
2. Cool it to approx. 60°C, then pour it into casting tray containing a sample comb. Now allow it to get solidified at room temperature. Remove the comb after the gel is solidified with utmost care without tearing the bottom of the wells.
3. Insert the gel into electrophoresis chamber horizontally. Now add buffer and load the sample into sample wells. Pass the current through it.
4. Biomolecules migrate towards the positive electrode, which are usually red colored. After the adequate migration of biomolecules, stain it with ethidium bromide for visualization of the sample.

5. Fluorescent dye is intercalated between bases of DNA during electrophoresis. But gel can also be stained after electrophoresis by soaking it in dilute solution of dye.

For visualization of biomolecules like DNA, RNA gel is kept on ultraviolet trans-illuminator. It is noted that DNA will diffuse within the gel over time, so take photograph after ending of electrophoresis.

Applications

1. Molecular cloning is the most important application of agarose gel electrophoresis. It creates various organism with genetic modification. The purpose of this modification is very specific like production of insulin in pharmaceutical industries.
2. It is a powerful technique to identify the individual on the basis of genetic code. This technique is known as fingerprinting frequently used in criminal investigation.
3. It also finds application for diagnostic purpose such as detection of gene abnormality, hence find helps in medical treatments.
4. It is also used for PCR products analysis after polymerase chain reaction.
5. Quantitative and qualitative analysis of DNA.

Advantages

- It has advantage of that gel can easily be poured.
- It does not cause denaturing of samples.
- Recovery of sample

Disadvantages

- Melting of gel during electrophoresis process.
- Buffer gets exhausted, therefore different forms of genetic material may run in unpredictable manner.

(b) SDS-Page

Sodium Dodecyl Sulphate (SDS) is also called polyacrylamide gel electrophoresis. It is an analytical technique used to separate proteins on the basis of their molecular weight. It is considered to be the most powerful techniques.

Principle

In SDS-PAGE, use of Sodium Dodecyl Sulphate (SDS) also known as sodium lauryl sulphate and polyacrylamide gel greatly removes the influence of the structure and charge. Thus, proteins are separated solely on the basis of length of polypeptide chain. It is an anionic detergent having strong denaturing effect. SDS in the presence of reducing agent cleaves the disulfide bond that is essential for tertiary structure of protein. As a result, proteins unfold into linear polypeptide chains with a uniform negative charge all over it. These denatured proteins are then loaded from cathode end of the electrophoretic tank having polyacrylamide gel. Clear bands of proteins are observed in decreasing order of their molecular mass from cathode to anode on applying the electric field. The rate of the movement of molecules depends upon pore size of gel and strength of applied electric field. DNA and RNA molecules can also be separated by this technique.

Requirements

Electrophoretic Tank: It may be vertical or horizontal tank fitted with electrodes, gel cassettes, Teflon spacer pipettes comb.

Power Supply: Ideal power supply of 100-200 volts is needed for running and transferring protein resolving gels.

Staining Box: These are generally plastic trays for staining the gel. These trays are resistant to most organic dyes, silver and other stains.

Supporting Media: White crystalline powder of acrylamide is used as supporting medium. It undergoes polymerization when dissolve in water. Polymer looks similar to net like structure called polyacrylamide which can readily be cross linked. It is of two types.

Resolving Gel: It is also called separating gel or running gel. It is prepared by 5-10% of acrylamide and constitutes 2/3 of the length of gel plate. There are numerous pores in this gel. They are smaller in size hence impacts the sieving property of the gel.

Stacking Gel: It constitutes $1/3^{rd}$ of the gel plate and is the upper layer. It is poured on the top of resolving gel and gel comb is inserted to form well. The percentage of the acrylamide depends upon the size of the protein sample.

Buffer: There are 2 types of buffer used in SDS-PAGE. The lower reservoir has amine buffer which is adjusted by HCl. The upper reservoir also contains amine buffer but its pH is bit higher than lower and is adjusted by glycine.

Dissociating Agent: Anionic detergent SDS is used as dissociating agent. It denatures the protein into individual polypeptide chain. It wraps around the polypeptide chain on heating with protein at 100°C.

Stains: These are used for visualization of bands of separated proteins after electrophoresis process is over. Coomassie Brilliant Blue R-250 (CBB) is the most commonly used stain for proteins. It is a anionic dye that is attached with protein non-specifically. Proteins are appeared as blue bands on a clear background.

Procedure

1. At first, protein sample is mixed with SDS to denature secondary structure of protein. Sometimes protein sample is boiled with SDS in presence of reducing agent such as Di-thio-threitol (DTT) or 2-mercaproethanol which further denatures the proteins by reducing/breaking disulfide linkages, thus overcoming some forms of tertiary structure and breaking of quaternary structure of protein. This is called as reducing SDS-PAGE.
2. Sometimes, non-reducing SDS-PAGE is used when native structure is required for further analysis. These denature proteins are then loaded into the wells of stacking gel containing stacking buffer.
3. Then connect this end with cathode of power supply. On applying the electric current across the gel pushes the negatively charged protein molecules to move towards anode. After crossing the stacking gel, these denatured proteins then enter the running gel which has its own buffer system.
4. Mobility of different proteins is different depending upon their size. Smaller proteins move faster whereas larger proteins move slower.
5. When separation is over, gel is stained with staining dye (CBBR-250) in the staining box. Excess of stains can be removed by de-staining agent like acetic acid. Blue bands are observed and can be analysed.

Application

1. It is used for determining the protein size.
2. It is also used for protein identification and for determining the sample purity.
3. It helps in identification of disulfide bonds.
4. Other applications include quantification of proteins and blotting purposes.

Advantages

1. It is a fast process because of high mobility of molecules.
2. All proteins carry negative charge thus migrate towards anode.

3. The proteins treated with SDS fixed dyes are better than the native proteins.
4. Hydrophobic as well as denatured proteins are soluble in SDS.
5. It is an easy, inexpensive and accurate method.

3.8 Separation of Amino Acids by Paper Electrophoresis

This method is based on isoelectric point of amino acids and pH control. As we all know that amino acids differ in their isoelectric point thus differ in their electrical properties hence becomes the basis for their separation. In this method, mixture of amino acids is placed at the centre of paper at particular pH. The pH of the paper is maintained by dipping the paper in buffer solution of suitable pH. Then the paper is connected to two different electrodes and electric current is passed through it. On passing the current, amino acids start migrating towards respective electrode depending upon the net charge on them. Amino acid having isoelectric point higher than buffer pH will gain proton becomes positively charged and moves towards cathode. Similarly, amino acid having isoelectric point less than buffer pH will lose proton becomes negatively charged and moves towards anode. On the other hand, amino acid whose isoelectric point is comparable to buffer pH will neither gain nor lose proton thus does not move on passing the electric current.

3.9 Separation of Amino Acids by Paper Chromatography

Without electric current: We can also separate amino acid mixture by paper chromatography. In this, mixture of amino acids is placed on a Whatman paper (no. 1) strip (vertical/circular) and then paper is kept in a gas jar or petridish containing solvent system i.e. n-butanol: acetic acid: water (BAW) in a particular ratio. Water acts as stationary phase through which nonpolar solvent rises up. When the chromatogram is developed paper is taken out and dried. It is then sprayed with visualizing agent (ninhydrin). This method is based on solubility difference of amino acids in stationary phase and mobile phase. For e.g. amino acid having polar side chain will be more soluble in stationary (water) phase so travel slowly. Whereas amino acid having non polar side chain will be more soluble in non-polar mobile phase, so travel faster. Hence, separated out as spots/concentric rings.

3.10 Isoelectric Point

The pH at which amino acid exists as zwitterion is known as isoelectric point of that particular amino acid. At isoelectric point amino acid is electrically neutral and does not carry any charge. Thus, does not move at all on passing the electric current. At isoelectric point concentration of zwitterion is maximum.

Calculation of Isoelectric Point of Amino Acids

As we know that isoelectric point of amino acid is the pH at which amino acid exist as zwitterion. In neutral solution amino acid exists in equilibrium with three species.

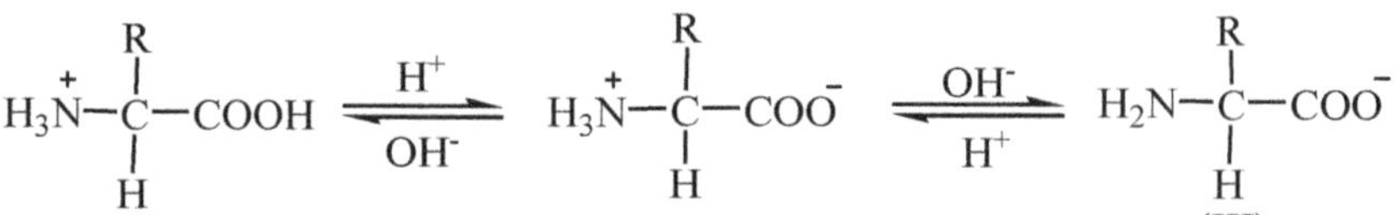

(I) Conjugate Acid (CA) — (II) Dipolar Ion (DI) — (III) Conjugate Base (CB)

The position of this equilibrium depends upon the pH of the solution. In acidic medium conjugate acid will predominate whereas in basic medium conjugate base will predominate.

At isoelectric point amino acid acts monoamino monocarboxylic acid and thus has two pKa values. pKa_1 is for carboxyl group whereas pKa_2 is for amino group (sign convention).

In the above equilibrium reaction two reactions are going on. In the first one when carboxylic acid loses proton to form dipolar ion and in the second one when amino (in ionic form) loses proton.

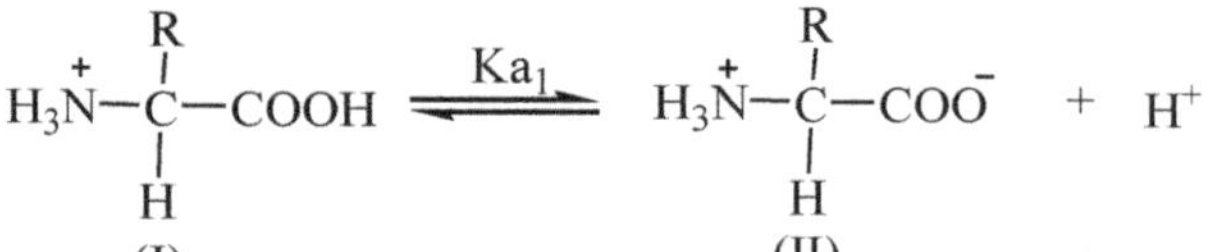

(I) Conjugate Acid (CA) — (II) Dipolar Ion (DI)

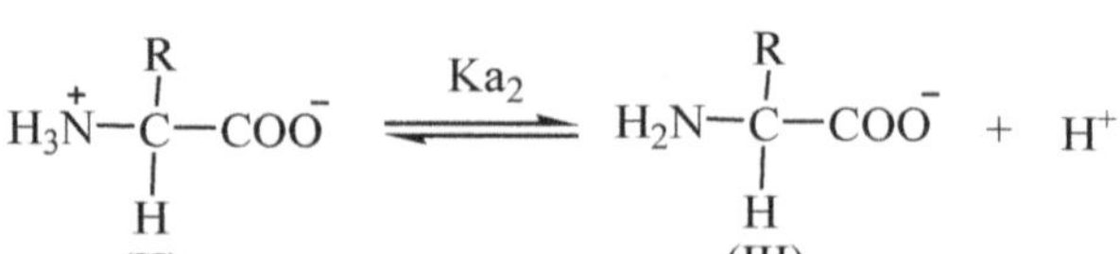

(II) Dipolar Ion (DI) — (III) Conjugate Base (CB)

Ka_1 and Ka_2 are the equilibrium constants.

$$Ka_1 = \frac{[D.I]\,[H^+]}{[CA]} \quad \text{......eq. 1} \qquad Ka_2 = \frac{[CB][H^+]}{[D.I]} \quad \text{......eq. 2}$$

at isoelectic point, [CA] = [CB]................eq. 3

from eq. 1, 2 and 3

$$\frac{\cancel{[D.I]}\,[H^+]}{Ka_1} = \frac{Ka_2\,\cancel{[D.I]}}{[H^+]}$$

$$Ka_1 \times Ka_2 = [H^+]^2$$

taking logrithum both sides

$$\log Ka_1 + \log Ka_2 = 2 \log [H^+]$$

$$pKa_1 + pKa_2 = 2\ pHi$$

$$pHi \text{ or } PI = \frac{pKa_1 + pKa_2}{2}$$

PI= isoelectric point of amino acid

For Example

1. Calculation of Isoelectric Point of Glycine

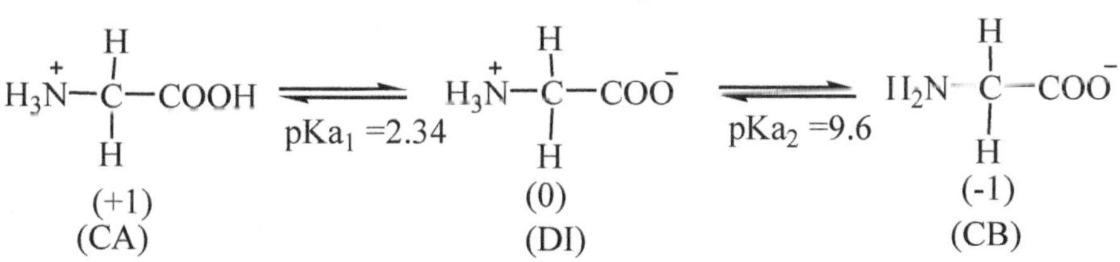

$$PI = \frac{pKa_1 + pKa_2}{2}$$

$$= \frac{2.34 + 9.6}{2}$$

$$PI = 5.97$$

2. Calculation of Isoelectric Point of Aspartic Acid

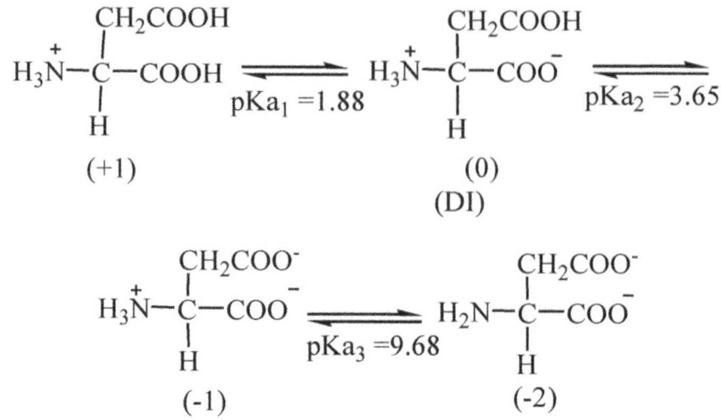

$$PI = \frac{pKa_1 + pKa_2}{2}$$

$$= \frac{1.88 + 3.65}{2}$$

$$PI = 2.77$$

3. Calculation of Isoelectric Point of Lysine

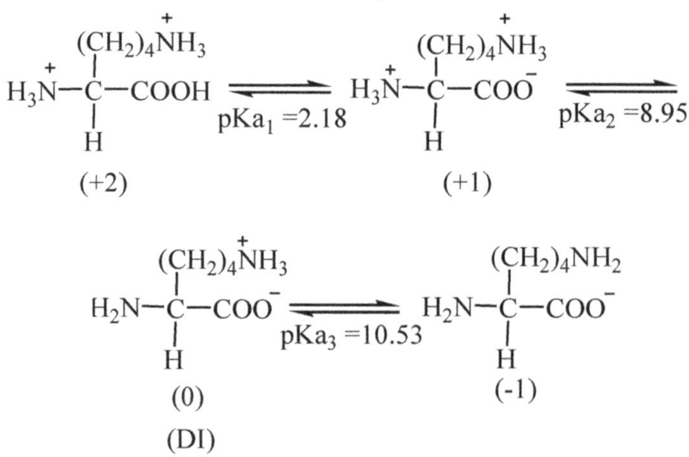

$$PI = \frac{pKa_2 + pKa_3}{2}$$

$$= \frac{8.95 + 10.53}{2}$$

$$PI = 9.74$$

Table 3.1: pKa values and isoelectric point of various amino acids

Amino acid	pKa1	pKa2	pKa3	PI
Glycine	2.34	9.60		5.97
Alanine	2.34	9.69		6.00
Valine	2.32	9.62		5.96
Leucine	2.36	9.60		5.98
Isoleucine	2.36	9.60		6.02
Phenylalanine	1.83	9.13		5.48
Tyrosine	2.20	9.11		5.66
Tryptophan	2.83	9.39		5.89
Serine	2.21	9.15		5.68
Threonine	2.09	9.10		5.60
Proline	1.99	10.60		6.30
Cysteine	1.96	8.18		5.07
Methionine	2.28	9.21		5.74
Aspartic acid	1.88	3.65	9.60	2.77
Glutamic acid	2.19	4.25	9.67	3.22
Lysine	2.18	8.95	10.53	9.74
Arginine	2.17	9.04	12.48	10.76
Histidine	1.82	6.00	9.17	7.59
Asparagine	2.02	8.80		5.41
Glutamine	2.17	9.13		5.65

3.11 Chemical Properties of Amino Acids

Amino acids contain amine group and carboxyl group. Therefore, they undergo three types of chemical reactions involving amine group, carboxylic group and both amine & carboxylic.

I. Reaction Involving Amino Group

(i) Reaction with Mineral Acids: Amino acids react with mineral acids like HCl, H_2SO_4 and get converted into corresponding salt. This reaction is called neutralization reaction.

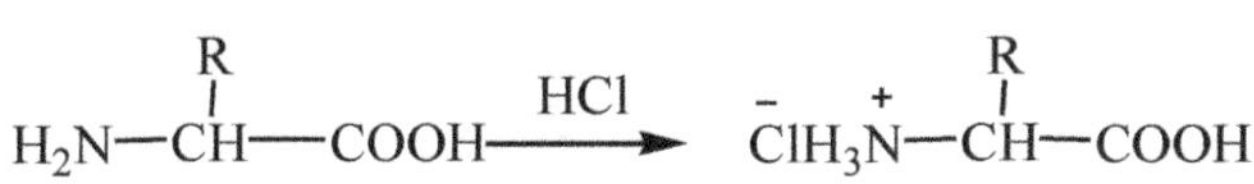

amino acid hydrochloride

(ii) Reaction with Acid Halide: Amino acid reacts with acid halides to give N-substituted amide or N-acylamino acid. This reaction is called acylation.

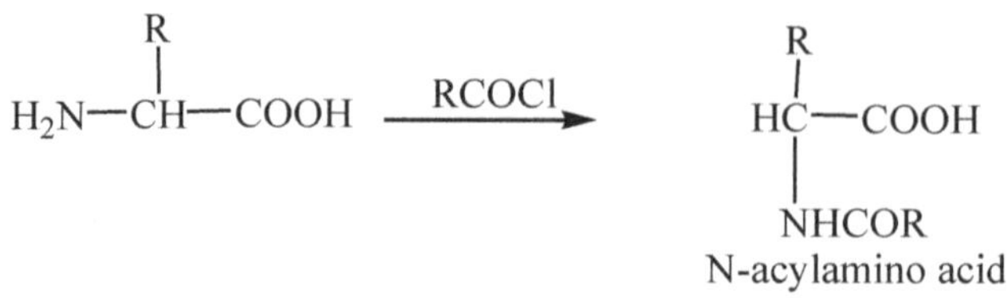

e.g.

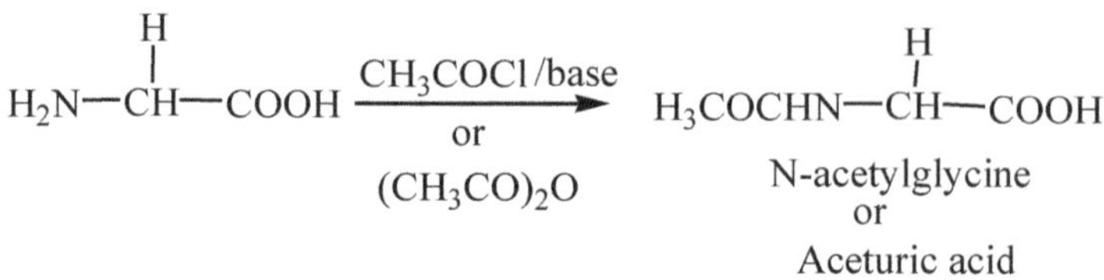

(iii) Reaction with Benzoyl Chloride: Amino acid reacts with benzoyl chloride to give N-substituted amide or N-benzoylamino acid. This reaction is called benzoylation.

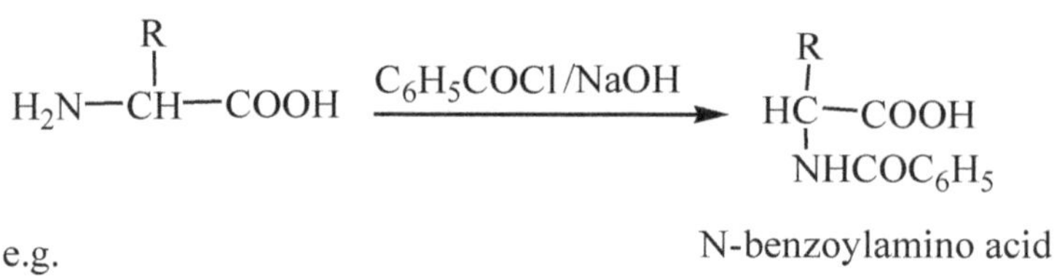

e.g.

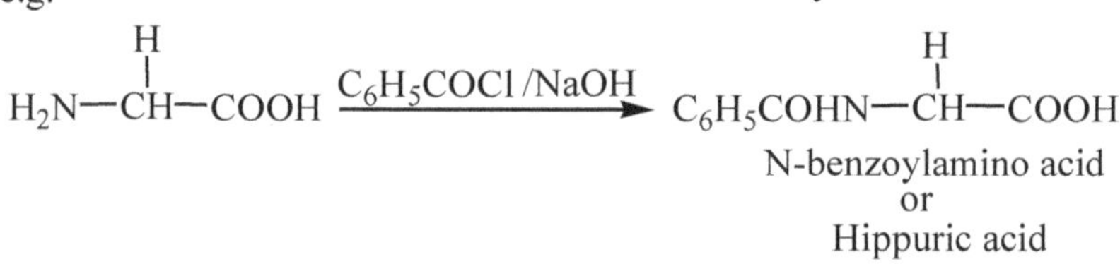

(iv) Reaction with Formaldehyde: Amino acids react with formaldehyde to form methylene amino acid.

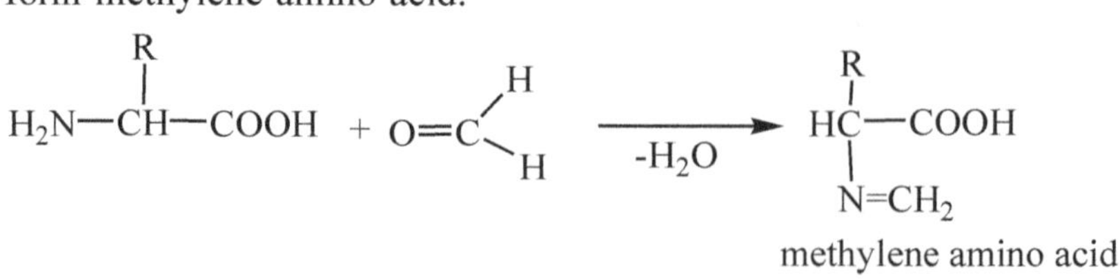

This reaction is used for estimation of amino acid. It is called Sorenson's formol titration method. In this method, amino acid cannot be titrated directly with alkali like NaOH because of the presence of free amino group which interferes in end point determination. Therefore, amino acid is treated with formaldehyde to convert it into methylene amino acid which is a strong acid thus can be titrated with alkali.

(v) Reaction with Nitrous Acid: Amino acids react with sodium nitrite in presence of hydrochloric acid to from aliphatic diazonium salt. This reaction is called diazotization. Aliphatic diazonium salt is unstable at room temperature hence, decomposed to form α-hydroxycarboxylic acid by liberating nitrogen gas.

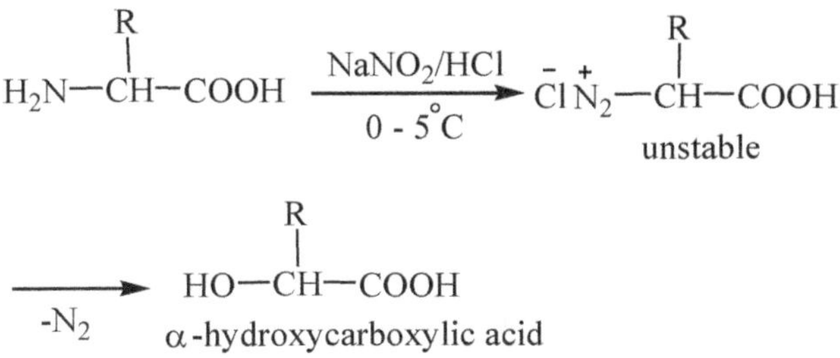

(vi) Oxidation: Amino acids react with alkaline solution of $KMnO_4$ or hydrogen peroxide and get oxidized to iminecarboxylic acid which on hydrolysis gives α-ketocarboxylic acid.

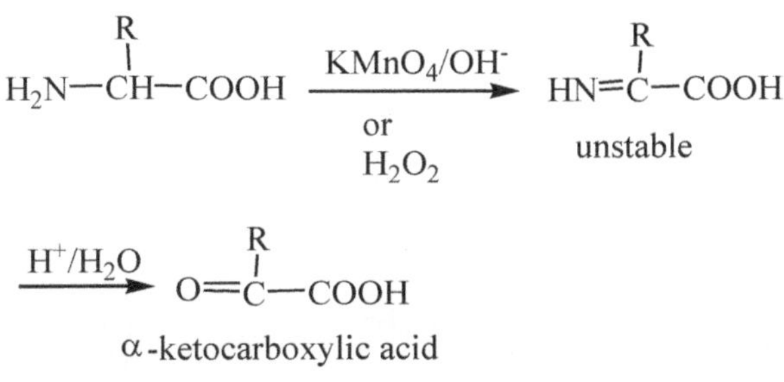

(vii) Reaction with HI: Amino acids on heating with hydriodic acid undergo deamination to form alkanoic acid.

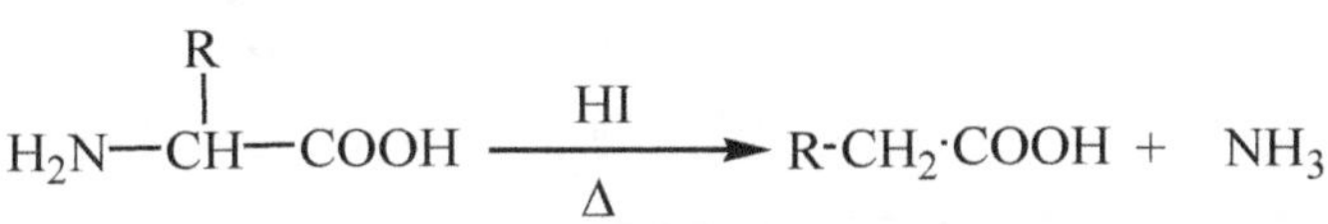

(viii) Reaction with Nitrosyl Chloride: Amino acids react with nitrosyl chloride in presence of light to form α-chloroalkanoic acid.

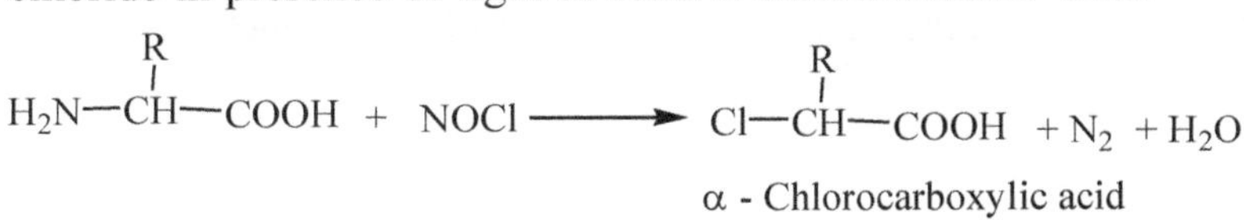

II. Reactions Involving Carboxylic Group

(i) Esterification and Amide Formation: Amino acids react with alcohol in presence of acid and get converted into salt of corresponding ester. Ester is liberated by treating with mild base. Ester when treated with ammonia it forms amino acid amide.

$$H_2N-\underset{}{\overset{R}{\overset{|}{C}}}H-COOH + R'OH \xrightarrow[-H_2O]{HCl} \overset{-}{Cl}\,H_3\overset{+}{N}-\overset{R}{\overset{|}{C}}H-COOR'$$

$$\xrightarrow{dil.\ Na_2CO_3} H_2N-\overset{R}{\overset{|}{C}}H-COOR' + NH_3$$

amino ester

$$\downarrow$$

$$H_2N-\underset{H}{\overset{R}{\overset{|}{C}}}-CONH_2$$

amino acid amide

(ii) Activation: Amino acids are activated by grinding with PCl_5 or heating with thionyl chloride in presence of pyridine to form corresponding amino acid chloride.

$$H_2N-\overset{R}{\overset{|}{C}}H-COOH \xrightarrow{PCl_5} H_2N-\overset{R}{\overset{|}{C}}H-COCl$$

(iii) Reduction: Amino acids react with reducing agents like lithium aluminium hydride or sodium borohydride to form β-amino alcohol.

$$H_2N-\overset{R}{\overset{|}{C}}H-COOH \xrightarrow{LiAlH_4} H_2N-\overset{R}{\overset{|}{C}}H-CH_2OH$$

β-aminoalcohol

(iv) Decarboxylation: Amino acids on dry distillation or on heating with barium oxide undergo decarboxylation to form amines.

$$H_2N-\overset{R}{\overset{|}{C}}H-COOH \xrightarrow[\Delta]{BaO} H_2N-CH_2-R + CO_2$$

amine

III. Reaction Involving Both Amino and Carboxylic Acid

(i) Effect of Heat: α-amino acids on heating undergo self-condensation reaction to form 2,5-diketopiperazine derivative.

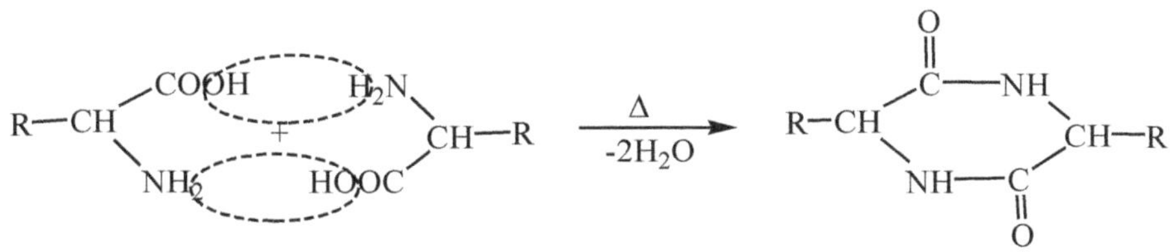

2,5-diketopiperazine

β-amino acid on heating loses ammonia to form α,β-unsaturated acid.

$$R{-}\underset{NH_2}{\underset{|}{CH}}{-}CH_2{-}COOH \xrightarrow[-NH_3]{\Delta} R{-}CH{=}CH{-}COOH$$

γ-amino acids on heating from lactam ring.

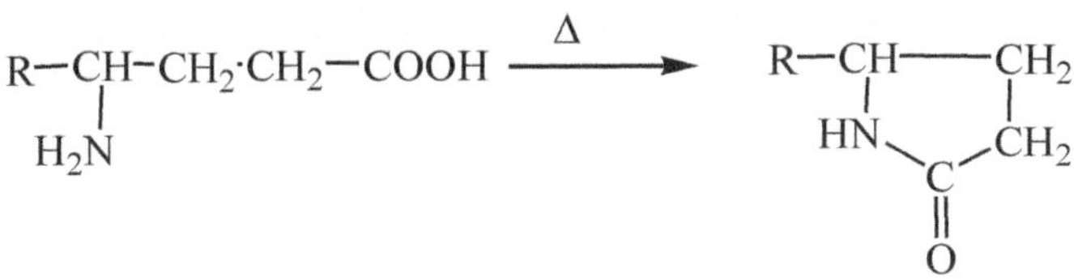

(ii) Ninhydrin Reaction: This reaction is used to analyse amino acids or proteins. It is called ninhydrin test. Ninhydrin reagent is prepared by dissolving indane-1,2,3-trione in alcohol or acetone. When amino acids are heated with ninhydrin reagent, they give bluish violet colored complex called Ruhemann's Purple complex except proline and hydroxyproline which give yellow color.

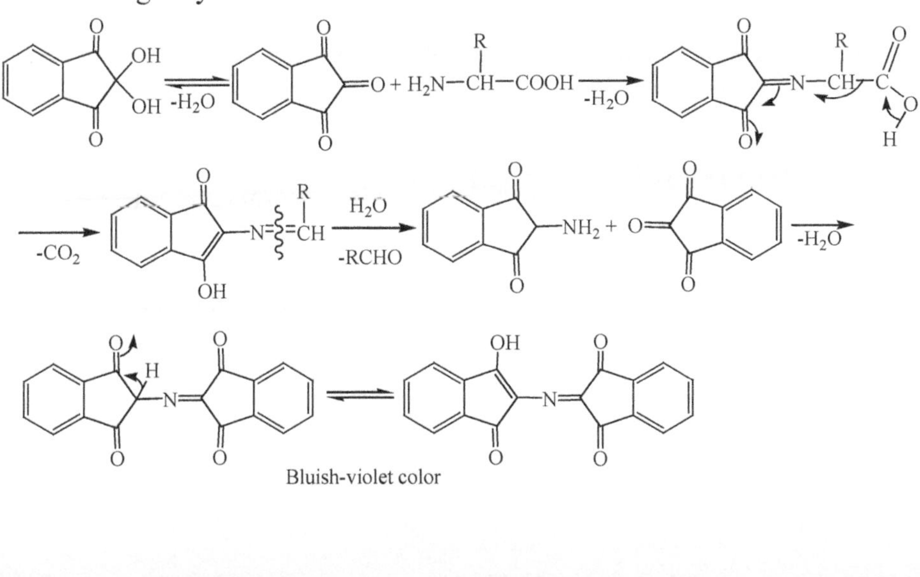

Bluish-violet color

(iii) Reaction with Isocyanate: Amino acids react with isocynate to form hydantoin derivative.

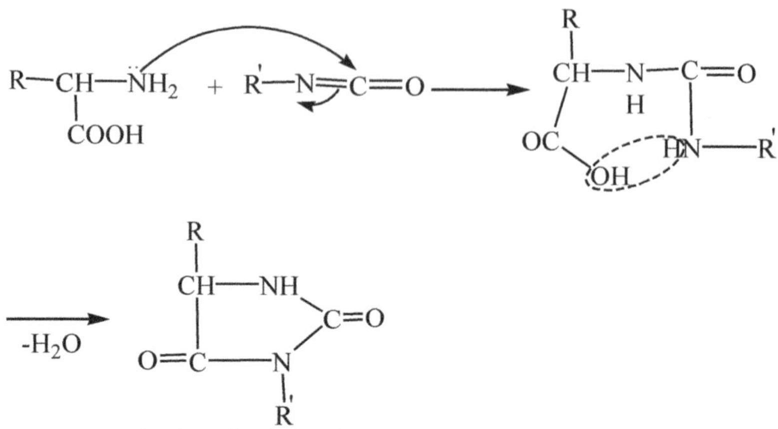

hydantoin derivative

(iv) Betaine Formation: Betaines are trialkyl derivatives of amino acids. They are prepared by treating amino acid with methyl iodide in presence of methanol. Betaine is a stable solid having basic nature. On reaction with HCl it forms betaine hydrochloride.

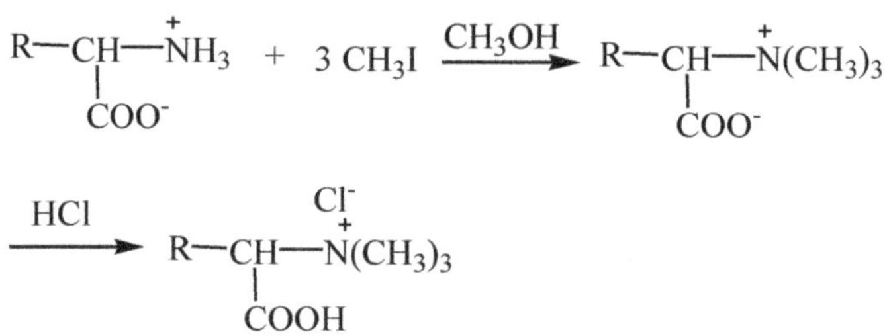

(v) Reaction with Metal: Amino acids form insoluble salt with heavy metal ions like Cu^{2+}, Ag^{+}, Hg^{2+}.

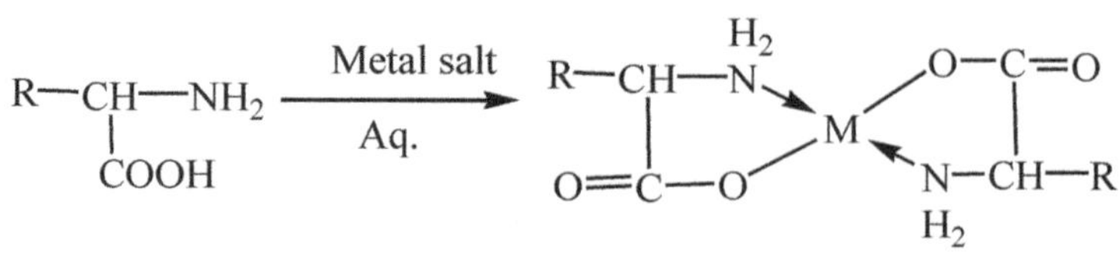

Insoluble complexes

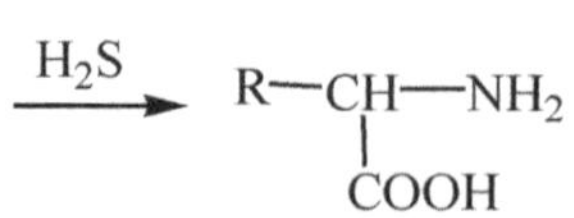

amino acid

This complex can be decomposed by passing hydrogen sulphide gas.

(vi) Dakin-West Reaction: Amino acids react with acetic anhydride in presence of pyridine forms methyl α-acetamido ketone. This reaction is called Dakin West reaction.

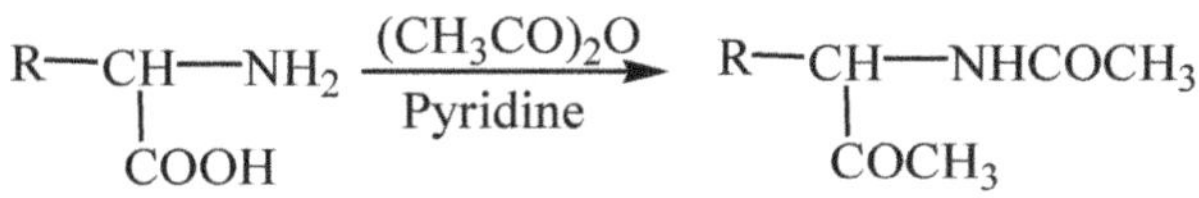

Methyl α- acetamido ketone

3.12 Peptides

Peptides are formed by joining together of α-amino acids. In peptide formation, amino group of one amino acid reacts with the carboxylic group of another amino acid. When two amino acids combine in this way the resulting product is called dipeptide.

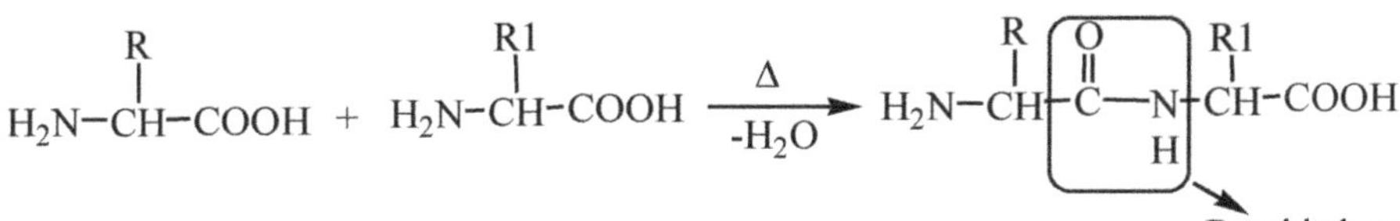

Peptide bond

Dipeptide

Similarly, combination of

3 amino acids = tripeptide

4 amino acids = tetrapeptide

5 amino acids = pentapeptide

6 amino acids = hexapeptide so on.......

n amino acids = polypeptides or proteins

So, proteins are polypeptides made up of at least 100 amino acids. Actually, there is a lot of variation found in protein structure inspite of containing the same number and kind of amino acids. The possibility of variation increases with increasing the molecular weight of protein. There are many ways in which different amino acids can combine together. When two amino acids are combined together, the resulting product is mixture of following two dipeptides. e.g.

Gly + Ala = Ala-Gly + Gly-Ala

In Ala-Gly, alanine is N-terminal and glycine is C-terminal residue. Whereas in Gly-Ala, glycine is N-terminal and alanine is C-terminal residue

Similarly when 3 amino acids combine together it gives a mixture of following six tripeptides.

Gly + Ala + Val = Gly-Ala-Val + Gly-Val-Ala + Ala-Gly-Val + Ala-Val-Gly + Val-Ala-Gly + Val-Gly-Ala

When we increase the molecular weight of the protein, this possibility of combination of amino acids also increases. That is the reason why there is a lots of variation found in proteins.

3.13 N-terminal Residue and C-terminal Residue

In a polypeptide chain amino acid containing free amino group is called N-terminal amino acid or residue and it is always written on the left hand side of polypeptide chain. Similarly, amino acid containing free carboxyl group is called C-terminal amino acid or residue and it is always written on the right hand side of polypeptide chain.

e.g. In a given polypeptide chain

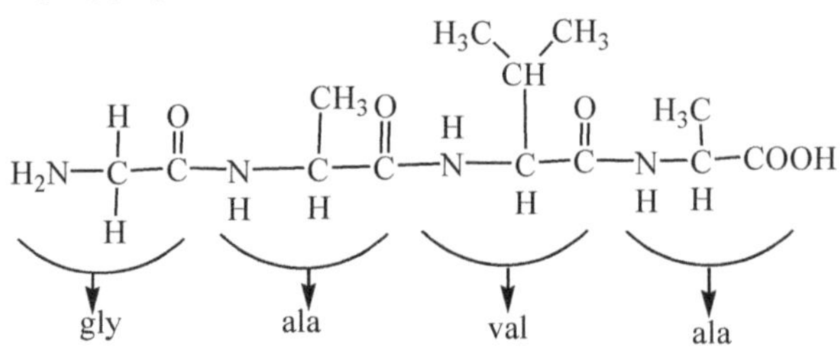

N-terminal residue = glycine ; C-terminal residue = alanine

3.14 Naming of Peptide

To name a peptide, write down a peptide chain with N-terminal amino acid on left hand side and C- terminal amino acid on right hand side. Replace -ine of all amino acids by -yl except C-terminal amino acid. e.g.

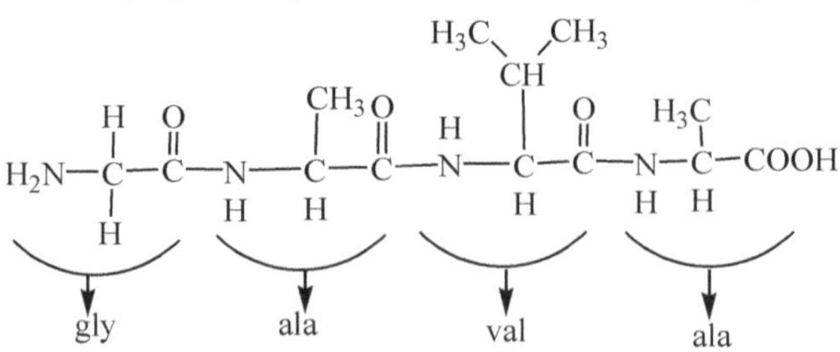

Glycylalanylvalylalanine

Or gly-ala-val-ala.

We can also use 3- letter abbreviations for each amino acid separated by hyphen to write the name of the peptide.

3.15 Structure Determination of Peptides or Proteins

Structure of protein can be determined by three simple steps which are as follows:

Step 1: Complete hydrolysis of peptide or protein

Step 2: End group of analysis of peptide or protein

Step 3: Partial hydrolysis of peptide or protein

3.16 Complete Hydrolysis of Protein or Peptide

In this, peptide is heated with mineral acid. On heating with dil. HCl, all peptide bonds are broken down and resulting mixture of α- amino acids can be separated by chromatography. This step gives information about the number and kind of amino acids present in a given peptide but no information about sequence of amino acids.

3.17 End Group of Analysis of Peptide or Protein

As we all know that there are two ends in a peptide i.e. N-terminal and C-terminal (Table 3.2).

Table 3.2: Methods for end group analysis

N-terminal Analysis	C-terminal Analysis
Edman's degradation method	Enzymatic method
Sanger's method	Thiohydantoin method
Dansyl method	Hydrazinolysis

3.18 N-terminal Analysis

Edman's Degradation Method: It is the best method to identify the N-terminal amino acid. In this method, peptide is treated with phenyl isothiocyanate in the presence of dil. alkali like NaOH and it gets converted into phenyl thiocarbamoyl derivative of protein (PTC-peptide). When PTC-peptide derivative is treated with dil. acid like HCl, CCl_3COOH, it is cleaved down to form phenyl thiohydantoin (PTH-peptide) derivative of N-terminal amino acid. The reaction takes place as follows (Figure 3.1).

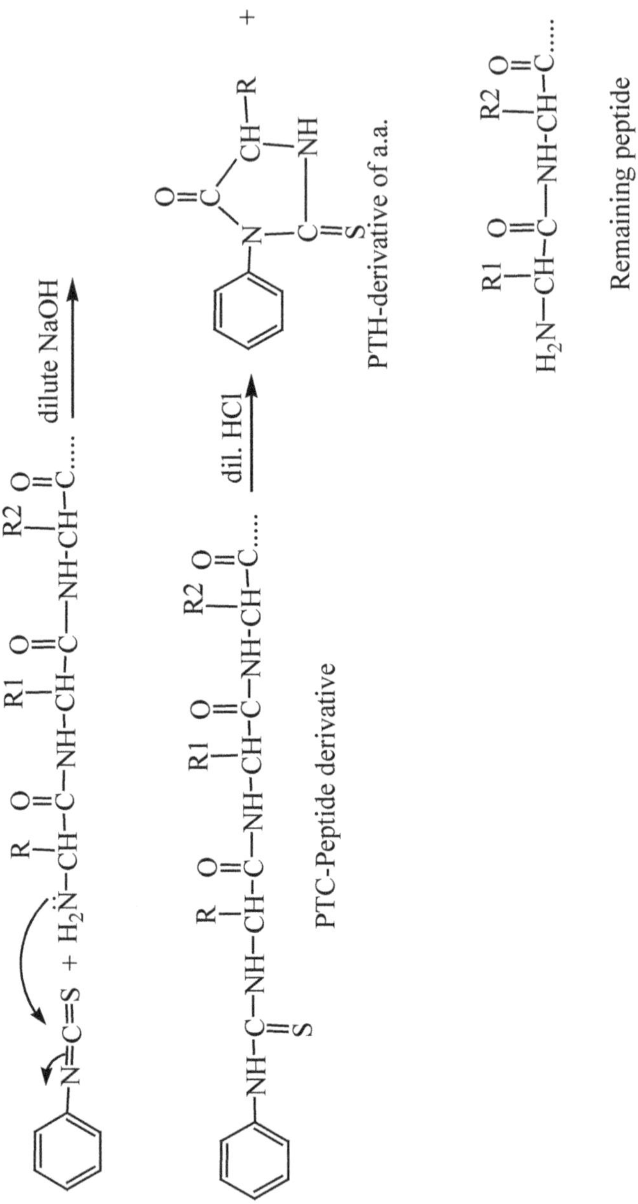

Figure 3.1: Edman's degradation method

Separate the PTH derivative of amino acid by suitable technique like chromatography and identify it. Repeat the same procedure on the remaining peptide to find out the sequence of amino acid. So, this method is not only used to find out the N-terminal amino acid but we can also determine the sequence of amino acids in a particular peptide or protein.

Sanger's Method: This method was introduced by Sanger in 1945. In this, peptide is treated with 2,4-Dinitrofluorobenzene (DNFB) or 1-fluro-2,4-dinitrobenzene (FDNB) in presence of mild alkali like $NaHCO_3$ and it gets converted into 2,4-DNP derivative of peptide. Hence, it is also called DNP method. It is then treated with dil. HCl to give DNP derivative of N-terminal amino acid and mixture of amino acids. Separate the DNP derivative of amino acid and identify it (Figure 3.2).

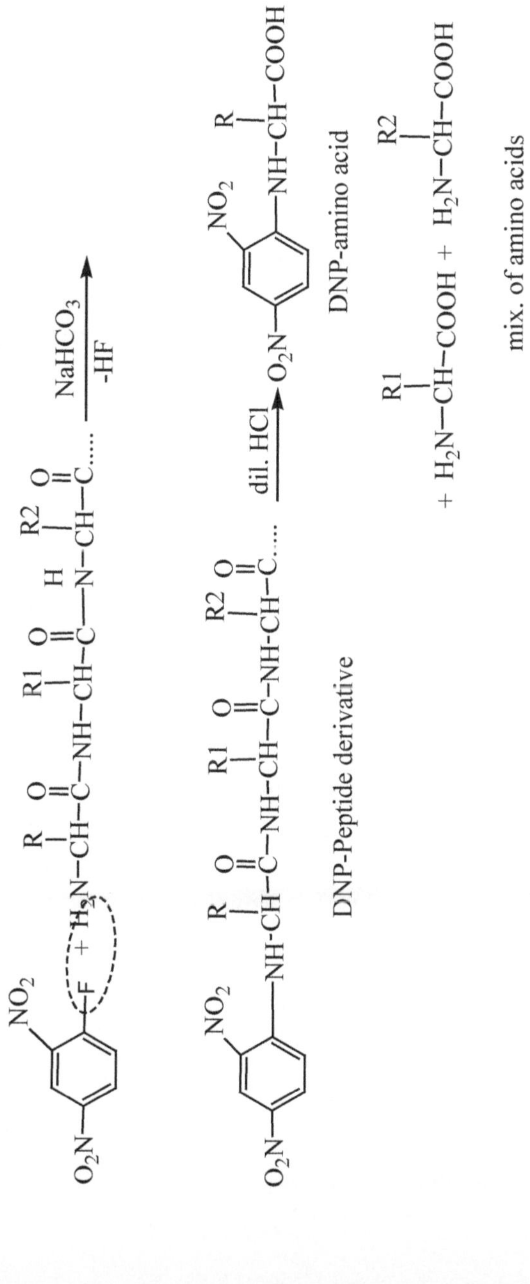

Figure 3.2: Sanger's method

Drawbacks of DNP Method:

1. This method is only used for determination of N-terminal amino acid not for determining the sequence of amino acids in a particular peptide due to the formation of a mixture of amino acids after the first procedure. Hence can't be repeated.
2. DNP reagent reacts with all the free amine groups present in the peptide. e.g. in lysine amine group is also present in the side chain that reacts with DNP reagent even if it is not present as N-terminal.
3. It also reacts with –OH group of tyrosine and –SH group of cysteine.

This problem can be solved in following way. If lysine forms mono-DNP derivative then it is not present as N-terminal. If it forms di-DNP derivative then it is present as N-terminal. Tyrosine and cysteine can also be distinguished in this way.

Dansyl Method: It is the most widely used method, also called modified DNP method. In this, peptide is treated with 5-dimethylamino-1-naphthalenesulphonyl chloride (commonly called Dansyl Chloride or DNS-Cl) in the presence of mild alkali and it gets converted into Dansyl derivative of peptide. It is then treated with dil. acid to give Dansyl derivative of N-terminal amino acid and a mixture of amino acids. Separate the DNP derivative of amino acid and identify it. Dansyl group is highly florescent in nature and therefore, can easily be detected by UV (Fluorimetric method). Even a minute quantity can be detected in few seconds.

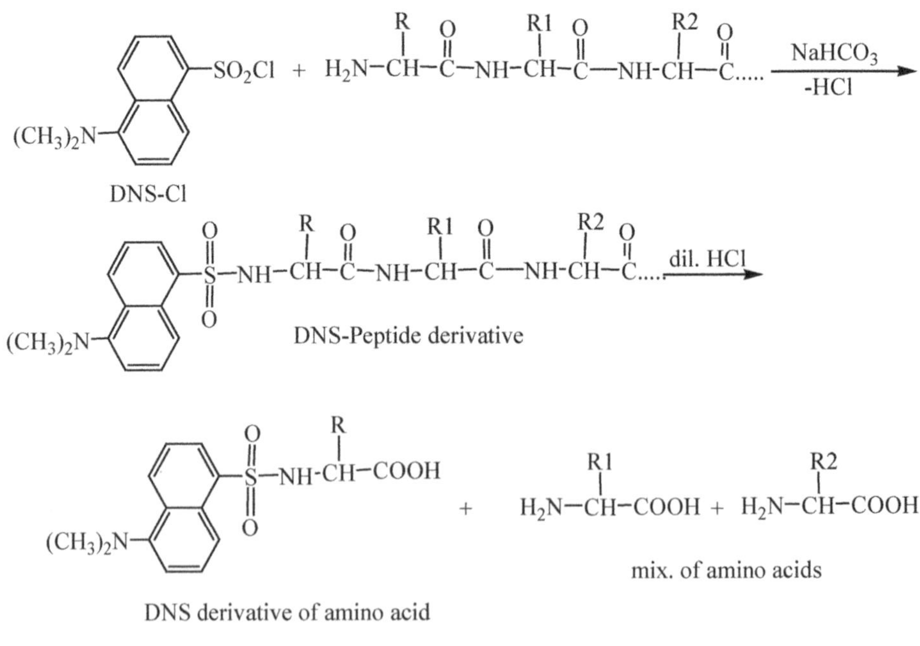

3.19 C-terminal Analysis

Enzymatic Method: In this method, peptide is incubated with carboxypeptidase the enzyme under suitable pH conditions. This enzyme selectively cleaves C-terminal amino acid that can be separated and identified.

Thiohydantoin Method: It is similar to Edman's degradation method. In this method, peptide is treated with ammonium isothiocyanate in the presence acetic anhydride. On reaction, it is converted into thiohydantoin derivative of peptide (Figure 3.3). This on treatment with cold solution of NaOH gets cleaved down to thiohydantoin derivative of C- terminal amino acid. Separate this and identify.

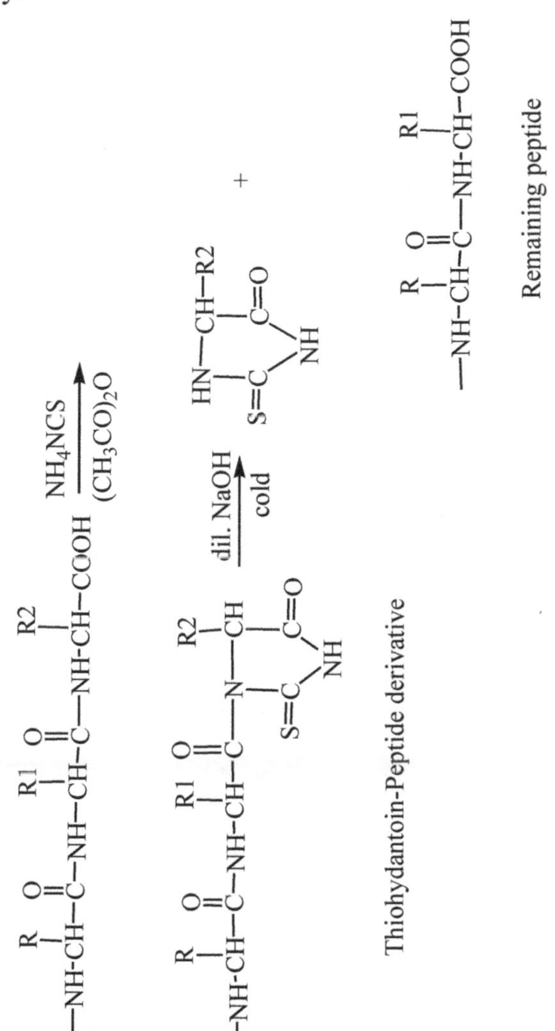

Figure 3.3: Thiohydantoin method

Hydrazinolysis Method: It is a widely used method. In this, peptide is heated with hydrazine at 100°C. On heating peptide bonds are broken and all amino acids except C-terminal are converted into corresponding hydrazones/hydrazides. This mixture is then separated using cation exchange resin, hydrazones/hydrazides being basic in nature are retained on the resin whereas C-terminal amino acid is eluted first and can be identified.

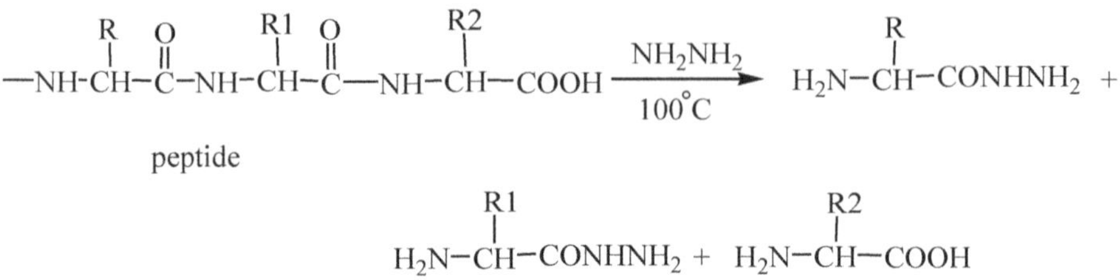

3.20 Partial Hydrolysis of Peptide

Partial hydrolysis of peptide with dil. HCl at varying temperatures gives a mixture of dipeptides, tripeptides, tetrapeptides etc. These fragments can be separated and arranged to get the sequence of amino acids. We can selectively cleave the peptide by other methods.

3.21 Other Methods for Selective Cleavage

1. Chemical Method

With 0.25 N Acetic Acid

When peptide is heated with 0.25N acetic acid, it selectively cleaves peptide bonds on both the sides of aspartic acid.

e.g. hydrolysis of following octapeptide gives a mixture of four fragments.

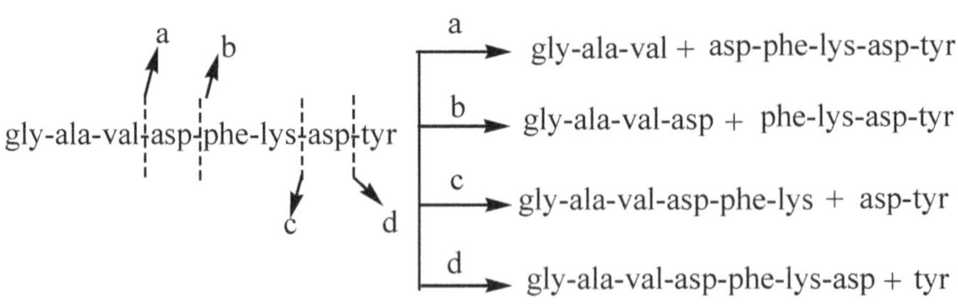

With Cyanogen Bromide (CNBr)

It is a another useful chemical to cleave peptide selectively. CNBr selectively cleaves the peptide bond on carboxyl side of methionine (Figure 3.4).

In this, peptide is treated with cyanogen bromide to form the intermediate that on hydrolysis gives homoserine lactone unit. Mechanism for this reaction is given as follows:

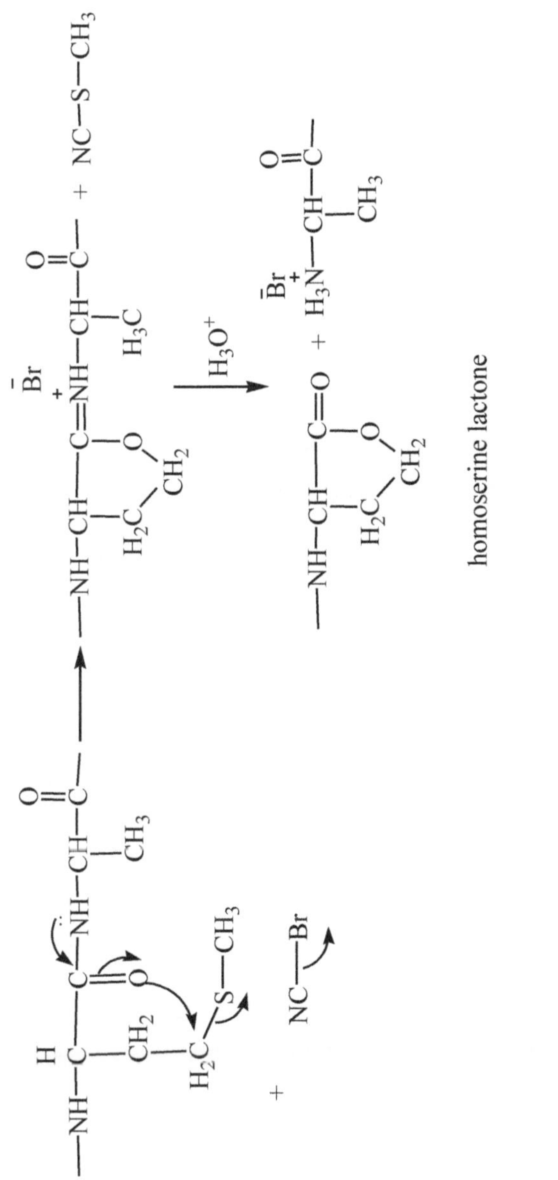

Figure 3.4: Cyanogen bromide method

With 10 N HCl

Peptide containing serine and threonine when heated with 10 N HCl gets hydrolyzed. Peptide bond on amino side of serine and threonine is cleaved. This reaction is believed to take place via the involvement of hydroxyl group present in the side chain of these amino acids. The reaction takes place as follows.

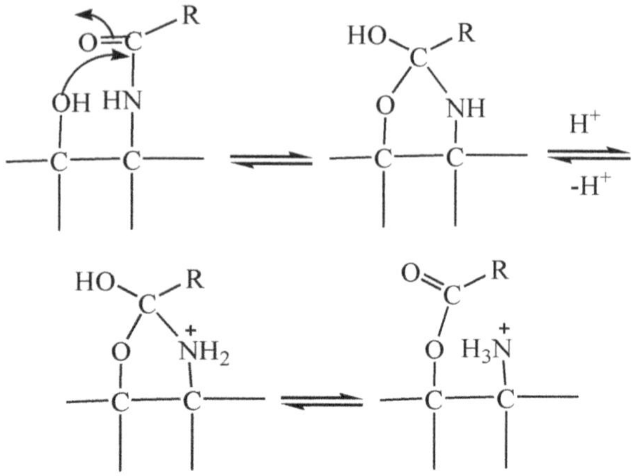

2. Selective Cleavage by Enzymes

Trypsin: This enzyme selectively cleaves the peptide bond at carboxyl group of basic amino acid. This enzyme works best in the pH range of 7-9.

Chymotrypsin: This enzyme selectively cleaves the peptide bond at carboxyl group of aromatic amino acid. This enzyme works best in the pH range of 7-9 but less specific than trypsin. It also hydrolyses the peptide bond at carboxyl group of leucine. Isoleucine, valine and methionine.

Pepsin: This enzyme selectively cleaves the peptide bond at amine group of aromatic amino acid. This enzyme works best in the pH range of 1-2. It is much less specific that leads to cleavage of leucine, aspartic acid, glutamic acid.

e.g.

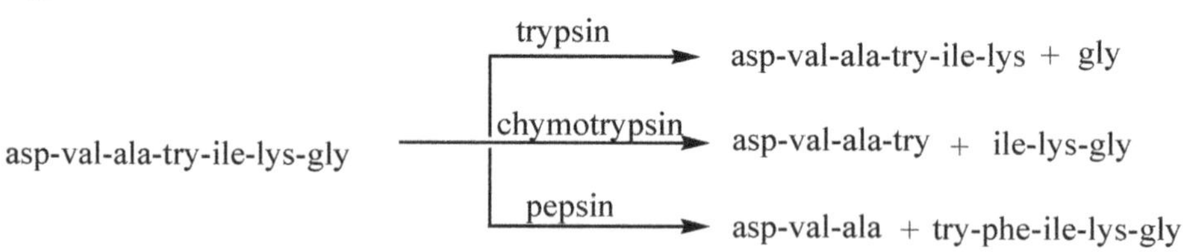

Few Examples:

Ques. No. 1: A heptapeptide "A" on complete hydrolysis gives a mixture Ala, Gly, Lys, Phe, Ile and Ser. On Edman's degradation "A" gives PTH derivative of Ala. Selective cleavage of peptide with carboxypeptidase enzyme cleaves down Ser. Partial hydrolysis of "A" yields Lys-Phe-Ala, Ala-Gly-Lys, Ala-Ile-Ser and Phe-Ala-Ile. Deduce the structure of "A".

Answer: Complete hydrolysis gives a mixture of six amino acids that shows one amino acid is present in twice amount. Edman's degradation of "A" gives PTH derivative of Ala that shows Alanine is N-terminal amino acid. Carboxypeptidase enzyme cleaves down Ser that shows Serine is C-terminal amino acid. Partial hydrolysis gives following fragments:-

Lys-Phe-Ala, Ala-Gly-Lys, Ala-Ile-Ser and Phe-Ala-Ile. On arranging these fragments, we get the sequence of amino acids in given peptide.

N — C

Ala ____ ____ ____ ____ ____ Ser

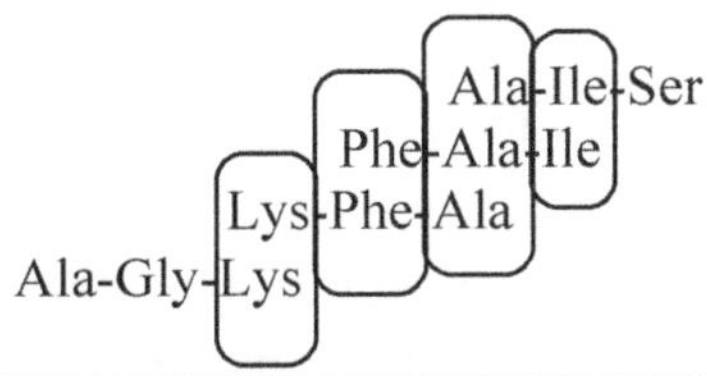

Sequence of amino acid = Ala-Gly-Lys-Phe-Ala-Ile-Ser

Ques No. 2: The reaction of nonapeptide "B" with dansyl chloride gives dansyl derivative of Cysteine. Peptide "B" on reaction with cyanogen bromide gives tripeptide containing Cys, Met, Lys and hexapeptide containing Try, Gly, Ala, Phe, Leu and Asp. Partial hydrolysis of "B" yields Lys-Met-Leu, Ala-Gly-Try, Cys-Lys-Met, Leu-Phe-Ala, Gly-Try-Asp and Met-Leu-Phe. Deduce the structure of "B".

Answer:

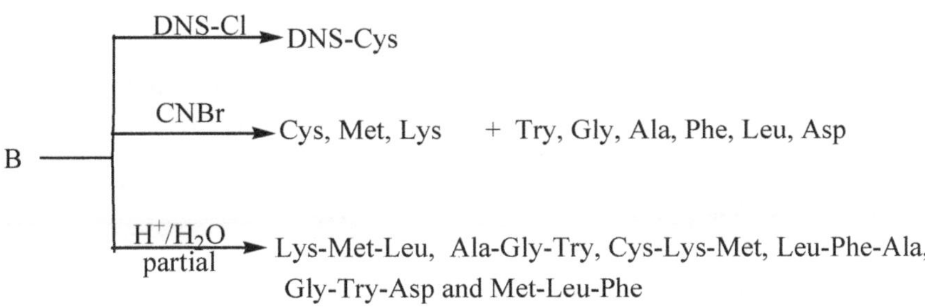

CNBr cleaves peptide bond at carboxyl side of methionine i.e.

Cys-Lys-Met___, ___,___,____,___,____

on arranging fragments

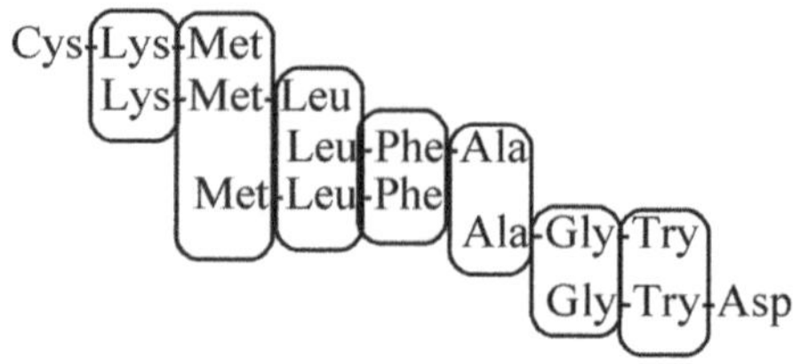

Cys-Lys-Met-Leu-Phe-Ala-Gly-Try-Asp

Ques No 3: A hexapeptide "C" on reaction with DNFB gives DNP derivative of Tyrosine. Hexapeptide "C" on hydrolysis with Carboxypeptidase enzyme give Valine. Partial hydrolysis of "C" with Trypsin gives two fragments one containing Val, Tyr, Lys and the other containing Phe, Val, Glu. Hydrolysis with Chymotrypsin gives dipeptide containing Glu, Val and tetrapeptide Tyr,Val, Phe and Lys. Deduce the structure of "C".

Answer:

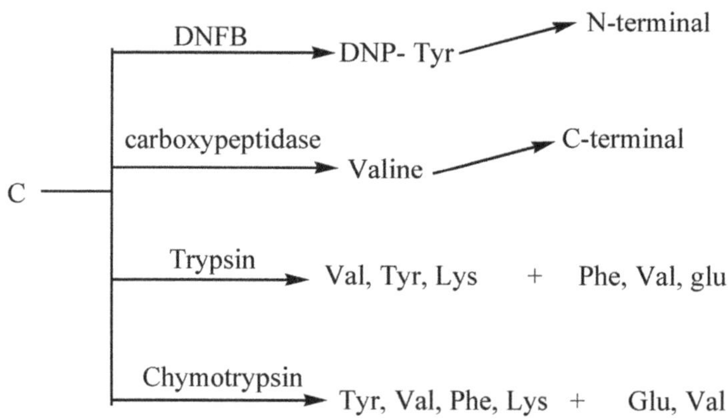

Trypsin cleave peptide bond at carboxyl side of basic amino acid i.e.
Tyr-Val-Lys ____ ____ Val

Chymotrypsin cleaves peptide bond at carboxyl group of aromatic amino acid i.e.
Tyr-Val-Lys-Phe Glu-Val

Sequence of C = Tyr-Val-Lys-Phe-Glu-Val

Ques No. 4: A tripeptide "D" is hydrolyzed completely to give two equivalents of Glu and one equivalent of Val and NH_3. D has only one free carboxylic acid and does not react with phenyl isothiocyanate. Valine is released when treated with carboxypeptidase enzyme. Deduce the structure of "D".

Answer:

1. Carboxypeptidase enzyme releases valine i.e. it is C-terminal.
2. Glutamic acid has two carboxylic acid groups, but D has only one free carboxylic acid that shows one of them is protected as amide since it liberates NH_3.
3. It does not react wih phenylisothiocyante that shows $-NH_2$ is not free, hence the structure is

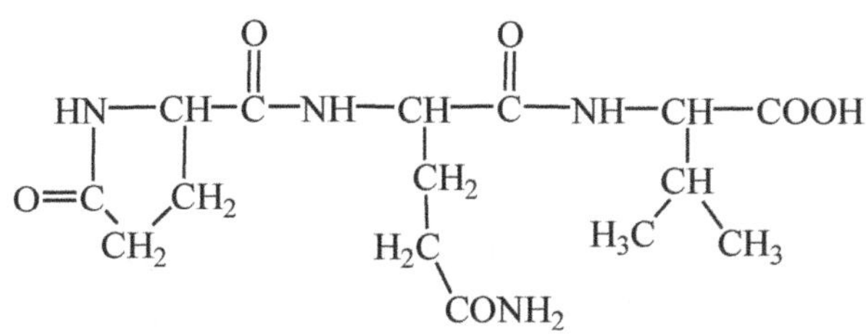

3.22 Synthesis of Peptide

Peptides cannot be synthesized by direct reaction of α-amino acids as reaction of two amino acids always lead to the formation of different dipeptides. Therefore, to synthesize a particular peptide we need to protect amino group of one amino acid and carboxyl group of another amino acid. There are two types of protecting groups i.e. N-protecting and C-protecting groups.

N-protecting Groups: The groups which are used for protection of free amino group of amino acid are known as N-protecting groups. There are five types of N-protecting groups and these are:

1. **Tertiary Butyloxycarbonyl Group or Carbo-t-butyloxy Group (t-BOC):** It is used in ester form generally p-nitrophenyl ester as halide form is not stable. It can be introduced by treating α-amino acid with ester from of t-BOC.

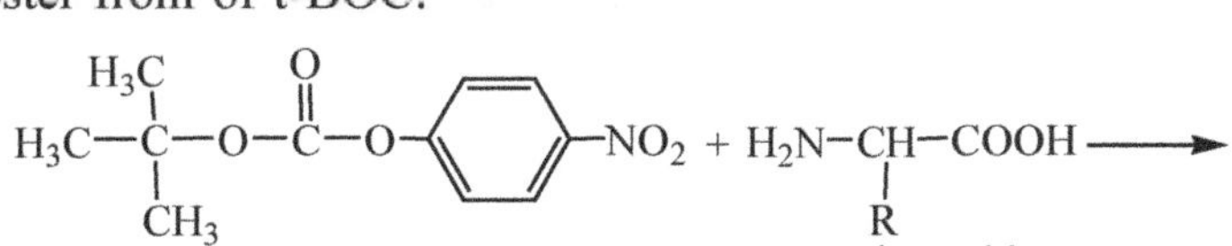

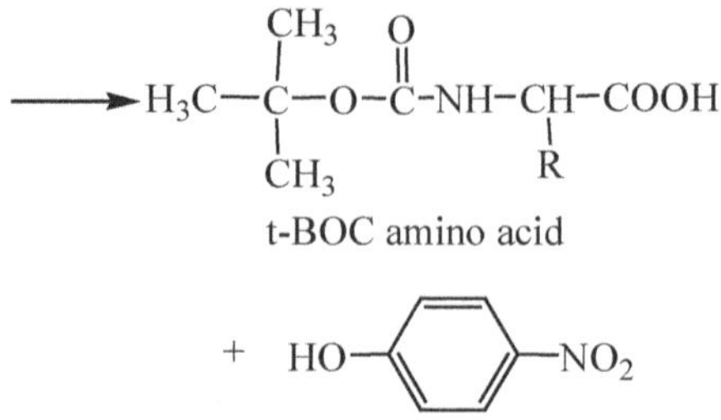

Removal: It can be removed by either of the following

(i) HBr: CH_3COOH (1:1)

(ii) HCl: CH_3COOH (1:1)

(iii) CF_3COOH (best reagent)

Synthesis of Gly-Ala-Ala

In this peptide, glycine is N-terminal and alanine is C-terminal amino acid. To synthesize above tripeptide, amino group of glycine needs to be protected followed by activation with either thionyl chloride or phosphorus pentachloride. Then it is condensed with second amino acid i.e. alanine in presence of DCC/THF (coupling solvent). Activation followed by condensation steps are repeated to get the desired peptide. Finally, N-protecting group is removed in deprotection step using either trifluroacetic acid or mixture of HCl and acetic acid (1:1 ratio). The reaction is given as follows:

1. Protection

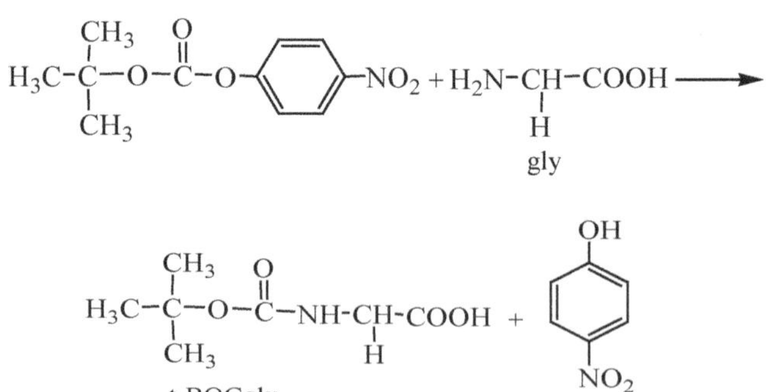

2. Activation

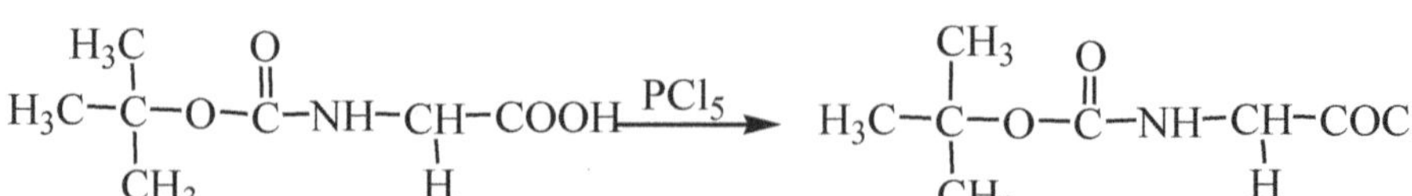

3. Condensation

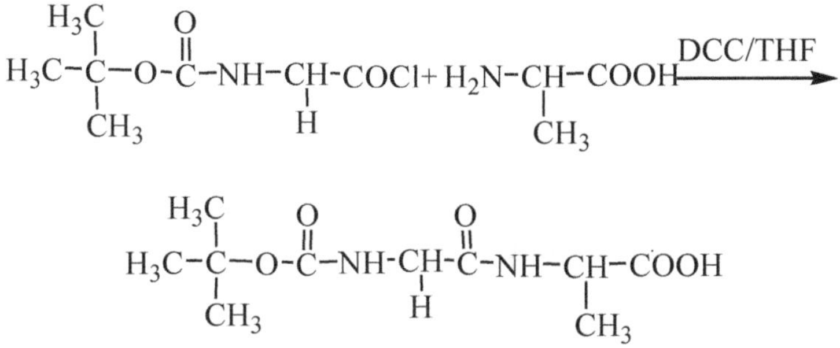

4. Activation

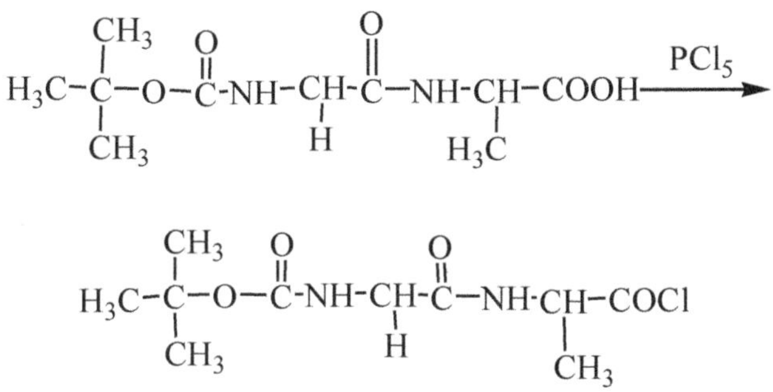

5. Condensation

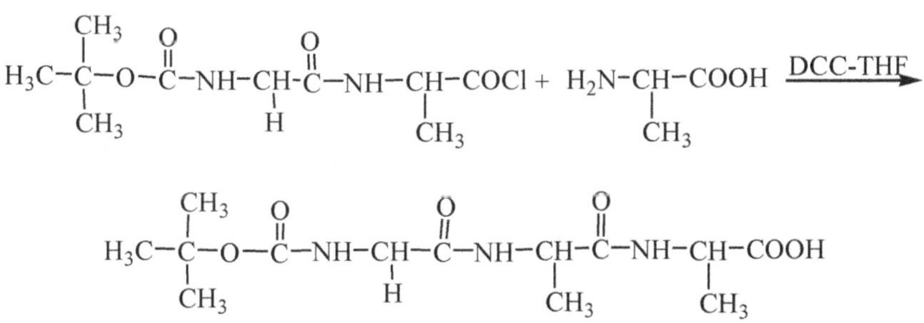

6. Deprotection

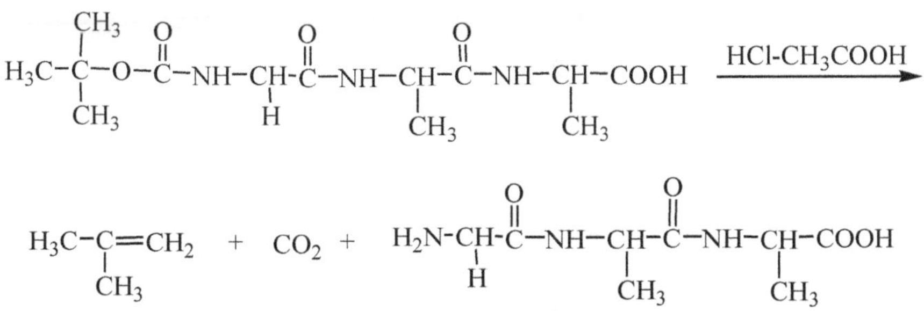

gly-ala-ala

2. Benzyloxycarbonyl Group or Carbobenzyloxy (CBZ)

Carbobenzyloxy group is introduced as halide. It is prepared by treating benzylalcohol with phosgene gas in situ.

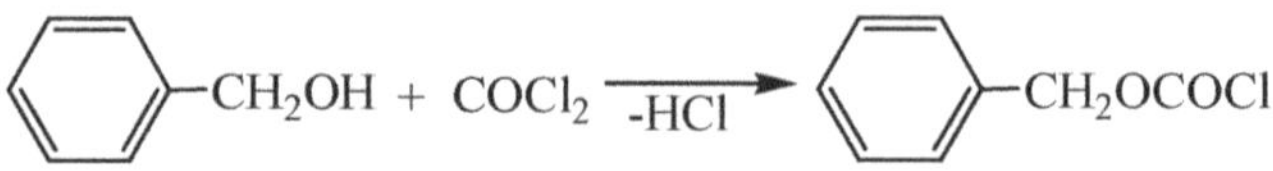

NOTE: Phosgene is a poisonous gas, therefore the reaction must be performed in fuming cupboard.

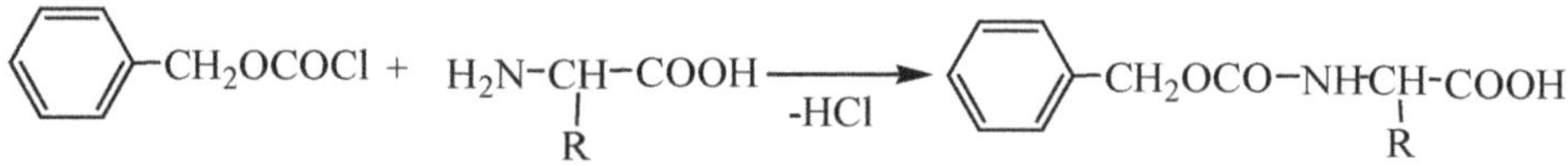

Removal

1. Pd/C
2. Na/liq. NH_3 (Birch reduction)

When amino acid contains sulphur in the side chain then it is removed by birch reduction (Na/liq. NH_3) because sulphur poisons the palladium catalyst.

Synthesis of Ala-Gly

In this peptide, alanine is N-terminal and glycine is C-terminal amino acid. To synthesize above dipeptide, amino group of alanine needs to be protected using benzyloxycarbonyl group followed by activation with thionyl chloride. Then it is condensed with second amino acid i.e. glycine in the presence of NaOH to neutralize the HCl produced. Finally N-protecting group is removed in deprotection step using either Birch reduction or palladium/activated charcoal. The reaction is given as follows:

1. Protection

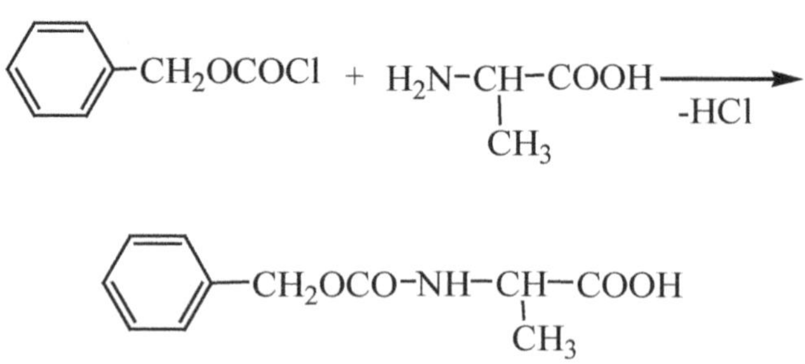

2. Activation

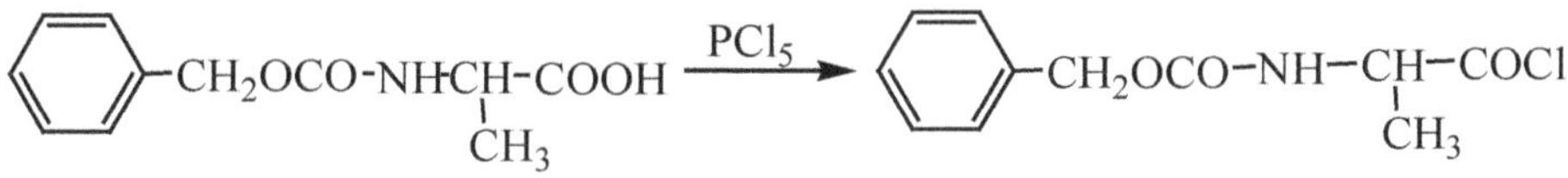

3. Condensation

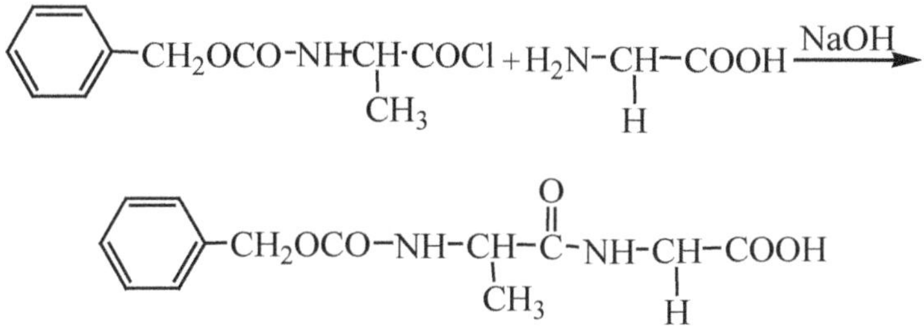

4. Deprotection

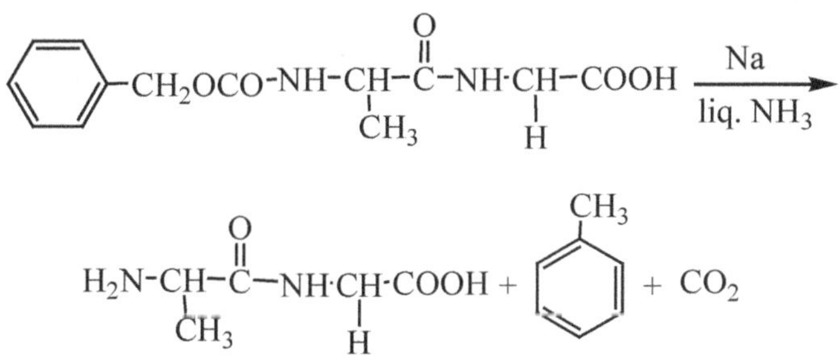

Advantage of Benzyloxycarbonyl Group

Use of this group proceeds with less or no racemization.

3. Triphenylmethyl Group (Trityl Group)

It is used as halide. It is introduced by treating N-terminal amino acid (whose carboxylic group is protected) with triphenylmethyl chloride.

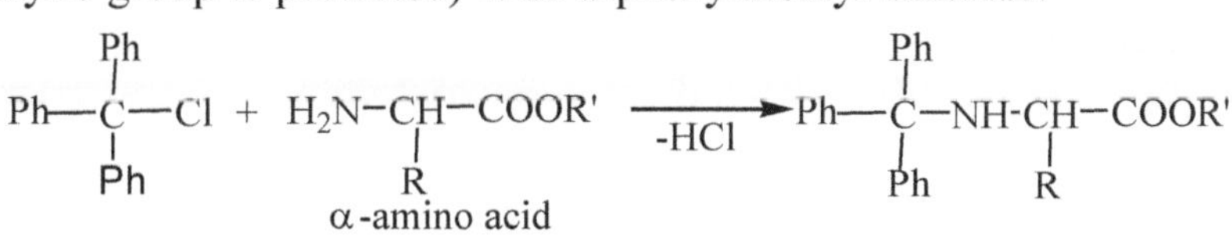

Removal

1. Pd/activated charcoal
2. Acetic acid

General Synthesis of Peptide

1. Protection

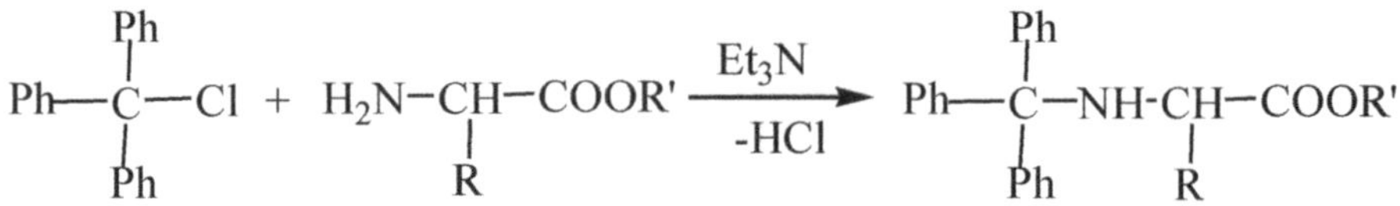

2. Removal of C-protecting Group

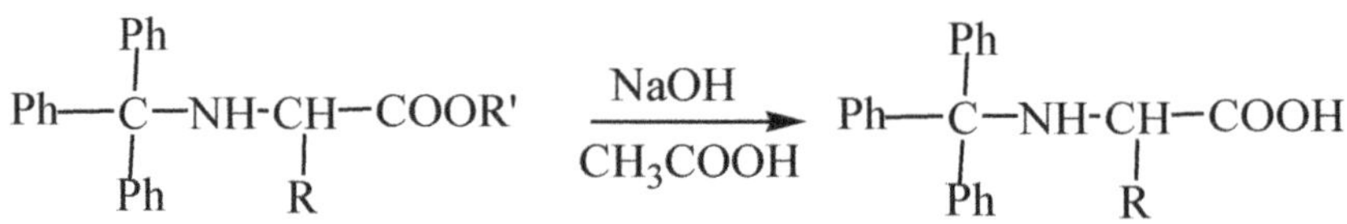

3. Activation

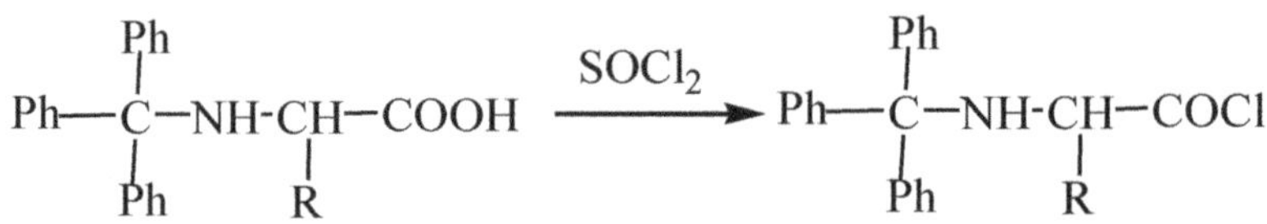

4. Condensation

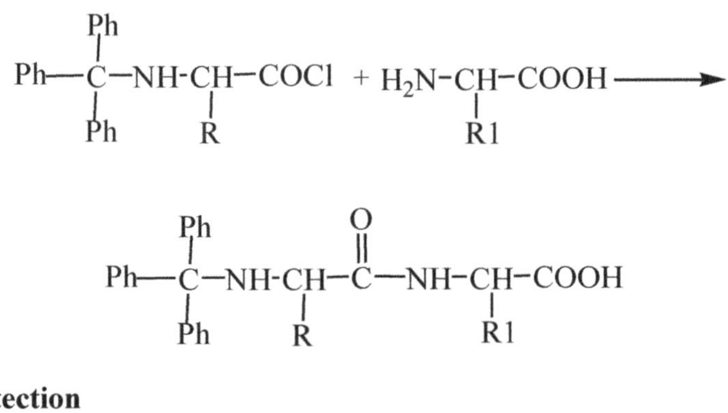

5. Deprotection

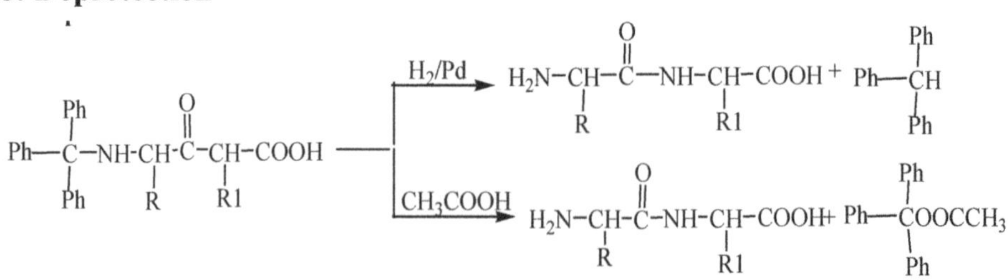

4. p-Toluenesulphonyl Group (Tosyl)

It is used as halide. It is introduced by the reaction of amino acid with tosyl chloride in presence of base.

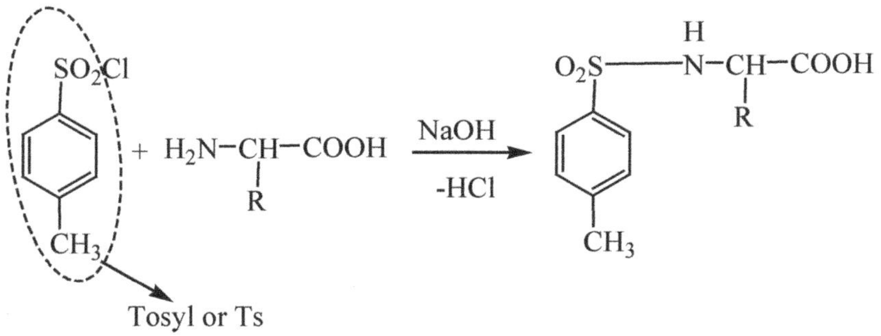

Removal

It is removed by Birch reduction (Na/liq. NH_3)

Advantage

Peptide synthesis proceeds without racemization.

5. Phthaloyl Group

It is introduced by heating amino acid with phthalic acid at optimum temperature (150°C). Heating above this temperature may lead to racemization of peptide.

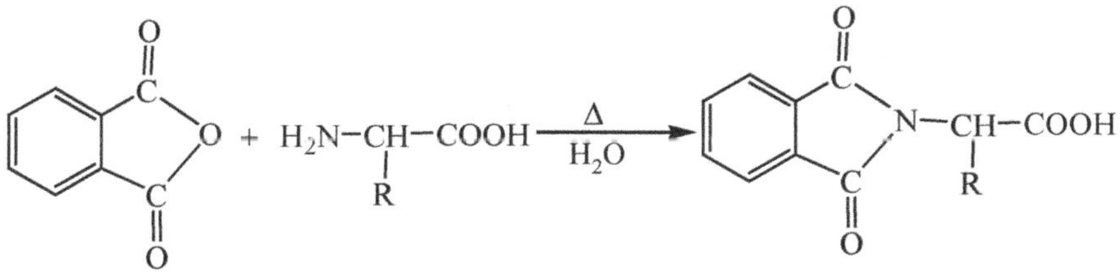

We can also introduce phthaloyl group by Gabriel's phthalimide method. In this method potassium salt of phthalimide is treated with α-haloester followed by reaction with amino acid.

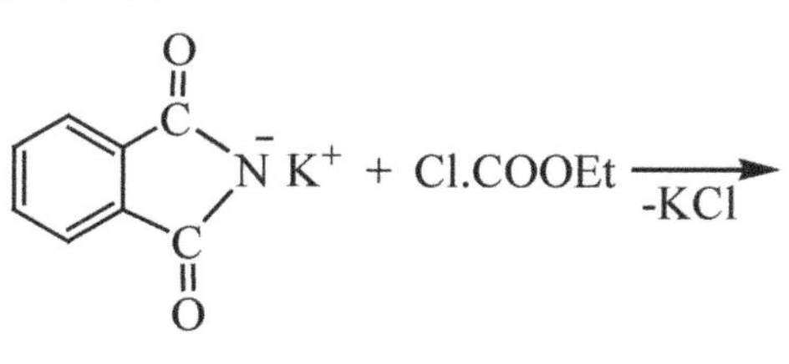

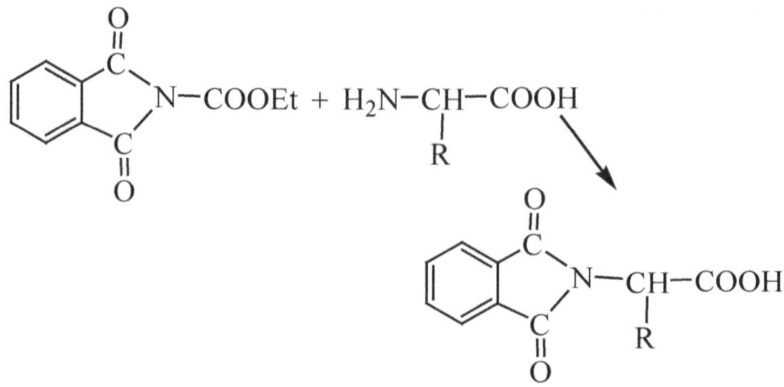

Removal

It is removed by heating with ethanolic solution of hydrazine followed by treatment with HCl.

Synthesis of Peptide

1. Protection

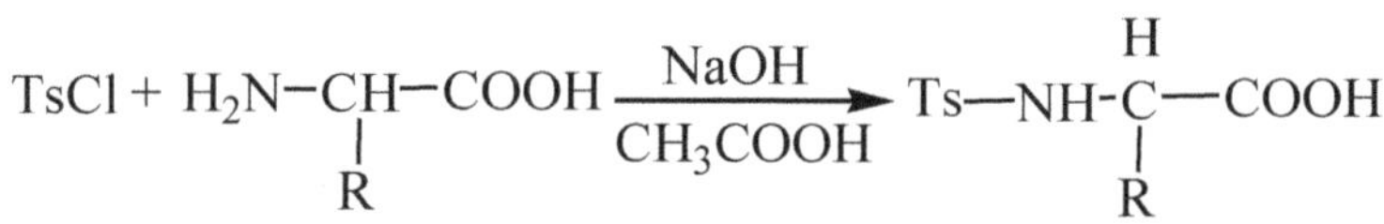

2. Activation

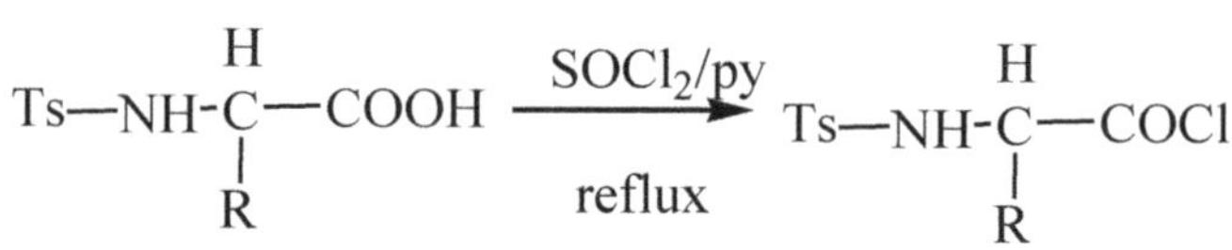

3. Condensation

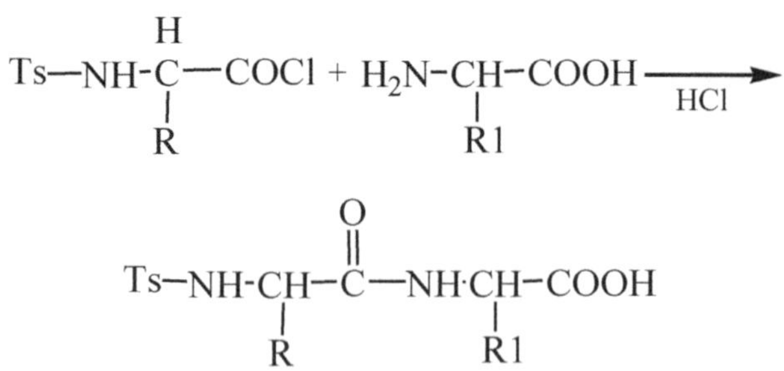

4. Deprotection

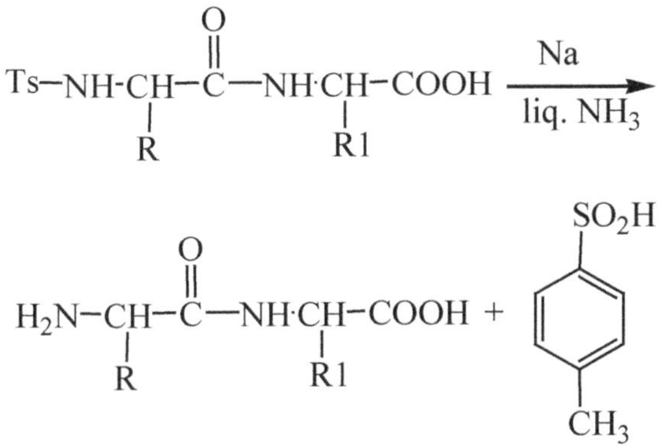

Advantage

The reaction proceeds without racemization.

3.23 C-protecting Groups

The groups which are used for protection of free carboxyl group of amino acid are known as C-protecting groups. Generally carboxyl group is protected by esterification reaction. It is protected as methyl, ethyl, tert-butyl ester, benzyl ester etc. It can be done by treating amino acid with alcohol in presence of dry HCl gas.

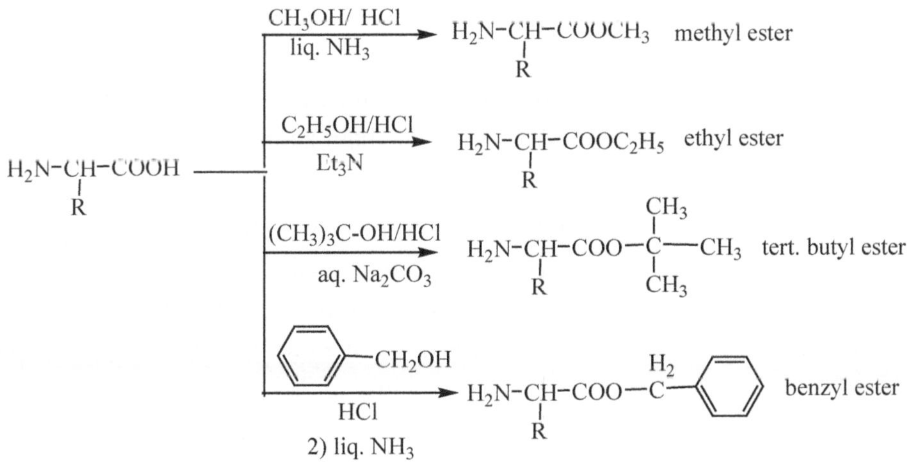

Alternatively, it is introduced by following reaction

$$H_2N\text{-}\underset{\displaystyle R}{CH}\text{-}COOH \xrightarrow{SOCl_2} H_2N\text{-}\underset{\displaystyle R}{CH}\text{-}COCl \xrightarrow{R'OH} H_2N\text{-}\underset{\displaystyle R}{CH}\text{-}COOR'$$

Where R′ = Me, Et, tert.butyl, benzyl

Removal

Ester can be removed by mild alkaline hydrolysis during peptide synthesis. Methyl, ethyl and tert.butyl group can be removed by this method. But this method leads to racemization. Benzyl group can be removed by mild alkaline hydrolysis as well as by catalytic reduction with palladium in presence of activated charcoal. Deprotection with Pd/C proceeds without racemization. Tert.-Butyl and benzyl groups are best C-protecting groups.

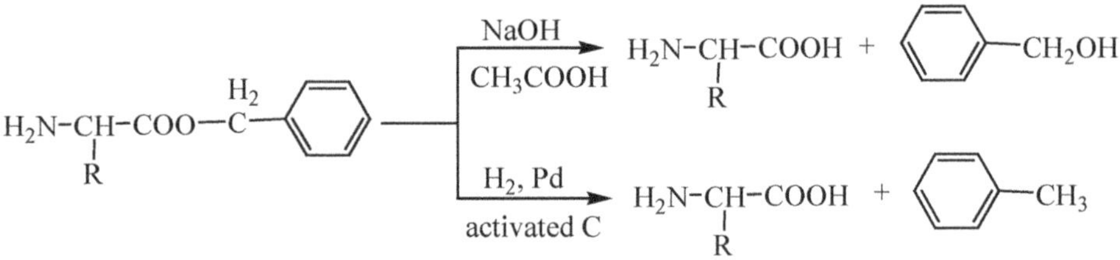

3.24 Protection of Side Chain

There are 4 types of functional groups present in the side chain of amino acids that need protection.

- **Amine Group (-NH_2):** Amino group can be protected by using any of the above N-protecting groups. But these groups must be different from the one used for protection of α-amino group. As different N-protecting groups have different stability under different reaction conditions. Sometimes, we have to deprotect the α-amino group for the formation of peptide bond without affecting the side chain amine group. For example, basic amino acids such as lysine, arginine, histidine.
- **Carboxyl Group (-COOH):** As we know that carboxyl group is protected by esterification reaction such as methyl ester, ethyl ester, tert. butyl ester etc. Here also the group used for protection of carboxyl group present in the side must be different from the one used for the protection of α-carboxyl group. For example, acidic amino acids such as aspartic acid, glutamic acid.
- **Hydroxyl Group (-OH):** Hydroxyl group of serine, threonine, tyrosine can be protected as ether like tert.butyl ether or benzyl ether. It can be introduced by treating amino acid containing hydroxyl group with tert. butyl chloride or benzyl chloride.

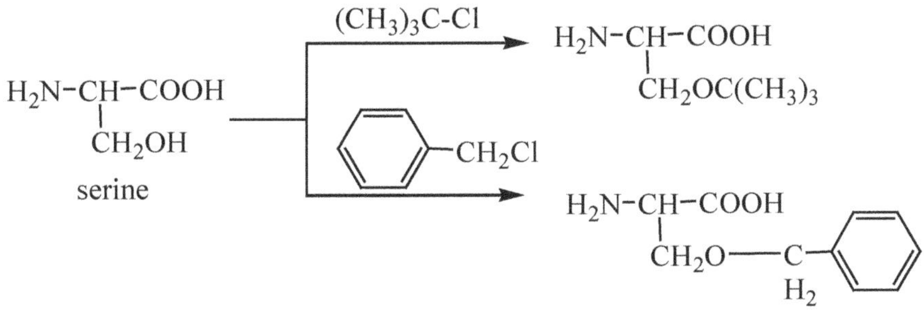

Removal: They can be deprotected by heating with HI

The better group used for protection of –OH group is tetrahydropyranyl (THP). It is introduced by treating amino acid with dihydropyran (DHP). THP ether has many advantages like low cost, easily introduced, stabililty under various reaction conditions and ease of removal under mild acidic condition.

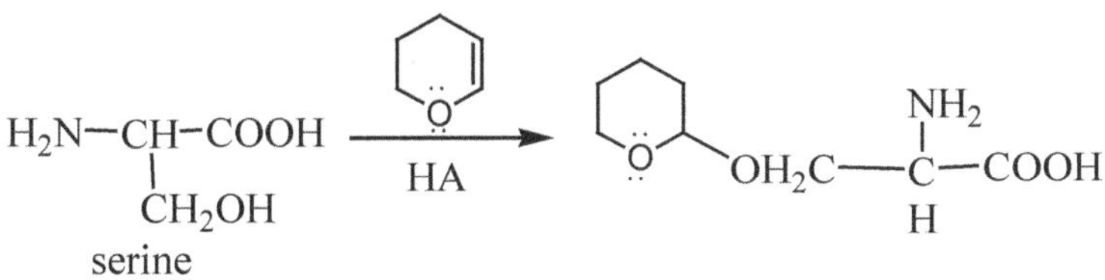

Removal: Acetic acid

- **Thiol Group (-SH):** It is present in cysteine and thiol group of cysteine is protected as s-benzylcysteine. It can be done by dissolving cysteine in liquid ammonia followed by treatment with sodium metal to form anion. This anion when treated with benzyl chloride it is converted into S-benzylcysteine.

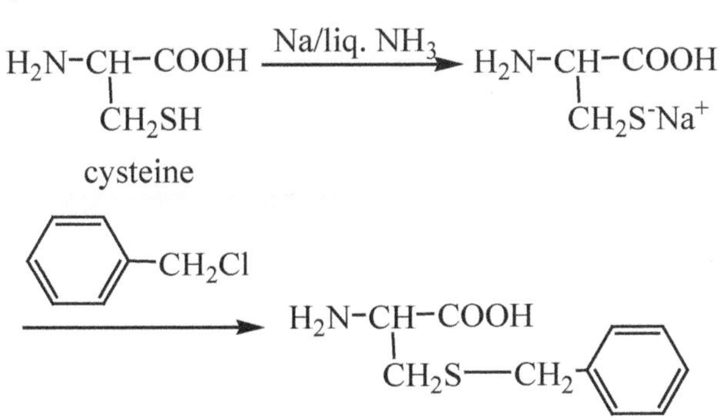

Removal:

Benzyl group is removed by Birch reduction (Na/liq. NH_3).

3.25 Activation of Carboxylic Group

Carboxyl group of amino acid is activated by heating it with thionyl chloride in the presence of pyridine or grinding with phosphorus pentachloride. This results in the formation of corresponding acid chloride.

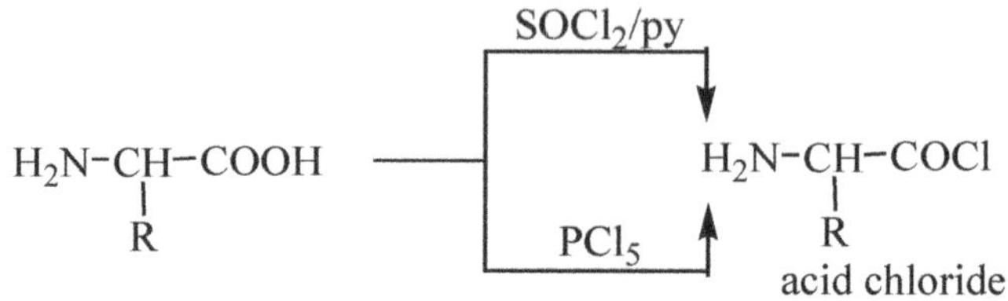

COOH group is also activated by converting it into p-nitrophenyl ester which is very reactive as compared to ethyl ester, hence called activated ester **(Activated Ester Method).**

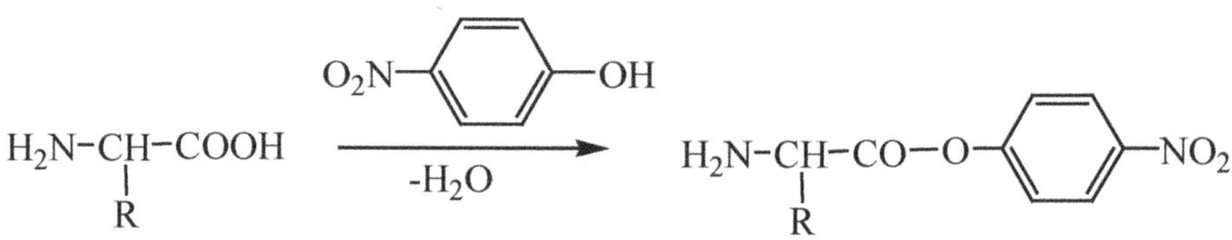

3.26 Activated Ester Method

It is similar to general method of peptide synthesis. In this, first of all amino group of N-terminal amino acid is protected. It is then activated by the reaction with p-nitrophenol instead of PCl_5 or $SOCl_2$. Activated p-nitrophenyl ester is then condensed with another amino acid to form protected dipeptide. Repeat the steps of activation and condensation till the desired peptide is obtained. When the synthesis is over, deprotect it by suitable reagent.

1. Protection

$$\text{Ar-CH}_2\text{OCOCl} + \text{H}_2\text{N-}\underset{\text{R}}{\underset{|}{\text{CH}}}\text{-COOH} \xrightarrow[\text{-HCl}]{} \text{Ar-CH}_2\text{OCO-NH-}\underset{\text{R}}{\underset{|}{\text{CH}}}\text{-COOH}$$

2. Activation

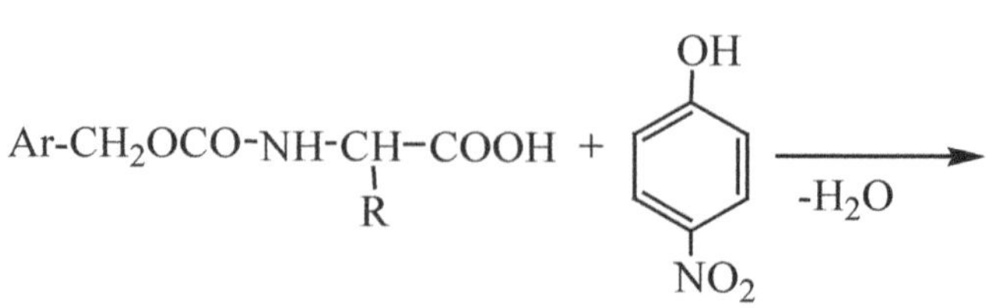

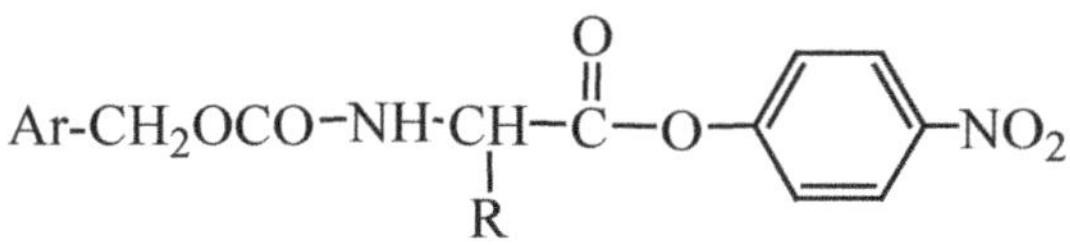

3. Condensation

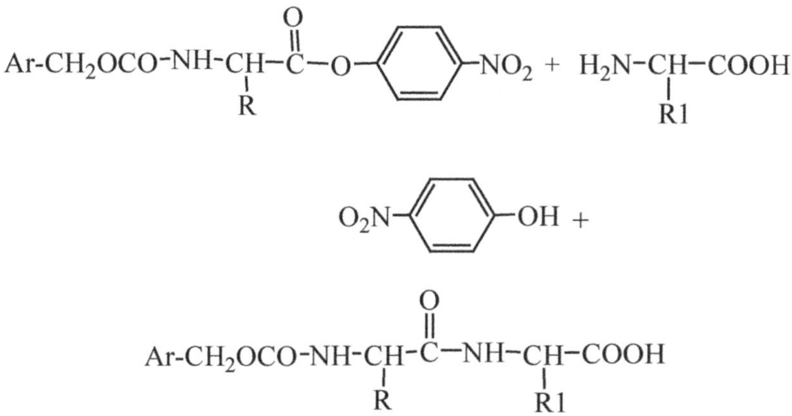

4. Deprotection

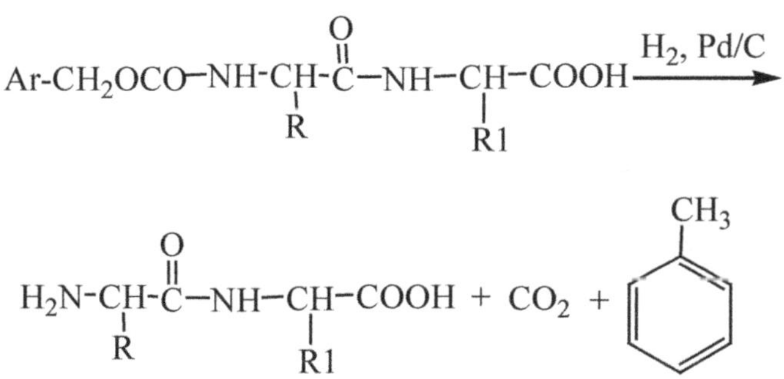

3.27 Azide Method

It is also called Curtius azide method. In this, first protection of amino group of N-terminal amino acid (C-protected) is carried out followed by reaction with hydrazine to form corresponding acid hydrazone. This acid hydrazone on reaction with nitrous acid is converted into corresponding acid azide. To this, another amino acid is added whose carboxylic group is protected and reaction mixture is stirred overnight for the condensation reaction to take place. When the desired peptide is obtained, remove the C-protecting and N-protecting group by suitable reagent/s.

$(N \longrightarrow C)$

1. N-protection of Amino Acid

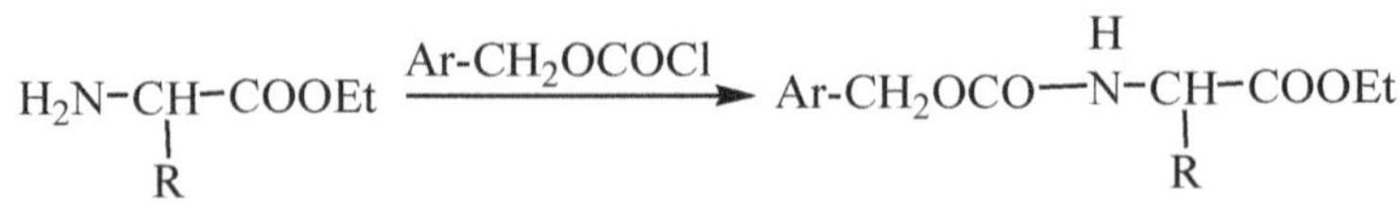

2. Reaction with Hydrazine

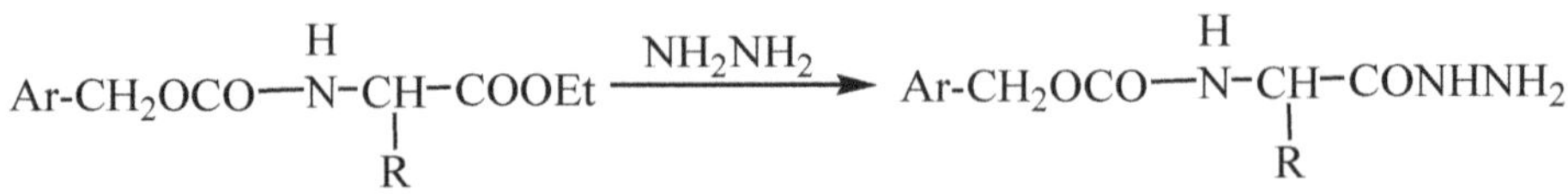

3. Reaction with Nitrous Acid

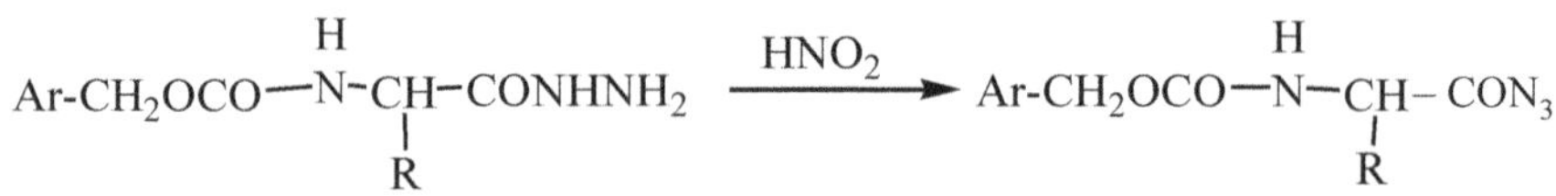

4. Condensation

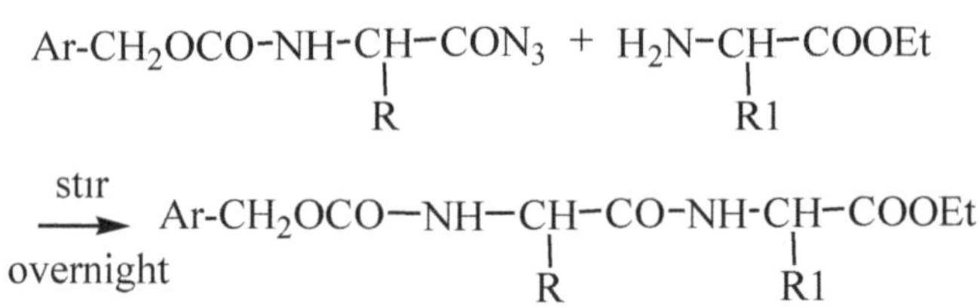

5. Deprotection of Carboxylic Acid

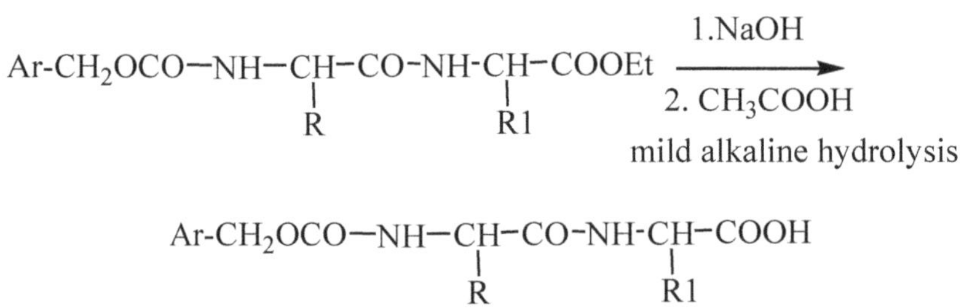

6. Deprotection of Amino Group

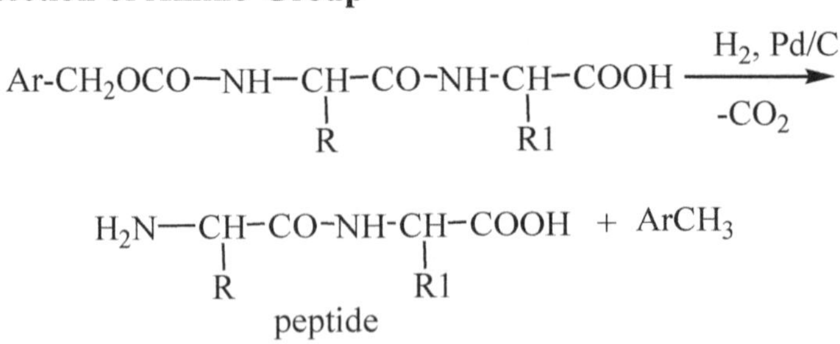

Advantage

- This process proceeds without racemization.

Disadvantages

- Large numbers of steps are involved.
- Acid azide formed may undergo Curtius rearrangement to form isocyanate. Hence, needs purification.
- Less economical.

3.28 Anhydride Methods

There are 2 types of anhydride methods

- N-Carboxyanhydride method (NCA)
- Mixed anhydride method

N-Carboxyanhydride Method (NCA)

NCA is a cyclic anhydride. It is prepared by treating N-protected amino acid (protected with N-benzyloxycarbonyl) with PCl_5 followed by heating under vacuum.

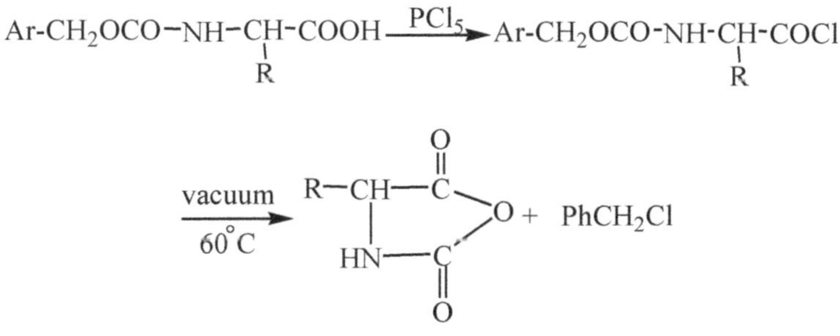

NCA

This NCA is a unit for polymerization. When NCA is treated with small amount of water or amines in inert solvent like DMF, it gets polymerized to give a peptide/protein containing same amino acid (polymer).

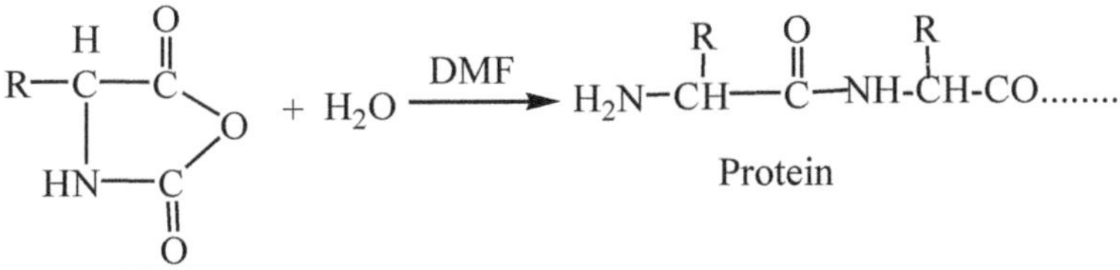

NCA

Protein

Peptide containing different amino acids can also be synthesized by NCA method. When NCA is treated with another amino acid in presence of alkaline medium followed by acidification it gives a dipeptide. Repeat the procedure to get the desired peptide.

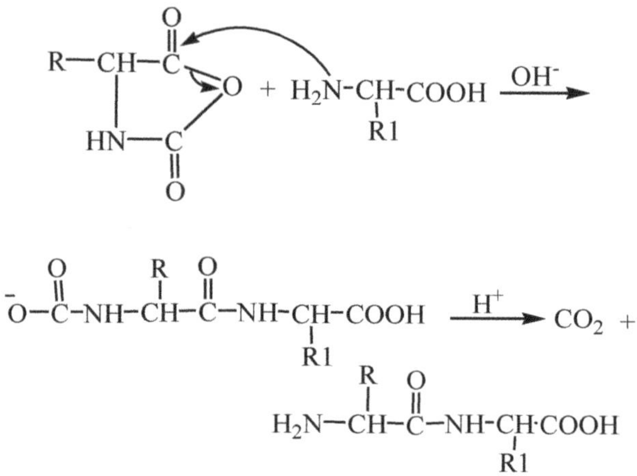

Mixed Anhydride Method

In this, N-protected amino acid is converted into mixed anhydride by reaction with ethyl chloroformate in the presence of triethylamine. Mixed anhydride is then condensed with another C-protected amino acid to form dipeptide. These steps are repeated to get the desired peptide. In the end, C-protecting and N-protecting group are removed by suitable reagents.

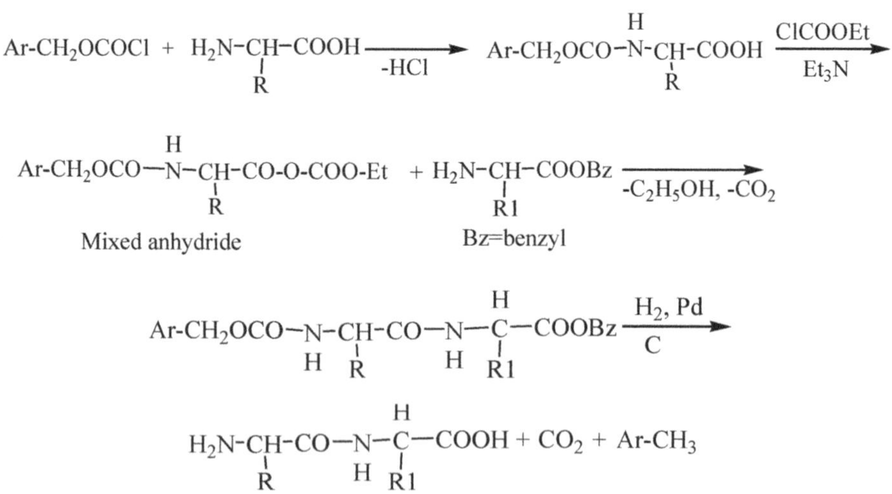

Protection of carboxylic group as methyl ester or ethyl ester leads to racemization of peptide due to alkaline hydrolysis. It is better to protect it as benzyl ester because it can also be removed by catalytic reduction with Pd/C.

3.29 DCC Method

Full form of DCC is dicyclohexylcarbodi-imide. Its structure is as follows-

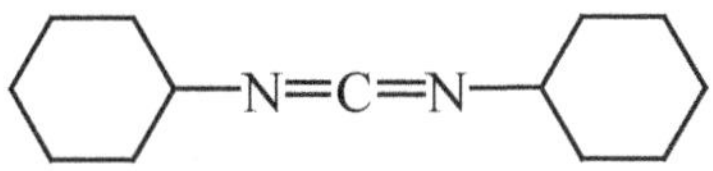

It is a coupling solvent, used to form an amide bond between carboxyl group of one amino acid and amine group of another amino acid and is removed as dicyclohexylurea after the reaction is over.

Peptides can also be prepared by DCC method. In this method, N-protected and C-protected amino acids are condensed together in the presence of coupling solvent, DCC. Resulting dipeptide is then subjected to mild alkaline hydrolysis or catalytic reduction to remove C-protected group. Now condense it with another N-protected amino acid in the presence of DCC to get tripeptide. In this way the process goes on to get the desired peptide. Finally deprotection of both the protecting groups gives the peptide.

1. Protection of Amino Group

$$\text{Ar-CH}_2\text{OCOCl} + \text{H}_2\text{N-}\underset{\text{R}}{\underset{|}{\text{CH}}}\text{-COOH} \xrightarrow[\text{-HCl}]{} \text{Ar-CH}_2\text{OCO-NH-}\underset{\text{R}}{\underset{|}{\text{CH}}}\text{-COOH}$$

2. Protection of Carboxylic Group

$$\text{H}_2\text{N-}\underset{\text{R1}}{\underset{|}{\text{CH}}}\text{-COOH} \xrightarrow[\text{HCl}]{\text{C}_6\text{H}_5\text{CH}_2\text{OH}} \text{H}_2\text{N-}\underset{\text{R1}}{\underset{|}{\text{CH}}}\text{-COOCH}_2\text{C}_6\text{H}_5$$

3. Condensation

$$\text{Ar-CH}_2\text{OCO-NH-}\underset{\text{R}}{\underset{|}{\text{CH}}}\text{-COOH} + \text{H}_2\text{N-}\underset{\text{R1}}{\underset{|}{\text{CH}}}\text{-COOCH}_2\text{C}_6\text{H}_5 \xrightarrow[\text{-H}_2\text{O}]{\text{DCC}}$$

$$\text{Ar-CH}_2\text{OCO-NH-}\underset{\text{R}}{\underset{|}{\text{CH}}}\text{-CO-NH-}\underset{\text{R1}}{\underset{|}{\text{CH}}}\text{-COOCH}_2\text{C}_6\text{H}_5$$

4. Deprotection

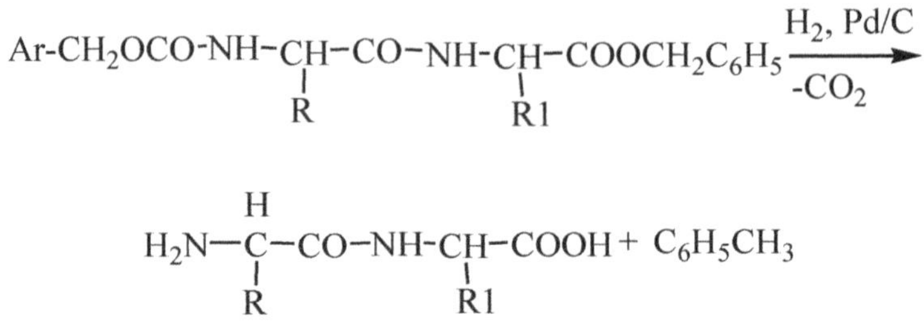

Mechanism of DCC Method

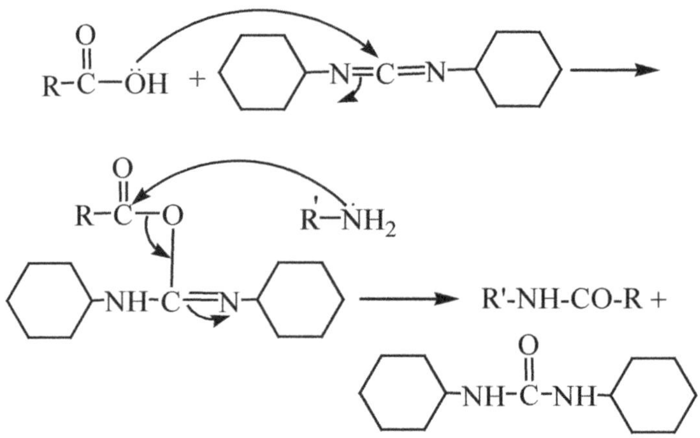

Dicyclohexylurea

3.30 Merrifield Solid Phase Peptide Synthesis

Solid Phase Peptide Synthesis (SPPS) was developed by R. B. Merrifield in 1963. It was a major breakthrough in the peptide synthesis. SPPS gives high yield of the pure products and is faster than classical method i.e. liquid phase peptide synthesis (LPPS). In this method, C-terminal amino acid is chemically attached to the insoluble resin i.e. by covalent bond then chain is built up. Solid support used should have following feature:

- It should be stable.
- It should permit the rapid filtration of liquids.
- It should be chemically inert to reagents and solvents used during peptide synthesis. Resin has the capability of swelling in solvents that allows for penetration of the reagents and attachment of amino acid.

Synthesis of Resin

The solid phase used in this method is a copolymer (resin). It is prepared by copolymerization of styrene and divinylbenzene.

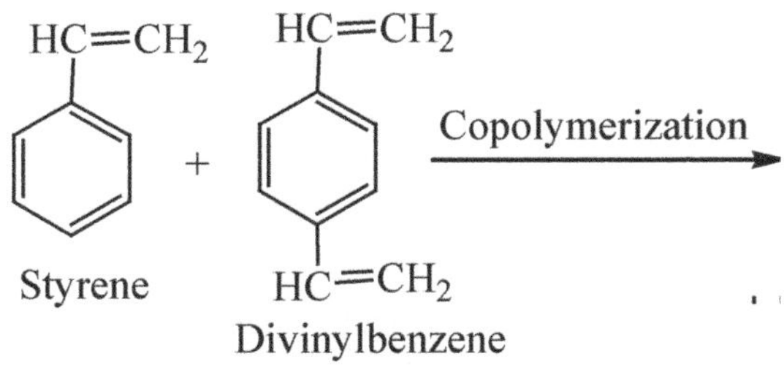

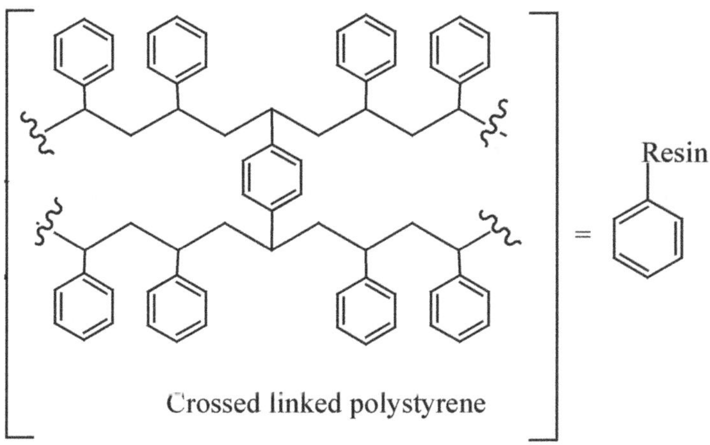

Then, it is activated by chloromethylation reaction with methoxymethyl chloride in the presence of stannous chloride.

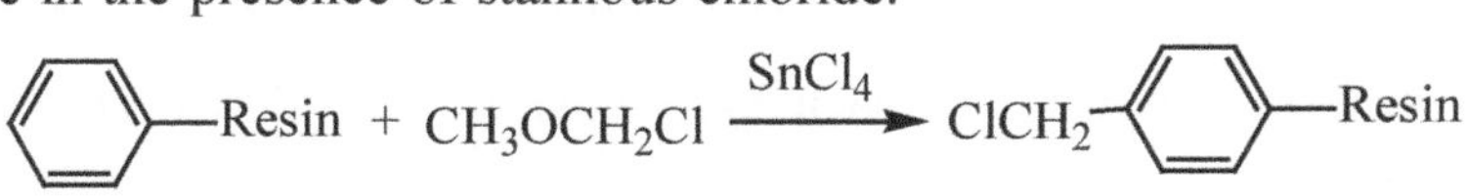

Synthesis of Peptide

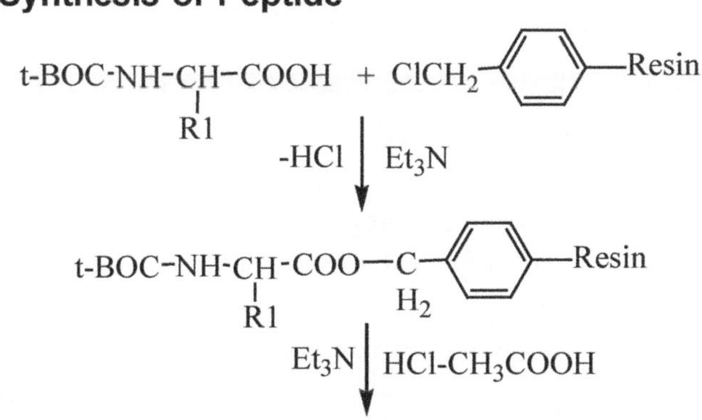

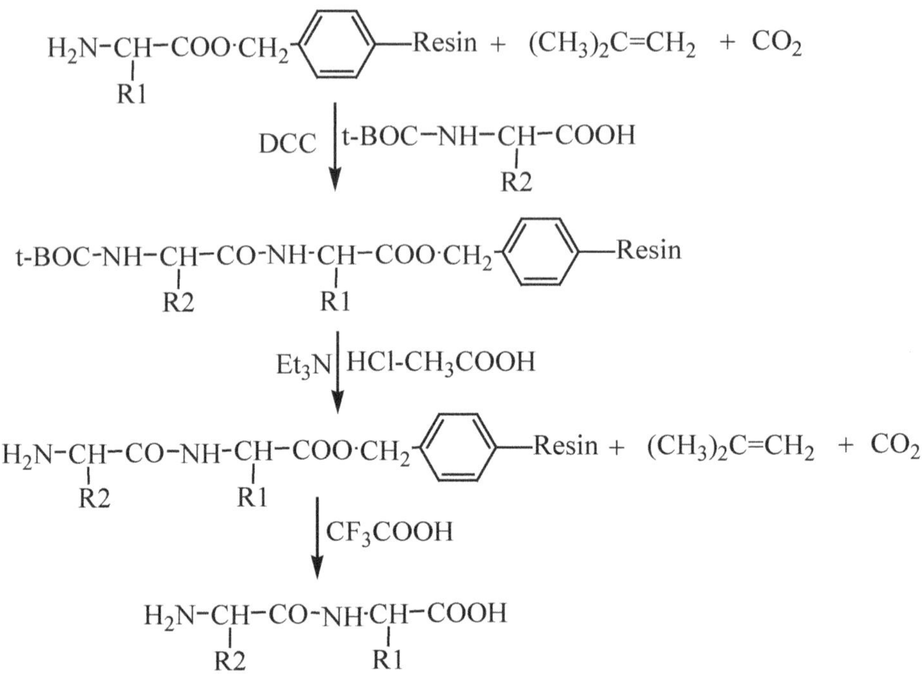

The Advantages of This Method are Very Considerable.

- Because of the use of solid support, no purification is required. Excess of the reagents and side products can be removed by washing and filtration.
- Through the replacement of a complicated purification procedure for each intermediate product with a simple washing procedure much time is saved.
- In addition, the yield in each individual step is 99.5% or better, So economical.

3.31 Proteins

Polypeptides are biomaterials composed of repeating amino acid units joined together by a peptide bond. They are formed by sequential reactions of protected amino acids. Polypeptides can acquire different three-dimensional conformations. It depends on their chemical composition such as versatility, coupled with their inherent biocompatibility and biological activity. Sometimes it also depends on making of ideal polypeptides suitable for drug and gene transfer applications and in the development of tissue scaffolds.

Short polypeptides containing less than 20–30 amino acids residues, are not considered to be proteins and are commonly called peptides or oligopeptides. The sequence of amino acid residues in a protein is defined by the sequence of a gene, which is encoded in the genetic code.

Proteins are macromolecules made up of amino acids joined through peptide bond. Most of the proteins contain more than 100 amino acids. The average molecular weight (based on the weight of a hydrogen atom as 1) of each amino acid is approximately 100 to 125. That means the molecular weights of proteins are usually in the range of 10,000 to 100,000. The mass of a protein is expressed in units of Daltons. One Dalton is equal to one atomic mass unit or weight of one hydrogen atom.

3.32 Average Molecular Weight of Protein

There are many methods by which molecular weight of proteins can be determined. This include osmotic pressure measurement, X-ray diffraction, mass spectrometry, ultracentrifugation, chemical method etc. Chemical methods are simple and can easily be carried out due to no specific requirements. Chemical method is based on the estimation of particular amino acid. If we know the percentage composition of the amino acid then we can calculate the mole fraction of each amino acid by dividing its percentage by its molecular weight. Suppose m is the molecular weight of amino acid, M is the molecular weight of protein and x is the percentage of that particular amino acid. then, by using following formula we can calculate molecular weight of protein.

$$x = \frac{m}{M} \times 100$$

By osmotic pressure method

$$M = \frac{C}{P} \times RT$$

Where M = molecular weight of protein, C = concentration of dry protein (gm per 100 cc solvent), P = osmotic pressure in cm H_2O of density 1, R = gas constant, 0.08207, T = temperature in Kelvin.

3.33 Classification of Proteins

We can classify proteins on the basis of the following:

- Constitution
- Shape

- Nature of molecules
- Function of proteins

I. Classification on The Basis of Constitution

There are Three Types of Proteins:

1. **Simple Proteins.** These proteins only give amino acids upon hydrolysis. They are of following types:
   - (i) **Albumins:** They are coagulated by heat and precipitated with solid ammonium sulphate from aqueous solution. e.g. serum albumin and lactalbumin and ovalbumin.
   - (ii) **Globulins:** These are insoluble in water but soluble in dilute salt solution. They are coagulated by heat and precipitated with half saturation with ammonium sulphate solution. e.g. ovoglobulin and serum globulins.
   - (iii) **Glutelins:** These are insoluble in water but soluble in dilute acid and alkali. They are coagulated by heat. e.g. glutelins from wheat.
   - (iv) **Histones:** These are soluble in water and dilute acids but insoluble in NH_4OH. They are not coagulated by heat. e.g. nucleoproteins.
   - (v) **Globins:** These are considered to be an example of histones but are different from histones. They are not basic like histones. They are not precipitated by NH_4OH. e.g. globins of hemoglobin.
   - (vi) **Prolamins:** These are insoluble in water but soluble in 70 percentage alcohol. These are rich in proline. e.g. zein of corn and gliadin of wheat.
   - (vii) **Protamines:** These are simplest proteins. These are polypeptide chains rich in basic amino acid. These are soluble in water and NH_4OH. They are not coagulated by heat. They are found in sperm cells.
   - (viii) **Albuminiod (Scleroproteins):** These are insoluble in all neutral solvents, dilute acids and alkalies. They are present in tissues like bone, cartilage, horn, hair.
2. **Conjugated Proteins or Complex Proteins:** The proteins which are made up of amino acids and some other non-protein material called as a prosthetic group are called conjugated proteins or complex proteins.

Depending upon the nature of prosthetic group, they are of following types:

(i) **Nucleoproteins:** In this, simple proteins like histones, protamines are attached to the nucleic acid (DNA, RNA). e.g. Chromatin of cell nuclei and virus.

(ii) **Glycoproteins:** These proteins are covalently attached to one or more carbohydrate units. They contain less than 4 % of carbohydrate. e.g. fibrinogen, transferring, immunoglobulins or antibodies.

(iii) **Mucoproteins:** These are glycoproteins containing more than 4% of carbohydrates. They are soluble in water. e.g. serum mucoprotein, mucin of saliva.

(iv) **Chromoproteins:** These proteins contain colored prosthetic groups. This includes hemoglobin and myoglobin, which bind to one and four heme groups respectively. For example, Chlorophylls which bind a porphyrin ring with a magnesium atom at its centre and Rhodopsins, which bind retinal.

(v) **Lipoproteins:** These proteins are attached to lipids like cephalin, lecithin. They are found in blood and cell membrane.

(vi) **Metalloprotein:** These proteins contain metal ions such as iron, cobalt, manganese, zinc. e.g. ceruloplasmin (Cu), heme (Fe), chromoproteins.

(vii) **Phosphoprotein:** In these proteins phosphoric acid is attached to serine and threonine residues. These proteins act as structural component like tooth dentin, reserve function such as milk caseins (alpha, beta, gamma and delta), and egg yolk.

3. Derived Proteins

- These proteins are the derivatives of either simple or complex protein that are obtained by the action of heat, enzymes and chemicals.
- Few artificially proteins are also included in this group.
- They are classified as primary derived protein and secondary derived protein.
- Examples are: denatured proteins and peptides.

II Classification on The Basis of Shape

Fibrous Proteins	Globular Proteins
• It has a fibre-like structure formed by the polypeptide chain. These proteins are held together by strong hydrogen and disulphide bonds. • Secondary structure is the most important functional structure in fibrous proteins. • They are usually insoluble in water. The presence of hydrophobic amino acids on their surface facilitates their packaging into very complex supramolecular structures. • Fibrous proteins are mainly used for structural purposes. • For example, collagen in tendons, keratin is present in nails and hair and myosin in muscles.	• They are made up of polypeptides that are coiled about themselves to form oval or spherical molecules. They have a compact and more or less spherical structure, more complex than fibrous proteins. • Tertiary structure is the most important functional structure in globular proteins. • They are generally soluble in water but can also be found inserted into biological membranes (transmembrane proteins), thus in a hydrophobic environment. • Globular proteins, unlike fibrous proteins that have structural and mechanical functions act as enzyme, hormones, receptors, transporter of fatty acid, oxygen, antibodies. • For example, myoglobin, hemoglobin, cytochrome c.

III. Classification on The Basis of Nature of Molecules

- **Acidic Proteins:** They exist as anion and contain acidic amino acids. e.g. blood groups.
- **Basic Proteins:** They exist as cations and are rich in basic amino acids e.g. lysine, arginine etc.

IV. Classification on The Basis of Function of Proteins

- **Structural Proteins:** Proteins are known as the building blocks of the body. They are an essential component of various structures in the cell and tissues like bone, tendon, cartilage, skin feather, hair, nail. Most of them are fibrous proteins and insoluble in water. e.g. collagen, keratin, elastin.
- **Defense Proteins:** They are the main constituent of antibodies that protect our body against antigens and pathogens thus preventing infections. e.g. antibodies.
- **Transport Proteins:** There are many proteins which act as carrier of many biological molecules such as serum albumin, hemoglobin.

- **Hormones:** Hormones are mainly composed of proteins. Hormones play a vital role in regulating muscle mass, sex hormones, growth and development. e.g. insulin, ACH, progesterone.
- **Enzymes:** Proteins are also called biological catalysts because they regulate many biochemical reactions that are occurring in the body. Most of them are globular proteins. e.g. DNA polymerase, lipase.
- **Storage Protein:** They store various molecules and ions in cells. They are found in seed, egg, milk etc. e.g. Ferritin store Iron, casein.
- **Contractile or Mobile Protein:** These proteins help in movement and locomotion of various body parts. e.g. Actin, myosin, tubulin etc.

3.34 Denaturation of Proteins

Denaturation is the change in the protein shape or structure by some external stress in such a way that it loses its ability to carry out its cellular function. i.e. protein loses its biological activity. This happens due to the loss of its three-dimensional structure responsible for its specificity i.e. its function after denaturation. Denaturation occurs when many weak bonds such as hydrogen bonds are broken with in protein that are responsible for the stability of protein structure, mostly tertiary structure i.e. its natural structure. Denatured proteins have a more random and looser structure that is insoluble in water. Denaturation can be brought about in many ways—*e.g.*, by heating, by treatment with alkali, acid, urea, detergents, and sometimes by vigorous shaking.

Factors

- **Effect of Heating:** The change of temperature brings changes in intermolecular attractions like H-bonding, covalent bonding etc. hence it gets denatured.
- **Effect of Acid or Alkali:** The change of pH changes the charge on many of the side chain. This disrupts various electrostatic interactions and hydrogen bonding.
- **Effect of Chemicals:** Certain reagents such as urea and guanidine hydrochloride denature proteins by forming hydrogen bond to the proteins that are stronger than the hydrogen bonds formed within the molecule.
- **Effect of Detergent:** Detergent such as sodium dodecyl sulphate denatures proteins by interacting with non-polar groups of the proteins.

Thus interfering in hydrophobic interaction. Some organic solvents like acetone or alcohol also interfere in hydrophobic interaction that also lead to denaturation of proteins.

- **Vigorous Shaking:** Sometimes protein is denatured on vigorous shaking as it changes the kinetic energy of the protein molecules overcoming/disrupting the weak interactions.

Few Examples of Denaturation

(a) Egg white contains egg albumins which is soluble in water. On heating it becomes hard mass and water insoluble.

(b) Similarly, altering the pH of milk by adding acids such as citric acid from lemon juice denatures milk proteins and curdles the milk. The solid white portion that separates from the whey is denatured protein. We can also observe when milk curdles naturally due to bacterial colonization. Bacteria produce lactic acid as a by product of metabolism. When properly controlled, this process of denaturation is used to make yogurt and fresh cheese.

(c) Denatured proteins are found to be the root cause of number of diseases e.g. Parkinson's disorder, Alzheimer etc. People suffering from these illnesses have abnormally folded proteins in some part of their body particularly associated with the brain and nervous system.

(d) Presence of denatured proteins in the lens of the eye are one of the major causes of blindness. In such cases, denaturation is said to arise due to ageing or from excess exposure to UV radiation.

Sometimes, protein regains its original structure upon removal of denaturing agent and restoration of temperature and pH conditions that favors the native state. This reversible process is called **Renaturation.** For example serum albumin of blood, hemoglobin and ribonuclease enzyme.

The denaturation of many proteins, such as egg white, is irreversible. Most important consequence of denaturation is the loss of biological activity. For example, loss of catalytic activity of enzyme.

3.35 Structure of Proteins

There are four levels of protein structures:

1. **Primary Structure:** It is the simple and basic structure of protein but important one as whole activity of protein depends upon it. Primary structure of protein is formed by joining together of α-amino acid by

peptide bond which is formed between amine group of one amino acid and carboxyl group of another amino acid. The linear sequence of amino acids within a protein is considered the primary structure of the protein (Figure 3.5).

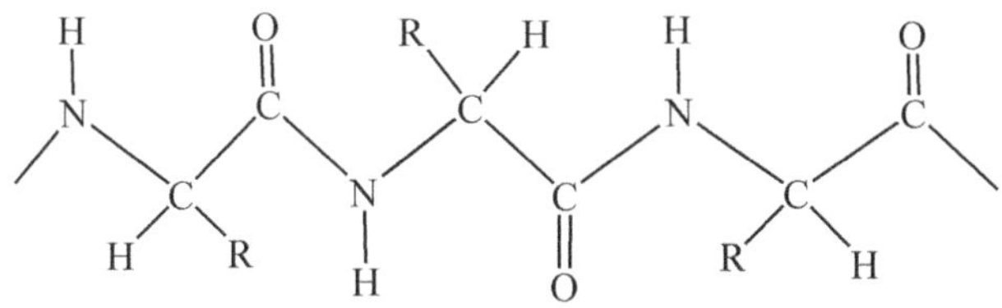

Figure 3.5: Primary structure of protein

The sequence of protein is determined by the DNA of the gene that encodes the protein. A change in the gene's DNA sequence may result in the change in the sequence of amino acid of the protein. Even a single change of amino acid affects the overall structure of protein and hence its function.

For example, a single amino acid change is associated with sickle cell anemia. It is an inherited disease that affects red blood cells. In sickle cell anemia, RBCs become sickle shaped due to the change in the sequence of amino acid. Hemoglobin molecule is made up of two α chains and two β chains, each chain consisting of about 150 amino acids. That means app. 600 amino acids are present in the whole protein. The difference between a normal hemoglobin molecule and a sickle cell molecule is just 2 amino acids out of the approximately 600. In sickle cell anemia, there is a slight change in the sequence of one of the polypeptide chains that make up hemoglobin that carries oxygen in the blood. The glutamic acid that is normally the sixth amino acid of the hemoglobin β chain is replaced by a valine. This change makes the hemoglobin molecules assemble into long fibers that leads to disruption of disc-shaped red blood cells into crescent shapes. The sickle cells get stuck while passing through blood vessels. Hence blood flow gets impaired that can cause serious health problems for people with sickle cell anemia, including breathlessness, dizziness, headaches and abdominal pain.

2. **Secondary Structure:** The next level of protein structure is the secondary structure that refers to local folded structures formed within a polypeptide due to the formation of intramolecular H-bonds. In secondary structure, no interaction takes place between side group "R" of amino acids. It is of two types: α-helical and β-pleated structures.

- **α-Helical Structure:** This structure was proposed by Linus Pauling and Robert Corey by X-ray analysis. Secondary structure of a protein is shape in which linear polypeptide chain exists. Sometimes, it acquires the shape of spiral staircase or curled ribbon (Figure 3.6). This shape is stabilized by the formation of intramolecular H-bond that is formed between the carbonyl (C=O) oxygen of one amino acid and amine hydrogen (N-H) of another amino acid that is four down the chain. It means the carbonyl of amino acid 1 would form a hydrogen bond to the N-H of amino acid 5. There is app. 3.6 residues/turn in an α-helix. The R groups of the amino acids is oriented outward from the α- helix and free to interact with others.

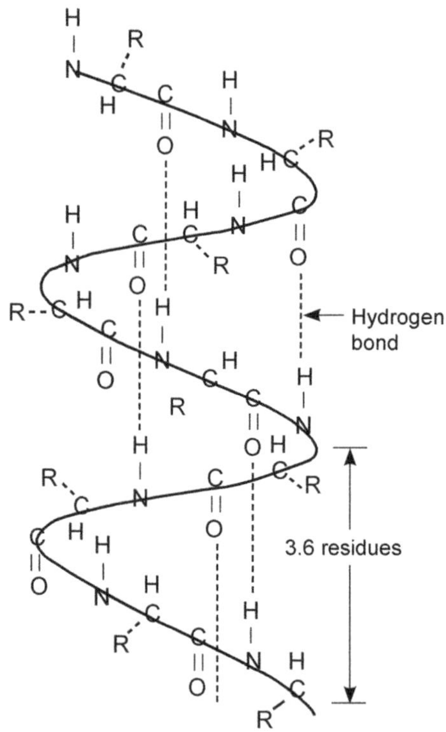

Figure 3.6: α-Helical structure of protein

Source: https://www.sarthaks.com/1730861/draw-a-neat-labelled-diagram-for-the-secondary-structure-of-protein

- **β-pleated Sheet:** In this structure, two or more linear polypeptide chains line up next to each other forming a sheet-like structure. This structure is held together by intermolecular hydrogen bonds formed between N-H group of one polypeptide chain with C=O of another polypeptide chain. While the R groups extend above and below the plane of the sheet structure (Figure 3.7). β- pleated structure is of two types:

- **Parallel** β-pleated structure in which two strands may be parallel, pointing in the same direction i.e. their N- and C-terminals match up.
- **Antiparallel** β-pleated structure in which two strands pointing in opposite directions i.e. N-terminal of one strand is positioned next to the C-terminal of the other.

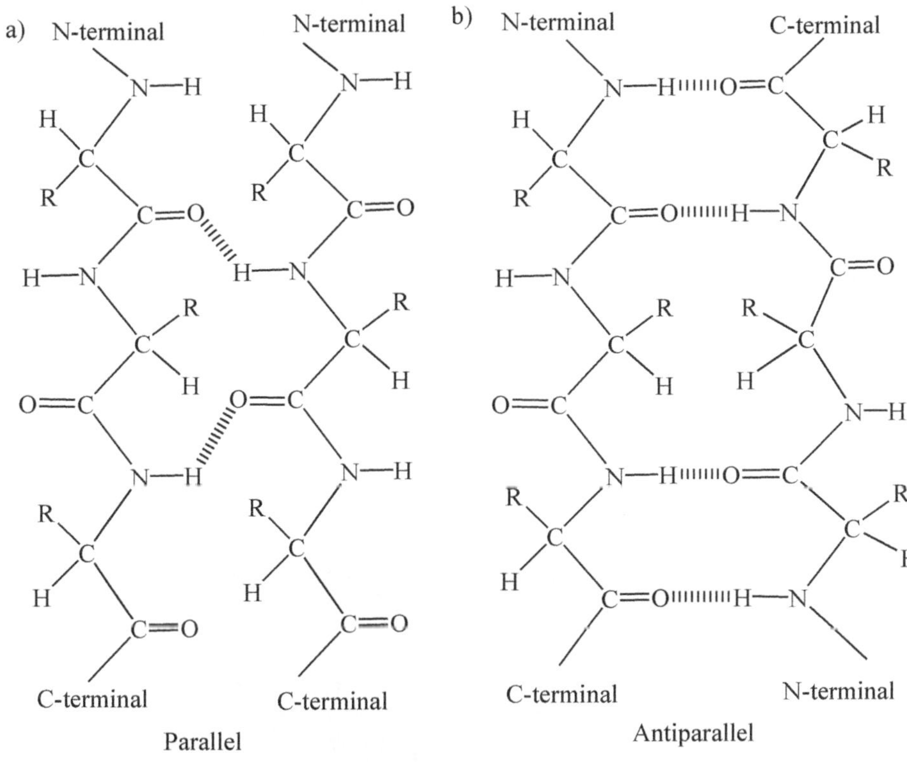

Figure 3.7: β-pleated structure of protein

3. **Tertiary Structure:** The overall three-dimensional structure of a polypeptide is called its **tertiary structure**. It is obtained by folding, twisting and bending of a polypeptide chain that gives a characteristics shape to the protein which is responsible for its functional properties. The tertiary structure is primarily due to interactions between the R groups of the amino acids that make up the protein. Various interactions responsible for stabilization of tertiary structure include hydrogen bonding, ionic bonding, dipole-dipole interactions and London dispersion forces,

the whole range of non-covalent bonds. For example, the side groups (R) containing like charges repel one another, while those which contain opposite charges form an ionic bond. Similarly, if side chain (R) contains polar group then it can form hydrogen bonds and other dipole-dipole interactions. Tertiary structure of protein is also stabilized by hydrophobic interaction that takes between non-polar side groups of an amino acid. It is present inside of the protein. This interaction is also called Vander wall's interaction and it is a weak interaction but play important role in its stability. Whereas hydrophilic groups of the amino acid are oriented outside and free to interact with surrounding water molecules (Figure 3.8).

Apart from above, a special type of covalent bond is also present in the tertiary structure i.e. the disulphide bond. Disulphide bond is the only covalent bond that is formed between the sulphur containing amino acids such as cysteine. It is much stronger as compared to other bonds that are responsible for the stability of tertiary structure. These bonds act like "safety pins," keeping parts of the polypeptide firmly attached to one another.

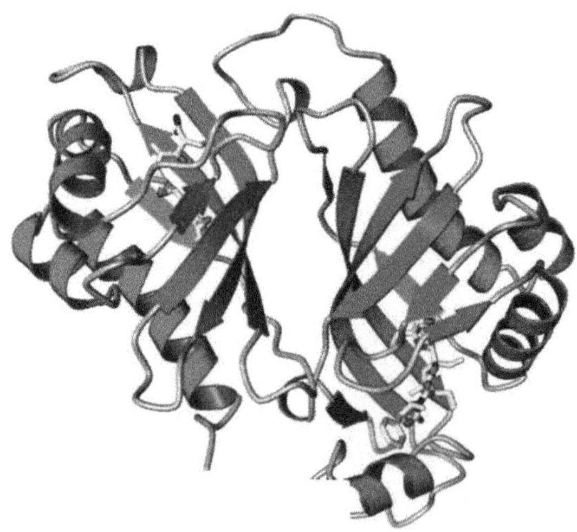

Figure 3.8: Tertiary structure of protein

Source: https://comis.med.uvm.edu/VIC/coursefiles/MD540/MD540-Protein_Organization_10400_574581210/Protein-org/Protein_Organization_print.html

4. **Quaternary Structure:** Many proteins are made up of a single polypeptide chain and have only three levels of structure. However, there are some proteins made up of multiple polypeptide chains, also known as subunits. When these subunits join together, they give the protein its **quaternary structure**. Quaternary structure is stabilized by same type of interactions that contribute to tertiary structure such as hydrogen bonding, London forces, hydrophobic interaction. There is no covalent interaction in this structure. For example hemoglobin. As discussed, hemoglobin carries oxygen in the blood and is made up of four subunits, two each of α and β types. Another example is DNA polymerase. This enzyme is made up of ten subunits and helps in synthesis of new strand of DNA.

3.36 Test of Proteins

1. **Biuret Test:** The compounds containing peptide linkage undergo this test. In this test proteins are treated with alkaline solution of copper sulfate, formation of a bluish violet color confirms the presence of protein. This test is given by protein containing 2 or more peptide bonds.

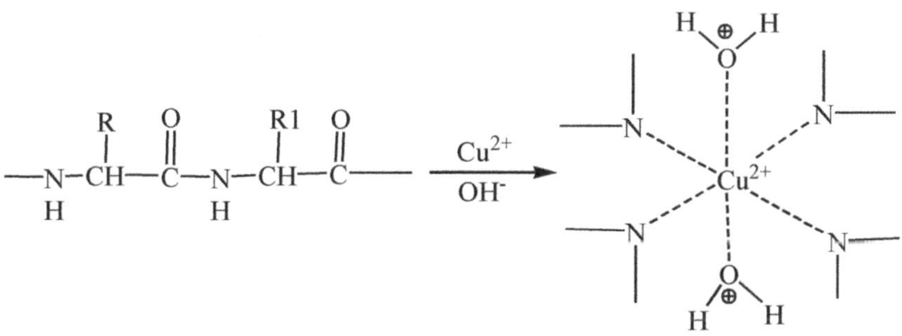

2. **Xanthoproteic Test:** When proteins are heated with Conc. nitric acid, formation of yellow color or precipitate which turns to orange color on treatment with alkali. The yellow substance obtained is called xanthoprotein, hence this test named xanthoproteic test. This test is given by proteins containing aromatic amino acids.

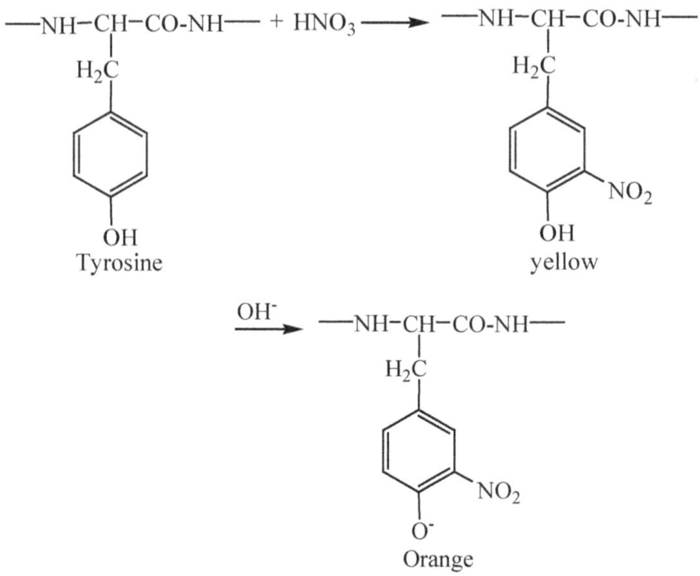

3. **Millions Test:** This test is given by protein containing tyrosine. When proteins are treated with millions reagent (mercury dissolved in nitric acid) a white precipitate is formed which turns to brick red color on boiling.
4. **Ninhydrin Test:** When Proteins or amino acids are heated with ninhydrin reagent, a deep blue to violet-pink color is obtained except proline and hydroxyproline which give yellow color.

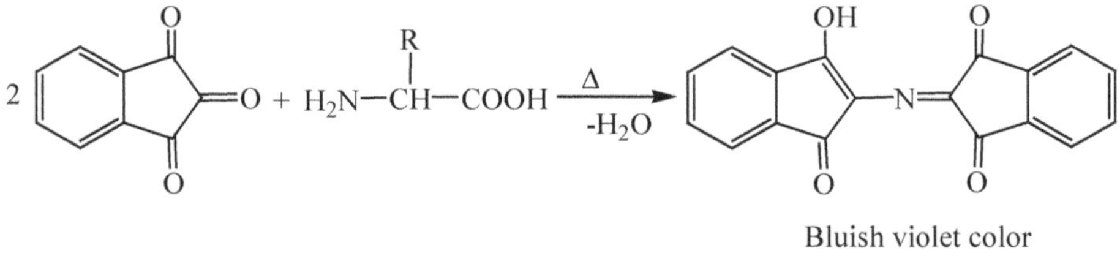

Questions

1. Define amino acids.
2. Classify amino acids on the basis of their nature. Give example of each.
3. What is the difference between essential and conditionally essential amino acids?

4. Write the configuration of naturally occurring amino acids.
5. Name the amino acid that has R configuration.
6. How would you synthesize alanine from gabriels phthalimide method?
7. Synthesize tryptophan by Erlenmeyer azlactone method.
8. Synthesis serine by acetamidomalonic ester method.
9. Synthesize proline by phalimide-malonic ester method.
10. Synthesize aspartic acid by Strecker's method.
11. What do you mean by N-terminal and C-terminal residue?
12. How would you detect the N-terminal end of peptide by Edmans degradation method?
13. Write down drawbacks of Sanger's method.
14. What is the significance of hydrazinolysis method?
15. How can you selectively cleave peptide using carbooxypeptidase, trypsin, chymotrypsin and pepsin?
16. Write down mechanism of cleavage of peptide at methionine end by cyanogen bromide.
17. How would you synthesize Gly-Leu using benzyloxycarbonyl group?
18. How would you synthesize Ala-Ala-Ala...by NCA method?
19. What are N-protecting and C protecting groups?
20. How would you synthesize Ala-Val-Gly by activated ester method?
21. How would you synthesize Ala-Val-Ser by Merrifield solid phase method? Give its advantage over general method.
22. Give difference between simple and complex proteins.
23. Write a short note on denaturation.
24. Explain electrophoresis.
25. Define isoelectric point. Give its significance.
26. Calculate isoelectric point of alanine. Pka1= 2.34 and Pka2= 9.69.

27. Synthesis Ala-Gla by DCC method.
28. Write down four levels of protein structure.
29. Draw structure of α-helical and β-pleated structure.
30. Write down difference between globular and fibrous structure of protein.
31. What is the significance of ninhydrin reaction?
32. Write down various test of proteins.

❑❑❑

4

Enzymes and Correlation with Drug Action

4.0 INTRODUCTION

Enzymes are those proteins which are biological in origin and regulate the specific function by catalyzing certain chemical reactions in every living things, from virus to man. So enzymes triggers the rate of metabolism, though is not itself changed in the course and is regenerated at the end of the course rather than changing the equilibrium constant of the reaction, it enhanced the rate of the reaction at which the equilibrium is approached. Breathing digestion, formation of the body tissues, heart action, movement of muscles and many more depend on the enzymes. In short we could say that without enzymes there is no life.

Another property of enzyme is their specificity i.e. the enzymes are not substrate specific only but also operate at specific condition of temperature, pH, etc. Maximum of the enzymes formed via a cell function within that cell and hence are called **endoenzymes** means act inside the similar cell where these are synthesized and so are also called metabolic enzymes e.g. respiratory enzymes, But some enzymes are liberated by living cell and catalyze reaction within the cell's environment, such enzymes are called as **exoenzymes** in other way you could say that operate outside the sites of biosynthesis e.g. - digestive enzymes

If an enzyme is secreted in a form which acts upon substrate as such i.e. without undergoing prior modification in structure. It is well-known as **Zymase** or active form of enzymes. e.g. Intracellular enzymes.

DOI: 10.1201/9781003727453-4

Whereas, in some cases the enzyme exist as inactive form instead of pressure called **Zymogens** (also known as pre-enzyme or pro-enzymes), which are activated on coming in contact with an activating agent.e.g. pancrease and pancreatic juices contains the zymogen. Trypsinogen which become active trypsin when coming in contact with previously formed trypsin or with the *enteropeptidase, enterokinase* of duodenal mucosa.

Enzymes are necessary in biological system because these allow the reaction to undergo at regulated pH and temperature. Secondly the reaction might occur too slow to support the metabolic processes needed to support life e.g. **G.R Stark** calculated that single molecule of urease enzyme hydrolyses about 30,000 urea molecules into CO_2 and ammonia in one second whereas in the lack of enzyme, same reaction will be completed in 3 million years.

The existence of enzyme as first reported by **W.Kuhne** in 1878 but it was **Edward Buchner** (1897), a German Chemist, who first isolated the enzyme from the yeast cells. He noticed that during the synthesis of yeast extract for medicinal purposes, the extract turned bad. He added sugar in the yeast extract to check the spoilage. He found that fermentation occurred and alcohol was produced. So, Buchner accidentally opened the door of modern enzymology. He reported that the yeast extract consist of an enzyme complex called Zymase which bring about enzymatic breakdown of hexoses.

Buchner gave the term enzyme from two Greek words. *En* = in, *Zyme* = yeast, so enzyme means inside the yeast.

There exist more than 1 lakhs of variable enzymes most of which being soluble in water or dilute solution; colorless solids, few are green, greenish brown or blue have been identified.

Proteinous nature of enzyme was reported by **J.B.Sumner** in 1926. He successfully isolated the urease enzyme in the pure crystalline form from the Jack bean extract. His work was further supported by **Northrop** *et al.* (1930-36) who isolated crystalline form of pepsin, trypsin and chymotrypsin.

A biological system needs the enzyme for number of metabolic processes some of which are catabolic while other are anabolic. A biological system needs the enzymes for various reactions.

- Metabolic actions occur at controlled temperature as these reactions might normally occur outside the body cells at very high temperature which might burn the cells.

- These act as biocatalysts to speeden the rate of biological functions which otherwise might have been too slow to meet the energy needs of body; or body temperature might have been too high to be fatal for the body

So enzymes make the essential biochemical process possible which makes the life possible.

4.1 Definition

Enzymes are colloidal organic macromolecules soluble in water, that are partially or wholly proteinous in nature; biological in origin; produced in micro amounts and having the capacity of catalyzing the specific biochemical reactions at given pH and temperature.

4.2 Chemical Nature

Enzymes, without any exception, are wholly or partially proteinous in nature so have specific 3-dimensional tertiary structure. But not necessarily all proteins are enzymes. Even their activity depends upon the integrity of their tertiary proteinous structure. The proteinous nature of proteins was first shown by **J. B. Sumner** (1926) and was further supported by **Northrop** et. al. (1930-36). The proteinous nature of enzymes is evident from:

- Both enzymes and proteins, when boiled with strong acid or base, undergo coagulation and lose their biological function.
- Enzymes are macromolecules having high molecular weight ranging from 12,000 to 1 million Daltons.
- Enzymes, on hydrolysis, yield a combination of amino acids.
- Enzymes and proteins both do no pass through dialyzing membrane.
- Both are colloidal in nature.
- On the basis of their chemistry, the enzymes are divided into two categories:
  - (i) **Pure Proteinous Enzymes:** These enzymes are formed of only one/more than one polypeptide chains formed of the amino acids interlinked through peptide bonds e.g. Pancreatic RNAase, Protease, Amylase, Pepsin, Trypsin, etc.

(ii) Conjugated, or Compound Proteinous Enzymes: A conjugated enzyme is formed of 2 components (Figure 4.1):

(a) Apoenzyme: It is the proteinous part of conjugated enzyme.

(b) Prosthetic Group: This is the non-proteinous fragment of a conjugated enzyme. Prosthetic group is called *cofactor* when it's a simple metallic ion or is called *coenzyme* when it's a complex organic molecule. B-Complex vitamins mainly act as components of coenzymes (Table 4.1). The cofactor or coenzyme is essential for the biological activity of enzyme. It is believed that the activators combine with the apoenzyme to orient it to form proper ES-complex. In some cases, the cofactor behave as a bridge between the substrate and enzyme forming a fragile ES-complex e.g. Mn^{++} for glycyl peptidase. So apoenzyme is hence catalytically inactive. Hence, the deficiency of a specific vitamin causes metabolic disorders in the individual. So the prosthetic group is either a vitamin or a metallic ion. The prosthetic group helps in biochemical functioning of cells. These probably mold the apoenzyme for forming a proper E-S complex both apoenzyme and prosthetic group are collectively called enzyme system or holoenzyme. So a holoenzyme is a conjugated protein.

$$\text{Holoenzyme} \rightleftharpoons \text{Apoenzyme} + \text{Coenzyme}$$

or

$$\text{Conjugated protein} \rightleftharpoons \text{Protein} + \text{Prosthetic group}$$

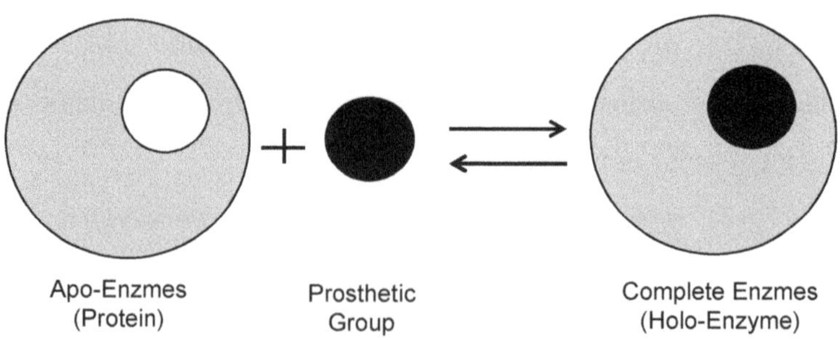

Figure 4.1: A Conjugated enzyme (Holoenzyme)

Table 4.1: Enzymes and their prosthetic groups.

Enzyme	Cofactor	Enzyme	Coenzyme
Catalase	Fe^{2+}	Succinate DHase	FAD
Hexokinase	Mg^{2+}		
ATPase	Na^{2+}	Malate DHase	NAD
Carbonic anhydrase	Zn^{2+}	Pyruvate DHase	NAD
Arginase	Co^{2+}		
Nitrogenase	Mo^{2+}		
α-Amylase	Cl^{-}		
Aconitase	Fe^{2+}	Decarboxylases	TPP
Lysine oxidase	Cu^{2+}		
Nitrate Reductase	Mn^{2+}		
Ascorbic acid oxidase	Cu^{2+}		

The master molecule DNA, consists of genetic info needed for the proteins synthesis. Biosynthesis of proteins or enzymes involves the unidirectional flows of genetic information commonly called **central dogma** as shown below (Figure 4.2).

Ribozymes and RNAase-P are two non-protein enzymes wherein RNA acts as catalyst.

Figure 4.2: Central dogma

4.3 Properties of The Enzymes

1. Enzymes are polymeric organic compounds with large molecular weight e.g. mol. wt. of Urease = 485,000 daltons. Smallest enzyme is peroxidase while largest being catalase found in peroxisomes.
2. Enzymes are wholly (e.g. Pepsin) or partially (e.g. Catalase with Fe^{2+} and Succinate DHase with FAD) proteinous in chemical composition.
3. Enzymes are needed in micro-amounts as these are used again and again.
4. Enzymes are biological in origin and have 3-dimensional tertiary structure maintained by H-bonds, ionic bonds, disulphide bonds and hydrophobic forces of attraction.
5. These act as biocatalysts so help in reaching the equilibrium constant faster by lowering the energy barrier and needed activation energy.

6. Enzymes neither initiate nor change the end products nature. These do not undergo any permanent change in the reactions being catalyzed by them.
7. Enzymes are substrate specific e.g. Maltase, Sucrase or Invertase and Lactase could hydrolyze only maltose (malt sugar), sucrose (cane sugar), lactose (milk sugar) respectively. This substrate specificity lies only at active sites of enzyme. On the basis of substrate specificity, enzymes show two sorts of substrate specificity:
   - **(a) Absolute specificity:** In this, the enzyme catalyzes only one type of substrate molecules and does not catalyze even substrate molecules related very closely. e.g. Aspartase reversibly converts fumaric acid into L-Aspartic acid but doesn't catalyze malic acid or the D-Aspartic acid Similarly, Urease (hydrolysis of urea into CO_2 and NH_3), Catalase ($2H_2O_2 \longrightarrow 2H_2O+O_2$) and Carbonic anhydrase ($CO_2+H_2O \longrightarrow H_2CO_3$) also show absolute specificity.
   - **(b) Relative Specificity:** Here, the enzyme catalyzes many substrate molecules having common structural features e.g. D-amino acid oxidases acting on all D-amino acids but cause oxidative deamination of tyrosine more swiftly than other D-amino acids.
8. Enzymes are temperature specific. Every enzyme is most active at a definite temperature called optimum temperature e.g. 37°C for Amylase Range optimum temperature is 25–40°C. For most of enzymes, the reaction rate doubles for a rise of 10°C (Q10) from 5° to 35°C.
9. Enzymes are pH-specific (pH means negative log of H^+ concentration) too, so every enzymes is most active at a definite pH called optimum pH e.g. 6.8 pH e.g. 6.8 pH for Amylse; 1.2–1.8 pH for Pepsin; 8.8–pH for Trypsin; 7.6 pH for Catalase; 7.2 pH for Glucose-6-phosphatase; etc. Average pH range is 6-7.5.
10. Enzymes are mostly endoenzymes or intracellular enzymes (act within same cell) e.g. digestive enzymes.
11. Enzymes have high turnover number or Katal (i.e. number of substrate micromoles acted upon by one micromole enzyme molecule in one minute) e.g. 5 million for catalase ($2H_2O_2 \longrightarrow 2H_2O + O_2$); 18 million for urease (Urea $\longrightarrow 2NH_3 + CO_2$); 1.9×10^4 for α-Amylase (Starch $\longrightarrow$ maltose); and 36 million for carbonic anhydrase ($CO_2 + 2H_2O \longrightarrow H_2CO_3$). Carbonic anhydrase is so far the fastest known enzyme while lysozymes have lowest turnover number.

High turnover numbers enzymes explain their remarkable effectiveness even though these are present in very lesser amounts.

12. In metabolism, enzymes act in a definite sequence e.g.

$$\text{Proteins} \xrightarrow[\text{(Gastric Juice)}]{\text{Pepsin}} \text{Proteoses Peptones}$$

$$\xrightarrow[\text{(Pancreatic Juice)}]{\text{Trypsin}} \text{Peptides} \xrightarrow[\text{(Intestinal Juice)}]{\text{Peptidase}} \text{Amino acids.}$$

So the final products formation is intermediated by a number of intermediate products.

13. Enzymes always catalyze reversible reactions e.g.

$$\text{Triglyceride} + 3H_2O \xrightarrow{\text{Lipase}} \text{Glycerol} + 3 \text{ Fatty acids}$$

In reversible reactions, the products of forward reaction become the substrates of backward reaction.

14. Enzyme activity could be increased by certain specific agents called activators which activate the inactive enzymes called proenzymes or zymogens. The synthesis of zymogens is a biological safeguard against autodigestion of the cell. These proenzymes might be activated at particular pH provided by H^+, or in presence of some specific enzymes. This kind of enzyme activation comprises the elimination of some peptide bond to produce catalytically active enzyme e.g.

(a) $\text{Pepsinogen} \xrightarrow[\text{(Provided by HCl of gastric juice)}]{H^+} \text{Pepsin} + \text{Inactive dipeptide}$

(Proenzyme secreted by Peptic cells of gastric glands) — Proteolytic (Endopeptidase)

(b) $\text{Trypsinogen} \xrightarrow[\text{(Intestinal juice)}]{\text{Enterokinase enzyme}} \text{Trypsin} + \text{Inactive dipeptide}$

(Proenzyme secreted by acinal cells of pancreas) — Proteolytic (Endopeptidase)

Pepsin or trypsin, once formed, and then start the activation of remaining proenzyme. This is called autocatalytic reaction or **autocatalysis**.

15. Enzyme activity is inhibited by certain chemical compounds called as inhibitors and the process of decrease in activity of the enzyme by inhibitors is called inhibition. This inhibition might be reversible (competitive or non-competitive) or irreversible. Some enzymes show feedback inhibition. The inhibitors consist of the metabolic poisons like cyanides inhibition. (Known to destroy respiratory cytochrome reductases of ETS and cause cyanide poisoning).

4.4 Classification of the Enzymes

As per the International Union of Biochemistry—1961, the enzymes are distinguished into six classes (Table 4.2). Every class is then divided further into subclasses which are divided further into sub-subclasses in which specific enzymes are being numbered. This system of classification is called IUB-system of enzymes classification. This system is grounded on the chemical reactions being catalyzed through the enzymes. In this system, every enzyme has a systematic code number, called E.C. Number, formed of four-digit number. First digit indicates class, second digit designates subclass, third digit indicates sub-subclass while fourth digit identifies a specific enzyme e.g. E.C. number 1.1.1.1 indicates NAD-oxidoreductase which transfers two H-atoms from an alcohol to NAD so trivial name of enzyme is alcohol dehydrogenase.

Class 1. Oxidoreductases: These enzymes catalyze the electrons transfer from one compound to the another, so these enzymes are indulged in oxidation (elimination of electrons or gaining of oxygen) and the reduction (gaining of electrons or loss of oxygen) which are together termed as redox reactions where reductant is electron-donor while oxidant is electron-acceptor.

Some necessary subclasses of oxidoreductases are:

(a) Dehydrogenases: These involve the elimination of 2-hydrogen atoms from the substrate molecule to a H-acceptor which is either FAD, FAD, NAD^+ or $NADP^+$. These enzymes act on functional groups like alcohol and –CHO, $-–CH_2-CH_2–,–CH_2-NH_2$ and $–CH=NH$

e.g.

$$C_2H_5OH + NAD^+ \xrightleftharpoons{\text{Alcohol DHase}} CH_3CHO + NADH + H^+$$

Ethanol (H-Acceptor) Acetaldehyde

(H-donor)

These include succinic dehydrogenase, lactate dehydrogenase, Malic dehydrogenase, etc.

(b) Enzymes Acting on CH–NH_2 Group as Electron-donor: These enzymes not only remove 2H-atoms from the substrate but also cause deamination (loss of –NH_2 group) eg. Glutamic dehydrogenase of animal liver.

$$\text{L-Glutamic acid} + H_2O + NAD^+ \xrightleftharpoons{\text{Glutamate DHase}} \alpha\text{-ketoglutaric acid} + NH_4 + NADH + H^+$$

(c) Enzymes Acting on H_2O_2 as Electron-acceptor: These are also termed as perodixases and catalyze the reactions where the hydrogen peroxide act as the oxidant while H-donor is reduced coenzyme e.g.

$$NADH + H+ + 'H_2O_2 \xrightleftharpoons{\text{NADH-Peoxidase}} NAD_+ + 2H_2O$$

Catalase is a peculiar peroxidase and catalyzes the reaction where H_2O_2 serves as both donor and acceptor and detoxifies the cellular hydrogen peroxide.

$$H_2O_2 + H_2O_2 \xrightleftharpoons{\text{Catalase}} 2H_2O + O_2$$

(d) Oxidases: These oxidoreductases transfer $2e^-$ from the donor to molecular oxygen, resulting, H_2O_2 formation e.g. Glucose oxidase or to nascent oxygen to form H_2O, e.g. cytochrome oxidases.

$$\beta\text{-D-Glucose} + O_2 \xrightleftharpoons{\text{Glucose Oxidase}} \delta - \text{Gluconolactone} + H_2O_2$$

These include ascorbic acid oxidase, oxalate oxidase, etc.

(e) Oxygenases: These oxidoreductases are catalyzing the assimilation of two oxygen atoms into the substrate e.g.

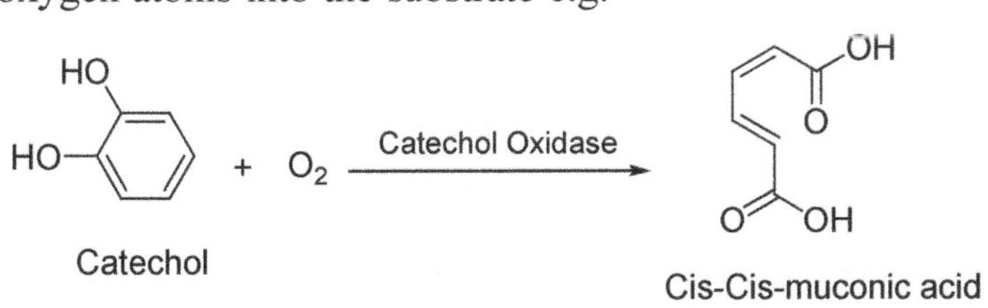

(f) Hydroxylases: These catalyze the assimilation of one oxygen atom of the molecular oxygen into the substrate whereas the second oxygen atom appears as water e.g.

$$\text{Progesterone} + O_2 + NADPH + H^+ \xrightleftharpoons[\text{Hydoxylase}]{\text{Progesterone}} \text{Deoxy-corticosterone} + NADP^+ + H_2O$$

Class 2. Transferases: The reactions involving transfer of groups like amino, acyl, phosphate, glycosyl group etc. from one compound to another compound, or from one position to another position in the similar molecule are catalyzed by these enzymes.

Main subclasses of transferases are:

(a) Acyltransferases: The transfer of acyle group e.g. Choline acyltransferase is catalyzed by these enzymes.

$$\underset{\text{(Donor)}}{\text{Acetyl Co. A + Choline}} \xrightleftharpoons[\text{Acyltransferase}]{\text{Choline}} \underset{\text{(Acceptor)}}{\text{Acetylcholine + Co. A}}$$

(b) Transaminases: These enzymes transfer the amino group from one amino acid to other keto acid acceptor so bring in a new amino acid.

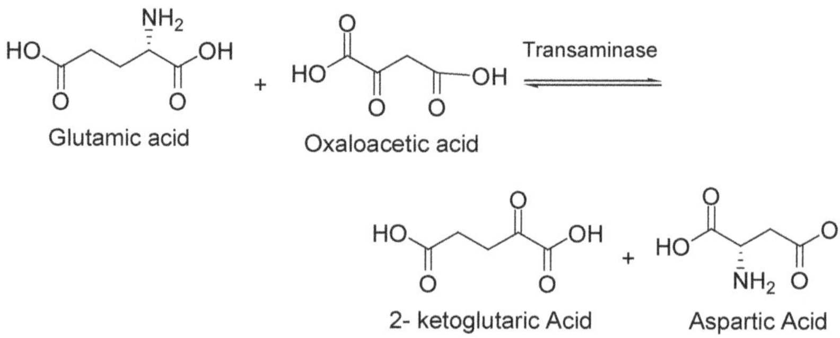

(c) Kinases: The transfer to phosphoryl group from the ATP or another nucleotide triphosphate, to amino or alcohol group acceptors so also called transphosphorylases e.g Glucokinase (Hexokinase) is catalyzed by these enzymes.

$$\text{D. Glucose + ATP} \xrightleftharpoons[Mg^{2+}]{\text{Hexokinase}} \text{Glucose-6-P + ADP}$$

$$\text{Creatine-phosphate + ADP} \xrightleftharpoons[\text{(Transphosphorylation}]{\text{Creatine kinase}} \text{Creatine + ATP}$$

(d) Glycosyl Transferases: The transferring of the activated glucosyl residue from uridine-diphosphoglucose (UDP-glucose) to the growing end of the glycogen primer during biosynthesis of glycogen e.g. is catalyzed by these enzymes

$$\text{UDP- Glucose + Glycogen primer} \rightarrow \underset{\text{by one glucosyl unit}}{\text{Glycogen extended}} + \text{UDP}$$

Class 3. Hydrolases: These enzymes catalyze the hydrolytic breakdown of complex polymeric compounds such as proteins, carbohydrates, fats and the nucleic acids into their monomeric units. Hydrolytic breakdown involves addition of water followed by cleavage of anhydro bonds formed by dehydration synthesis. The commonly broken bonds are $C-O, C-N, O-P$ and $C-S$ bonds. **Digestive enzymes are hydrolases.**

Depending on the nature of bonds hydrolyzed, hydrolases are classified into following categories:

(a) Esterases: These enzymes hydrolyze the ester bonds commonly present in the fats e.g. lipase, acetyl cholinesterase etc.

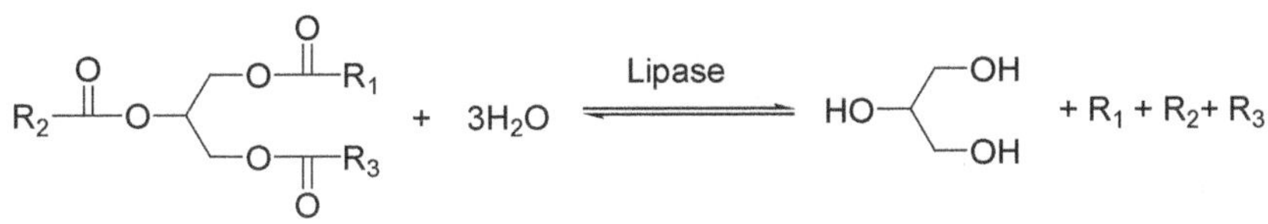

$$\text{Acetylcholine} + H_2O \xrightleftharpoons{\text{AchE}} \text{Acetic acid} + \text{choline}$$

(b) Glycosidases or Carbohydrolases: These hydrolyze the glycosidic bonds of the carbohydrates e.g. sucrose, maltase, lactase, amylase, etc.

$$\text{Sucrose} + H_2O \xrightleftharpoons[\text{(Invertase)}]{\text{Sucrase}} \text{Glucose} + \text{Fructose}$$

$$\text{Maltose} + H_2O \xrightleftharpoons{\text{Maltase}} 2\ \text{Glucose}$$

(c) Proteases or Peptidases: These hydrolyze the peptide bonds present between the amino acids in the proteins and might be endopeptidases (e.g. Pepsin, Trypsin) or **exopeptidases** (e.g. carboxypeptidase and aminopeptidase), e.g.

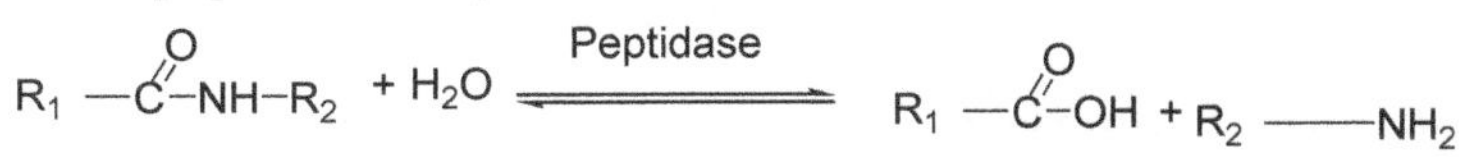

Dipeptide

$$\text{Protein} + H_2O \xrightleftharpoons{\text{Pepsin}} \text{Proteoses} + \text{Peptones}$$

(d) Nucleases: These hydrolyze the phosphodiester bonds present between thenucleotides of nucleic acids e.g. RNAase, DNAase, etc.

$$\text{RNA} + H_2O \xrightleftharpoons{\text{RNAase}} \text{Ribonucleotides}$$

$$\text{DNA} + H_2O \xrightleftharpoons{\text{RNAase}} \text{Deoxyribonucleotides}$$

Class 4. Lyases: These enzymes catalyze the elimination of groups like water, CO_2 or ammonia without hydrolysis, so these act on bonds like C–C,

C–O, C–N, C–S and C–halide bonds. The main subclasses are:

(a) Aldehyde Lyases: These enzymes catalyze the compounds having aldehyde group e.g. Aldolase.

Fructose-1, 6-Diphosphate $\xrightleftharpoons{\text{Aldolase}}$ 3-PGAL + Dihydroxyacetone-phosphate.

(b) Carbon-oxygen Lyases: These catalyze the C = O bonds and remove water molecule e.g.

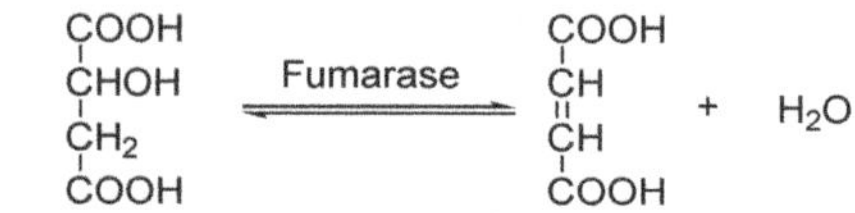

L.Malate Fumarate

Citric acid $\xrightleftharpoons[\text{dehydratase}]{\text{Citrate}}$ Cis-Aconitic acid

(c) Decarboxylases (C–C lyases): These enzymes remove CO_2 from α-and β-keto acids or amino acids e.g. Histidine decarboxylase.

Histidine $\xrightleftharpoons{\text{Histidine decarboxylase}}$ Histamine + CO_2

Class 5. Isomerases: These catalyze the rearrangement of molecular structure to induce cis-trans, keto-enol and aldose-ketose interconversions. These change the compounds into their isomers.

Main sub classes of isomerases are:

(a) Cis-trans Isomerases: This class catalyze cis-trans interconversions e.g. Retinene isomerase

Trans-Retinen $\xrightleftharpoons{\text{Retinene isomerase}}$ Cis-Retinene

(b) Aldose-ketose Isomerases: This class of enzymes catalyze the interconversion of aldose form to its isomeric ketose form and vice-versa e.g. Transphophoisomerase.

3-Phosphoglyceraldehyde $\xrightleftharpoons[\text{Isomerase}]{\text{Triosephospho}}$ Dihydroxy acetone-P

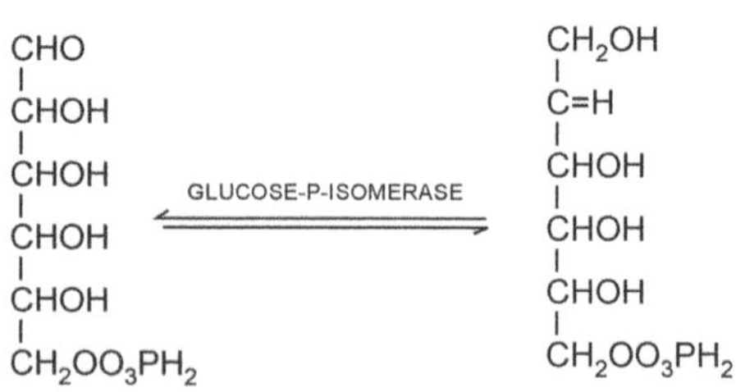

GLUCOSE-6-P FRUCTOSE-6-P

(c) Epimerases: These catalyze interconversion of asymmetric carbons and are also known as racemases.

$$\text{D-Xylulose-5-P} \xrightleftharpoons{\text{Epimerase}} \text{D-Ribulose-5-P}$$

$$\text{D-Lactic acid} \xrightleftharpoons{\text{Racemase}} \text{L-Lactic acid}$$

(d) Mutases: These catalyze interconversion of asymmetric carbons and are also known as racemases.

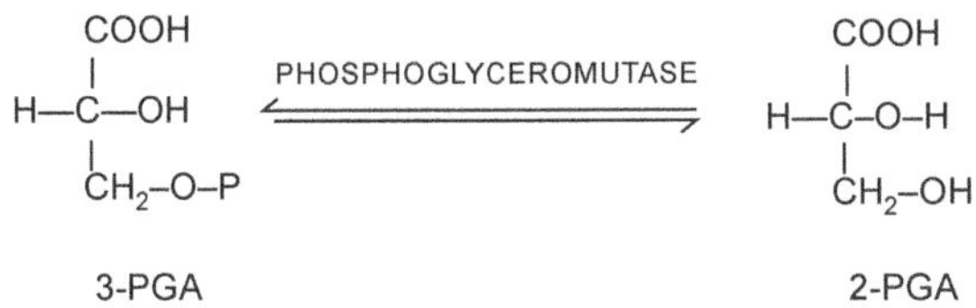

Class 6. Ligases (Ligare = To bind)**:** These enzymes catalyze the condensation reactions and involve the forming of bonds like C–C, C = O, $C \equiv N$, etc. due to coupling of two dissimilar molecules in the presence of energy provided by ATP cleavage. These are also called synthetases. It includes the following subclasses.

(a) C=O Synthetases: These catalyze the forming of C=O bonds in the presence of energy e.g. Pyruvate decarboxylase.

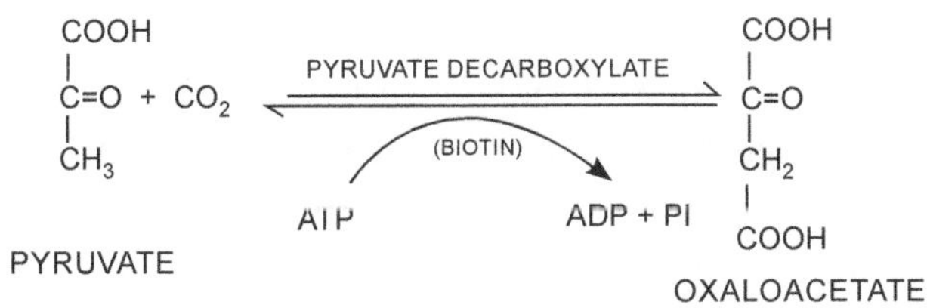

(b) $C \equiv N$ Synthetases: These catalyze the forming of $-C \equiv N$ bond due to addition of ammonia e.g. Glutamine synthetase.

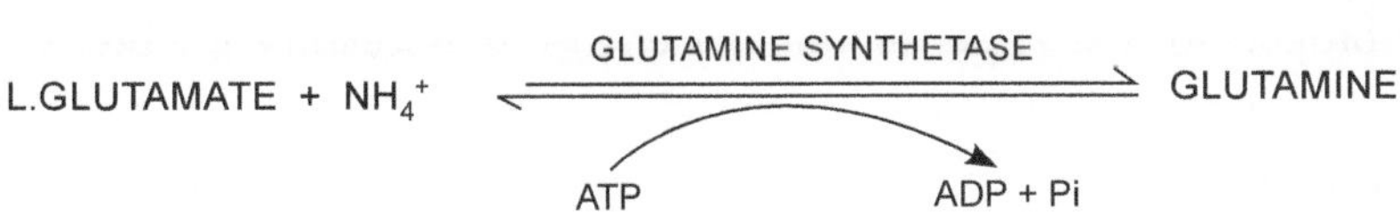

(c) C–C ligases: These catalyze the forming of C–C bond due to carboxylation e.g. Acetyl Co. A carboxylase.

$$\text{Acetyl Co.A} + CO_2 + \text{ATP} \xrightleftharpoons{\text{Acetyl Co. A. caboxylase}} \text{Malonyl Co.A} + \text{ADP} + \text{Pi}$$

Table 4.2 : Classes and major subclasses of enzymes

S. No.	Class	Major Subclasses
1.	**Oxidoreductases** (Oxidation reduction reactions)	Dehydrogenases, Oxidases, Reductases, Peroxidases, Catalase, Oxygenase, Hydroxylase.
2.	**Transferases**. (Transfer of functional groups)	Transaldolase and Ketolase. Acyl,-methyl, glucosyl-and phosphotransferases, Kinases, Phosphomutases.
3.	**Hydrolases**. (Hydrolytic reactions) Splitting of compounds)	Esterases Glycosidases, Peptidases, Phospholipases, Amidases, Deaminases.
4.	**Lyases**	Decarboxylases, Ketolases, Hydratases, Lyases.
5.	**Isomerases.** (Isomerization reactions)	Epimerases, Isomerases, Racemases, Mutases.
6.	**Lagases.** (Formation of bonds with ATP cleavage)	Synthetases, Carboxylases.

4.5 Mode of Enzyme Action

The detail of the mechanism of enzymatic reactions is still not known: many theories have been given for explaining the mechanism.

(a) Lock and Key Mechanism. (Figure 4.3): The bigger enzyme molecule coming in close contact with smaller substrate molecules leading to the intermediate complex formation. Enzymes bear distinct cavities or active sites or catalytic sites or active centers in which specific types of substrate molecules could form intimate association by their specific areas called reactive sites. Because these active sites are specifically designed, only specific substrate molecules having proper geometric shape could fit correctly. This combination occurs by electrostatic hydrogen and/or hydrophobic bonds and produces a strain on the bond which is to be broken so that lesser amount of energy is necessary for breaking it. The intermediate complex or enzyme substrate complex (also called Michaelis complex) quickly breaking for releasing the enzyme and the product of biochemical reaction. When substrate molecule attaches to the active site, its parts are detained together in such a way leading to the distortion of chemical bonds, i.e. the bonds are weakened. These weakened bonds accelerates the rate of reaction. Thus the overall process could be summarized by following reaction:

$$\text{Enzyme + Substrate} \rightarrow \text{Enzyme Substrate Complex} \downarrow \text{Enzyme + Product}$$

This simple model of enzyme activity indicates the biphasic nature of the enzyme mechanism. First step involves binding of substrate molecule with the active site of the enzyme while the second step involves the chemical modification of the substrate.

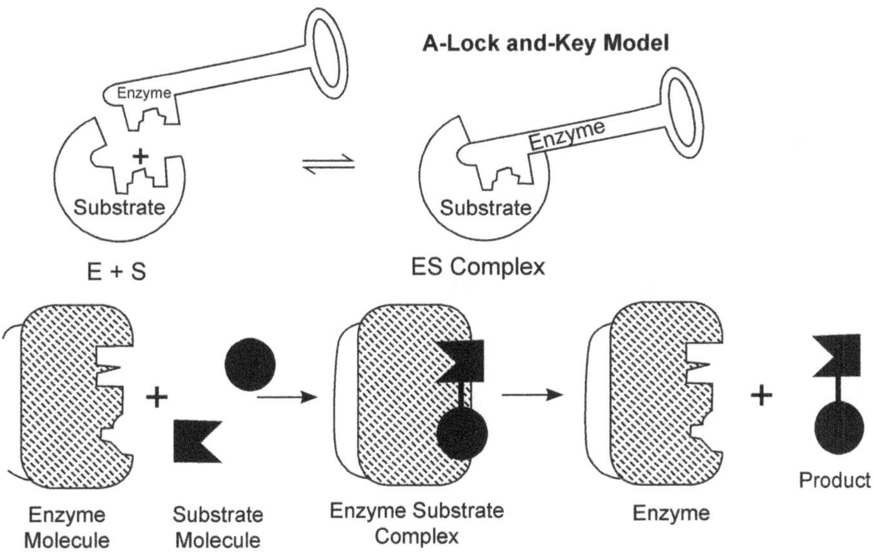

Figure 4.3: Lock and key theory of enzyme action

Nature of Active Sites

Active sites consist of special groups e.g. –OH/hydroxyl group of hydroxyamino acids e.g. serine; $-NH_2$ of basic amino acids e.g. lysine, –COOH of the amino acid; –SH of cysteine amino acid for establishing contact with substrate molecules. Sometimes greater than one side chains might be forming the active site. As particular lock could be unlocked by a particular type of key so particular enzyme could act on particular substrate due to their specificity, the name for the mechanism has been given as Lock and Key Mechanism (Fischer, 1989). He noticed that glycosidases which act on glycosides are highly specific w.r.t. their substrates which is because of the close similarity in the shape of enzyme and substrate molecule.

So the enzyme molecule provides a suitable platform or template on which the specific substrate molecules could bind and interact to form the end products. After the reaction, products leave the enzyme surface so that it's free to receive more substrate molecules. So enzyme bring the reactants together at the active sites so causing tremendous rise in the concentration of reactions locally on the enzyme molecule This approximation rises the reaction rate several thousand times.

$$E + S \underset{k_2}{\overset{k_1}{\rightleftharpoons}} \text{ES complex} \xrightarrow{k_3} E + \text{Products}$$

Where k_1= Rate constant of formation of ES complex.

k_2= Rate constant of dissociation of ES complex.

k_3 = Rate constant for the formation of products from ES complex.

(b) Induced Fit Mechanism (Figure 4.4): It was put forward by Koshland in 1959. According to this theory, the active site of enzyme bears two groups: **buttressing** and **catalytic.**

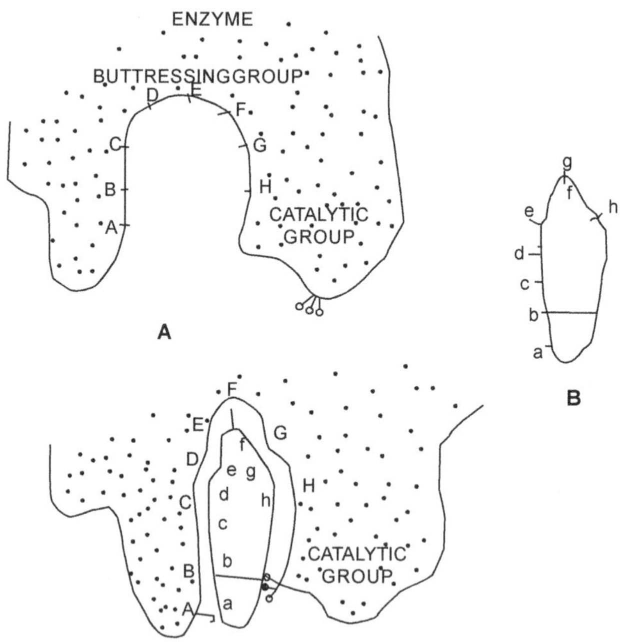

Figure 4.4: Induced fit theory of enzyme action A, active site of enzyme, substrate molecule C, enzyme substrate complex with change in confirmations so as to bring the catalytic group against the substance bonds to be broken

The buttressing group supports the substrate. The catalytic group weakens the bond of reactants by nucleophilic and electrophilic processes. As the substrate comes in contact with supporting group, the active site undergoes conformation changes. Consequently, catalytic group is brought opposite to substrate bonds, and is finally broken. In this mechanism variation in the shape of active site occurs. Induced fit is possible as of the flexibility of the protein molecules.

A single molecule of enzyme carbonic anhydrase, the fastest acting enzyme, is competent to hydrate 36 million (36×10^6) molecules of CO_2, in one minute. This catalysed reaction is 10 million times faster than non-catalysed reaction.

$$CO_2 + H_2O \xrightarrow{\text{Carbonic anhydrase}} H_2CO_3$$

(c) Michaelis-Menton Hypothesis (Figure 4.5): All chemical reactions possess a potential energy barrier the needed to be crossed before the reactants are converted into end products. So the reactant molecules must be given sufficient energy for entering in the reactive phase called transition state (TS) lying at the top of the energy barrier. Activation energy, Ea is the minimum energy needed for bringing the substrate molecules of one mole in the reactive form. In the transition state, the reactants are in the intermediate stage of the reaction pathway and are in their activated state. This transition state (T.S) complex could break down to products or reform the reactants (Figure 4.5).

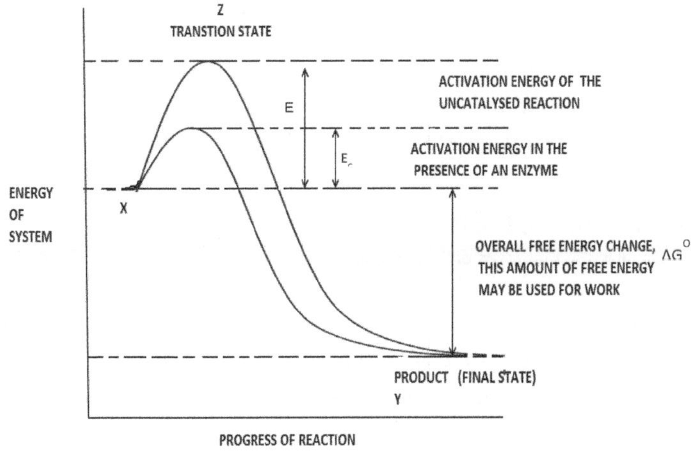

Figure 4.5: Energy profile for uncatalyzed and enzymic reactions

A comparison of the energy diagrams for catalyzed and non-catalyzed reaction (Figure 4.5) shows that there is no variation in the energy of the reactants and the products. This confirmed that enzymes don't change the thermodynamics of the system but do change the pathway for reaching the final state.

4.6 Michaelis-Menton

Michaelis-Menton (1913) explained the mode of action of enzymes on the basis of activation energy. They proposed that in an uncatalyzed reaction, the numbers of substrate molecules in transition state are less so the rate of reaction is low, but the presence of enzyme lowers the energy barrier. so more and more substrate molecules become reactive and react to produce the end-products so increasing the reaction rate.

So the enzyme provides an alternative pathway for the chemical reaction whose energy barrier is lower so needed activation energy is less.

A well-controlled body temperature of 37°C does not let temperature to be enhanced to speed up the reaction and 37°C is not warm sufficient to provide the reaction rates needed for metabolically active animals. It is so as the mammals are homeothermal. In such cases, the presence of enzymes enhances the reaction rates even at ordinary physiological temperatures by lowering the energy barrier so less activation energy is needed e.g. the energy of activation for the decomposition of H_2O, is 18 Kcal/mole, whereas in the enzyme (catalase) being there, needed activation energy is only 6.4 Kcal/mole. Thus, the activation energy in this case becomes about one-third when the enzymes is present. Similarly, casein is hydrolyzed by hydrogen ions at the energy level equivalent to 20.6 Kcal/mole, but when trypsin is used it is hydrolyzed at the energy level of only 12.0 Kcal/mole.

Energy-Kinetics of Enzymes (Michaelis-Menton Equation)

Michaelis and Menton (1913) proposed an equation which shows the interrelationship between the enzyme activity and substrate concentration and is known as Michaelis-Menton equation. The three basic assumptions of the Equation are:

1. E-S complex is in the steady state i.e. concentration of E-S complex remaining constant.

2. Under saturating conditions, all enzyme molecules are involved to form E-S complex and none is free.
3. Rate of formation of E-S complex is maximum under saturating conditions.

Under these assumptions, equilibrium expression for the formation and the breakdown of the E-S complex is expressed as:

$$E + S \underset{k_2}{\overset{k_1}{\rightleftharpoons}} ES\ complex \xrightarrow{k_3} E + Products$$

So
$$Km = \frac{k_2 + k_3}{k_1}$$

Rate of enzyme action (Vi) is expressed as:

$$Vi = \frac{V_{max}[S]}{Km + [S]}$$

S = Substrate concentration

Km = Michalis-Menton constant

V_{max} = Maximum rate of enzyme activity under a given setoff pH, temperature, ionic strength etc. and is constant for specific enzyme.

When substrate concentration [S] is equal to Km [Michaelis-Menton constant]

Then
$$Vi = \frac{V_{max}[S]}{Km + [S]}$$

$$Vi = \frac{V_{max}[S]}{[S] + [S]}$$

$$Vi = \frac{V_{max}[S]}{2[S]}$$

$$Vi = \frac{V_{max}}{2} = \frac{1}{2}V_{max}$$

This shows that when initial velocity is half-maximal, then the substrate concentration indicates Michalis-Menton constant, or Michaelis-Menton constant is equal to substrate concentration that provides half of the maximum velocity.

It could be derived in another way:

If the initial velocity is equal to half-maximal, then

$$Vi = \frac{V_{max}[S]}{Km + [S]}$$

$$Vi = \frac{1}{2}V_{max} = \frac{V_{max}[S]}{Km + [S]}$$

$$Km + [S] = \frac{2V_{max}[S]}{V_{max}}$$

$$Km = [S].$$

4.7 Chymotrypsin Catalyzed Hydrolysis of Peptide Bond

Chymotrypsin catalysing the hydrolysis of the peptide bond is symbolized by numerous steps as shown in (Figure 4.6).

Step 1: In the step I, the polypeptide substrate travels to the enzyme's active site.

The Polypeptide is binding to the active site due to the size, shape and sequence of the amino acid of chymotrypsin.

When the peptide is present in the site, H^+ travels from the serine amino acid positioned at 195 (Ser-195) of amino acid sequence of the enzyme's to the histidine positioned at 57 (His-57). Oxygen atom of the hydroxyl group of the serine's forming a covalent bond with the –C atom of one of the substrate's peptide leading to the shifting of the two electrons from one of the double bonds and forms a lone pair.

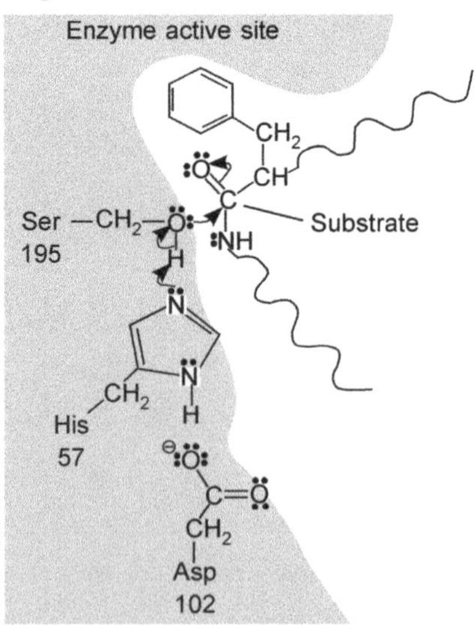

Stage 2: The positive charge present on the His-57 is pacified by the negatively charged aspartic acid positioned at 102 (Asp-102). Also the peptide bond in between the –C and –N is broken, as because the double bond is restructured between the carbon and the oxygen atom in the peptide bond,. By forming a bond with hydrogen atom available from His-57 the nitrogen-contaning group is steadied.

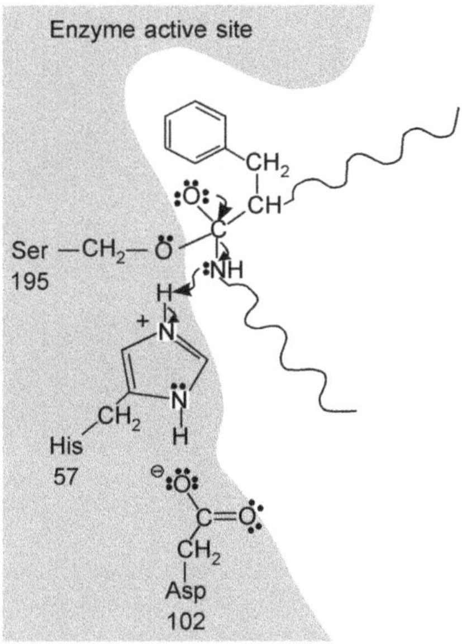

Stage 3: The part of the polypeptide having the "N" atom from the broken peptide bond comes from the active site.

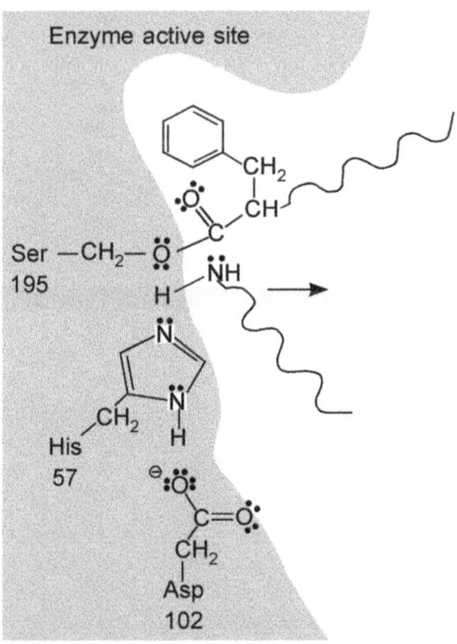

Stage 4: The H_2O molecule moves into the active site and the oxygen atom of the water molecule by losing an H^+ ion to nitrogen on His-57 facilitates the oxygen of the water for forming the bond with the remaining substrate through carbon atom. Alike in stage I, one bond in the double bond shifts up forming a lone pair.

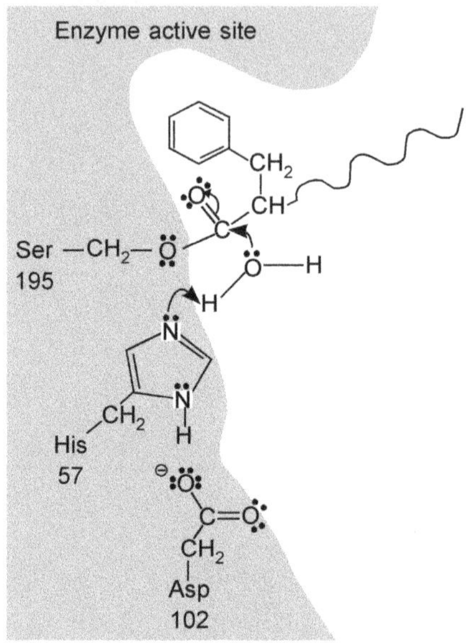

Stage 5: On formation of double bond again, the C-O bond of Ser-195 gets broken. The hydroxyl (–OH) on Ser-195 is reinstated by transferring an H^+

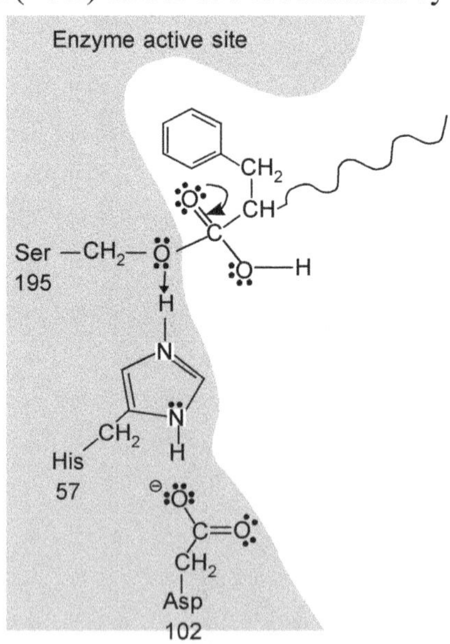

Figure 4.6: Mechanism of Chymotrypsin

from His-57. After this, both the His-57 and Ser-195 are reverted to their respective usual forms.

Stage 6: Along with leaving the active site in the novel form, which is ready for repeating stages 1-5 with an another polypeptide molecule, left over substrate portion also moves out from the active site.

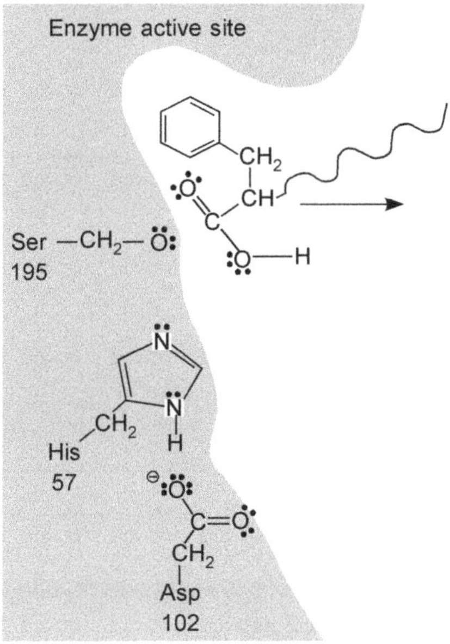

4.8 Factors Affecting Enzyme Activity

4.8.1 Substrate Concentration

Enzyme molecule being large in size is having several active sites over its surface keeping all the other factors constant, rise in substrate concentration enhances the reaction rate, since more molecules will be colliding with enzyme to fill more active sites. Hence, at constant concentration of enzyme, the enhancement in concentration of substrate speed up the reaction rate upto a level. Additional the rise in substrate concentration doesn't enhance the reaction rate.

When reaction rate is no longer changing with a rise in substrate concentration, we assume that enzyme active sites are completely filled or saturated with the substrate molecule. Here, all the active sites of the enzyme molecules get engaged by the substrate molecules and is called saturation of enzymes. Graph in Figure 4.7 represents that lower is the Michaelis constant (Km) value, higher is the substrate affinity of an enzyme. Km (Michaelis constant) of an enzyme is the concentration of the substrate at which the reaction attains half of its maximum velocity (V_{max}). It is the measure of the empathy of the enzyme for its substrate. Km value varies from substrate to substrate as because the different enzymes differ in their affinity towards different substrates. For most enzymes, the Km varies between 10 and 10^{-6}M and varies with the environmental conditions like ionic strength and the temperature.

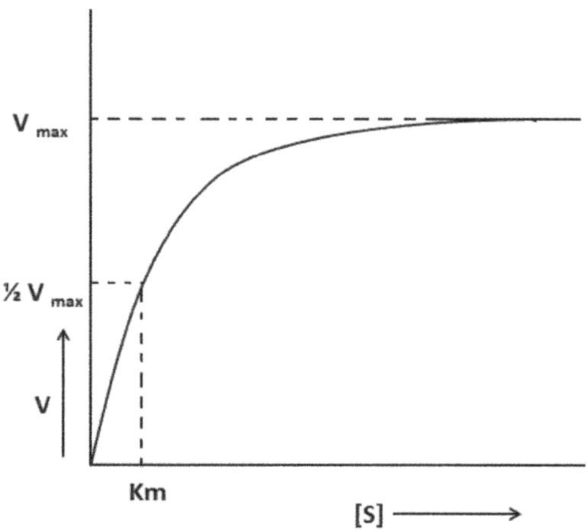

Figure 4.7: Influence of the substrate concentration versus velocity of enzymatic reaction

The concentration of the substrate affects the velocity of enzymatic reaction

Proteases act on different proteins; however, Km value differs in every case.

$$\text{The value of } V = \frac{V_{max}[S]}{Km + [S]}$$

Where V =Velocity of enzymatic reaction.

V_{max} = Limiting or maximum velocity at infinity substrate concentration

[S] = Substrate concentration in moles per litre of solution.

Km = Michaelis constant in mole per litre of solution.

4.8.2 Temperature

Enzymes are temperature specific as every enzyme is most active a specific temperature called optimum temperature e.g. 37 C for Amylase. Range of optimum temperature is 25-40°C. Any variation in the optimum temperature increases the enzyme activity (Figure 4.8). Velocity of enzymatic reaction increases upto certain optimum temperature. Enzyme activity rises from 0° to optimum temperature and doubles with every 10° rise in temperature. This is called temperature coefficient (Q 10). This is so as more and more substrate molecules become reactive by having needed activation energy. When the rate of reaction doubles by a rise of 10°C, then Q 10 is equal to 2. At lower temperature, the catalytic activity of enzyme dominates over the thermal denaturation of enzyme At very high temperature, as denaturation of enzymes predominates so Q 10 increases as high as 10-15 due to which stability of enzyme decreases rapidly with increase in temperature and is of weeks at 0°C, of days at 10°C, of hours at 30°C but only of fraction of seconds at 70°C. A very high temperature leads to sudden fall in enzymatic activity This sudden fall is due to denaturation of enzymes. At this stage a collapse of enzymatic structure occurs due to breaking of hydrogen bonds and disulphide bonds so causing loss of tertiary structure of enzymes.

Biological phenomena thus have critical temperatures al which they function Above a critical temperature enzyme is non-functional. If damaging temperature is maintained for too many minutes, the enzymes will be completely inactivated. The denaturation of enzymes is a permanent phenomenon and couldn't be renatured.

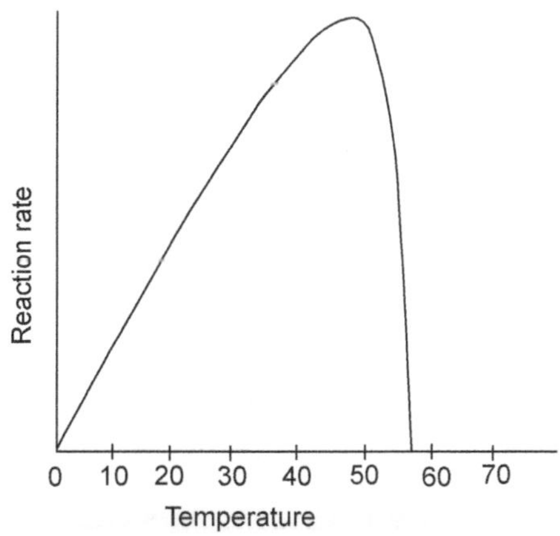

Figure 4.8: Effect of temperature

The thermal behavior of enzymes imposes serious limitations on organisms. Most cells lose their capacity to carry on metabolism at temperature above 45°C Only very few organisms which bear heat resistant enzymes are able to persist at very hot places e.g bacteria found in hot springs. The winter metabolism of most of the organisms subsides to point where dormancy is inevitable. Warm blooded animals such as man and few other vertebrates are only creatures that have evolved a method of regulating their body temperature. Very cold temperature thus preserves enzymes in a temporarily inactive stage. But the enzyme inactivation at low temperature is a reversible phenomenon so enzyme activity is regained on increase in temperature So the food and other edible things could be preserved for a longer period in frozen stage, because both microbial and food enzymes fails to act at very low temperature. The enzymes of plants and cold blooded animals of arctic areas show biochemical adaptability at very low temperature and denature even at moderate temperature.

4.8.3 pH

Enzymes normally function most efficiently at particular concentration of hydrogen ion called optimum pH which generally lies around 6.0-7.5 pH. Every enzyme reduces its activity outside its own optimum pH range (Figure 4.9). Enzymes like pepsin work at lower pH, whereas enzymes like trypsin are highly active at high pH. Urease acts at neutrality. Pepsin present in the stomach, digests proteins and works at pH value of 2.0 (acidic). Trypsin is found in duodenum and works at alkaline pH of 8.8.

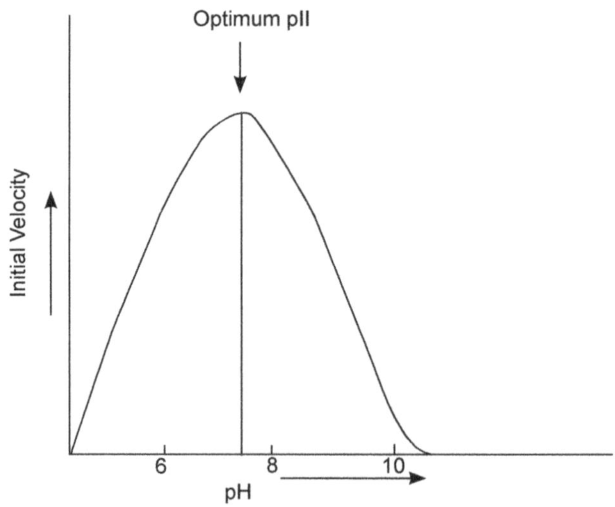

Figure 4.9: Effect of pH on enzyme activity

The concentration of hydrogen ion might change the ionization and solubility of enzymes, the inhibitors, substrates and even the activators. It might also alter the enzymes susceptibility to heat. This might be the possible explanation for different optimum pH values for different enzymes (Table 4.3).

Table 4.3 : Optimum pH of Some Enzymes

Enzyme	Optimum pH
Pepsin	2.0
Invertase (Sucrase)	4.5
Peroxidase	5.0
Cellobiase	5.0
Chlorophyllase	5.9
Maltase	7.0
Lipase	7.5–8.0
Cholinesterase	7.0
Amylase	6.8
Urease	7.0
Catalase	7.0
Trypsin	8.8
ATPase	9.0

But the optimum pH of an enzyme varies largely and is dependent on the absorbed ions, presence of other colloids, activating agents, inhibitors, history of enzyme, the age of cell and the ratio of concentration enzyme and substrate.

It has been witnessed that in case of most enzymes, reaction rate is low at strongly acidic and basic solutions as these are known for the denaturation of enzymes. Usually the enzyme is maximally active at or close to the isoelectric point. (pH where the net charge of the solution is zero e.g. optimum pH for trypsin is 8.0 while its isoelectric point is 10.1

4.8.4 Enzyme Concentration

With high substrate concentration, upsurge in enzyme concentration accelerates the velocity of enzyme-catalyzed reactions. Though in practice, it's not applicable. Figure 4.10 shows that at zero enzyme concentration, the rate of activity will be zero. This isn't correct because most of enzyme-catalyzed metabolic reactions go on even in the deficiency of the enzyme though at a very low rate. Similarly, under normal conditions, the enzyme concentration rarely exceeds 0.001M. Hence, we are restricted to work with very low concentration of enzyme. Within certain limits, the rate of enzyme activity and reaction is directly proportional to the enzyme concentration.

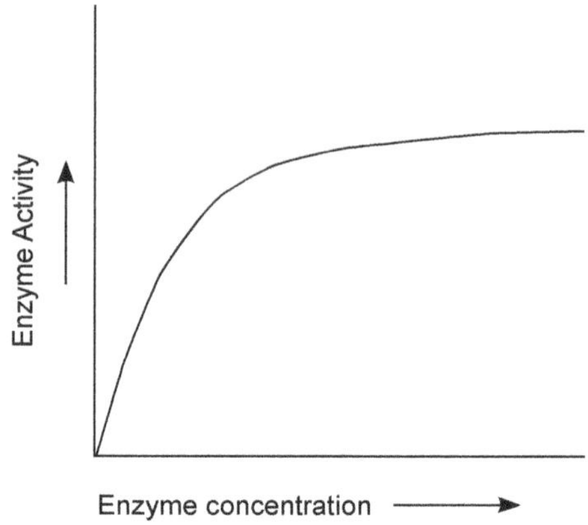

Figure 4.10: Effect of enzyme concentration on enzyme activity

This relationship could be expressed as:

$$V \propto [E] = V = \text{Velocity of reaction}$$

$$[E] = \text{Enzyme concentration}$$

$$V = K\,[E]$$

$$K = \text{Constant}$$

So $$[E] = \frac{V}{K} \qquad \because V = \frac{\text{Amount of product}}{\text{Time of reaction}}$$

So $$[E] = \frac{\text{Amount of product}}{\text{Time of reaction} \times K}$$

This relationship between the concentration of enzyme and the velocity of reaction could be used detect the strength of a specific enzyme in a combination of proteins.

4.9 Enzymes Specificity

There are mainly four kinds of enzyme specificity. The fact that every enzyme has its own set of specificities suggests that there must be a large no of different enzymes. In general the stereo and substrate specificities of enzymes attributed to the protein of the apoenzyme while the reaction specificity is more attributable to the co-enzyme prosthetic group of the apoenzyme. The specificity of enzymes could be observed in following manner.

(i) Reaction Specificity: Enzymes are specific in the sense that almost invariably one enzyme catalysis only one of the reactions which the substrate could undergo. Different enzymes bring out different reactions on same substrate. e.g.

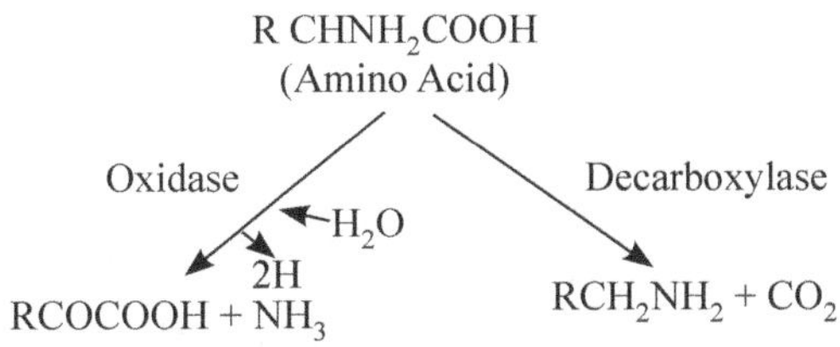

(ii) Substrate Specificity: The extent of substrate specifity varies from enzyme to enzymes. A few Hydrolase are relatively nonspecific, other require substrate containing certain groups, while still other enzymes react very specifically with one substrate only e.g. urease catalyzing the hydrolysis of urea.

$$NH_2CONH_2 \xrightarrow{H_2O} 2NH_3 + CO_2$$

(iii) Group Specificity: Several enzymes hydrolyze polypeptides, but every enzymes has preference for peptide bonds involving specific sites of the polypeptide chain or peptide bon involving specific amino acids e.g.

(a) Pepsin hydrolyzes preferably those peptide bonds lying on C-terminus of tyrosine, tryptophan, leucine and phenylalanine amino acids.

(b) Trypsin hydrolyzes those peptide bonds on the C-terminus side of arginine and lysine amino acids.

(c) Carboxypeptidase hydrolyzes the peptide bonds on the C-terminus of the polypeptide change and separates amino acids one by one.

(d) Aminopeptidase hydrolyzes the peptide bond on the N-terminus of the polypeptide change and separates amino acids one by one

(e) Similarly, phosphorylase acts only on the $1 \longrightarrow 4$ glyosidic linkages of glycogen, while branching enzyme acts on the $1 \longrightarrow 6$ glyosidic linkages of glycogen.

(iv) Stereo Specificity: Most of the biological reactions are extremely specific in the synthesis of stereoisomers. Some enzymes show optical specificity for a D or L optical isomer. thus the same amino acids has two oxidases ane for L-isomer and other for its D-isomer.

$$\text{L-amino acids} \xrightleftharpoons[\text{L–Amino Acid Oxidase}]{O_2} \text{D-Keto acid} + NH_3 + H_2O_2$$

$$\text{D-amino acids} \xrightleftharpoons[\text{D–Amino Acid Oxidase}]{O_2} \text{L-Keto acid} + NH_3 + H_2O_2$$

Although enzymes show optical specificity, still certain enzymes interchanges the optical isomers between L and D forms e.g. alanine racemase.

$$\text{L-Alanine} \xrightleftharpoons{\text{Alanine Racemase}} \text{D-Alanine}$$

Similarly, some enzymes expresses specificity towards the geometric or cis-trans forms e.g. fumarase.

$$\underset{\text{(trans-isomer)}}{\text{Fumaric acid}} + H_2O \xrightleftharpoons{\text{Fumarase}} \underset{\text{(cis-isomer)}}{\text{Malic acid}}$$

But fumarase does not react with cis-isomer (malic acid).

Though in most cases, the enzyme specificity is shown for only those compounds where the carbon was asymmetrical. However, enzyme specificity is sometimes found even in those compounds having symmetrical carbon eg. Glycerol and citric acid where group a_1 is identical with group a_2. But when enzymatic action takes place, the a_1 group of glycerol and citric acid is preferentially attacked while the a_2 group is not attacked. This phenomenon was reported by Ogston (1948). This shows that although the substrate might be symmetrical but the enzyme-substrate relationship is symmetrical. He proposed the three point attachment of the substrate to the enzyme e.g. in glycerol, the three sites attached were the groups, a_1, a_2, and H.

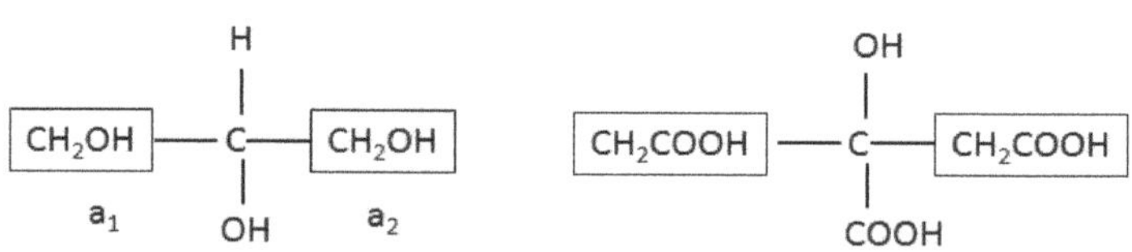

GLYCEROL CITRIC ACID

4.10 Inhibition of Enzymes

A. Competitive Inhibition

When competitive inhibition occurs, another organic molecule contests with the substrates for binding at the active site of the enzyme.

Generally such inhibitors exhibit a close structural similarity to the substrates of the enzyme they inhibit. For such a case, instead of enzyme- substrate complex, enzyme-inhibitor complex (EI- complex) is formed and the enzyme activity is inhibited. EI-complex is stable and does not dissociate to yield the products.

$$\text{Enzyme + Inhibitor} \longrightarrow \text{Enzyme Inhibitor Complex}$$

In the meantime the substrate and inhibitor are contending for the same active site so the Km value for the substrate enhances in the presence of inhibitor though V_{max} remains unchanged (Figure 4.11).

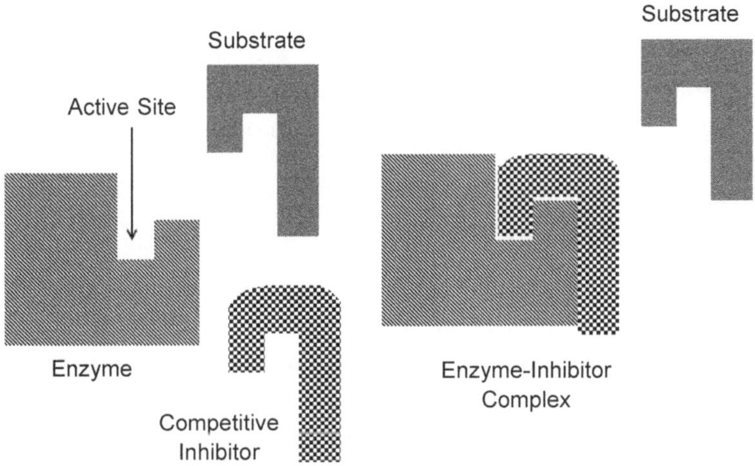

Figure 4.11: Competitive inhibition.

The characteristic example of the competitive inhibition is succinic acid dehydrogenases (SAD) that oxidize succinic acid to fumaric acid. When concentration of malonic acid, is additional, the activity of succinic action dehydrogenase drops rapidly. Hence malonic acid acts like competitive inhibitor because it has structural similarity to succinic acid. Other structurally similar competitive inhibitors of succinic dehydrogenase are oxalic acid and glutaric acid. Competitive inhibition could be reversed but enhancing the concentration of the substrate which dislodges the inhibitor molecules by the law of mass action.

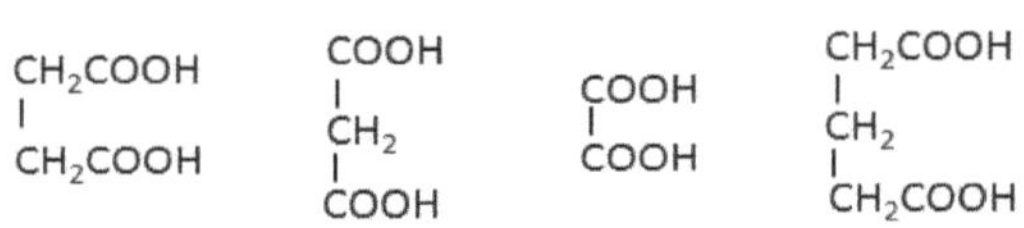

SUCCINIC ACID MALONIC ACID OXALIC ACID GLUTARIC ACID

Competitive inhibitors are used in the control of bacterial pathogens. Sulpha drugs (e.g. sulfanil amide) work as competitive inhibitors of folic acid synthesis in bacteria as they substitute for *p*-amino benzoic acid.

Other chemical compounds showing the competitive inhibition are cyanides and carbon monoxide.

Ephedrine and amphetamine act as competitive inhibitors of enzyme monoamine oxidase enzyme (which oxidizes amines like adrenaline and nor-adrenaline) because structurally similar to the amines.

Antimetabolites (e.g. antibiotics like penicillin, streptomycin, chloramphenicol and tetracycline) act as competitive inhibitors of certain metabolites like vitamins, amino acids, purine, pyrimidine, etc. as both are structurally similar. Similarly, flourouracil is an analog of thymine and 6-mercaptopurine is an analog of hypoxanthine so competes with adenine or guanine or their derivatives.

B. Non-competitive Inhibition

Sometimes, there is not competition for active sites of enzyme by inhibitor and it fixes itself with some other site of enzyme. As a result, the physical structure of enzyme is altered. This fails in forming enzyme-substrate complex and the reaction fails to occur. Non-competitive inhibitor could attach to free enzyme or to ES-complex or both. In the presence of non-competitive inhibitor, there is decrease in V_{max} while Km does not change.

In non-competitive inhibition, many enzymes could be poisoned. For example, cyanide poisoning results the inhibition of cytochrome oxidase vital for nearly all mammalian cells. The non-competitive inhibition could not be inverted by enhancing the substrate concentration. In the above case, cyanide (inhibitor) does not compete for active sites of enzyme with substrate because it has no similarity with substrate (cytochrome) but it acts at some other site of enzyme (Figure 4.12). Another example of non-competitive inhibition is secretion of two polypeptides secreted by Ascaris, one inhibits the pepsin and other inhibits trypsin. This prevents the worms from being digested by the action of these enzymes in the gut.

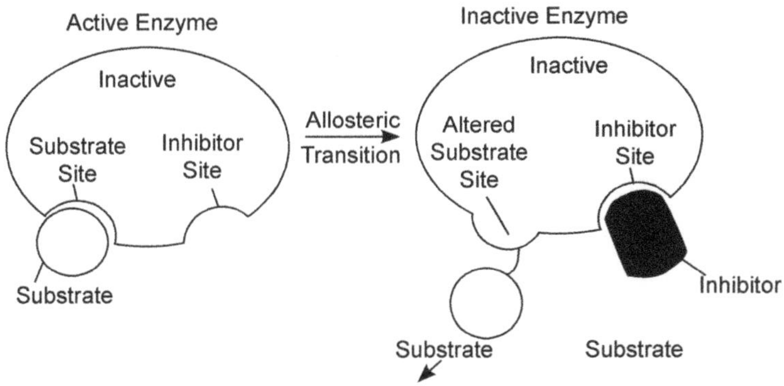

Figure 4.12: Non-competitive inhibition

Heavy metal ions such as Cu^{2+}, Hg^{2+}, and Ag^{+} have been known to act as non-competitive inhibitors of enzymes having–SH groups at active sites. These metallic ions inhibit the enzymes as these form mercaptides with –SH groups. Similarly, EDTA (Ethylene diamino tetraacetic acid) reversibly binds Mg^{2+} and other divalent cations and thus inhibits enzymes requiring divalent ions as cofactor for their enzymatic activity e.g. blood clotting enzymes, hexokinase, etc.

(c) Allosteric Modulation or Feedback Inhibition

It is reasonably useful mechanism as it checks the accumulation of products. Such type of feedback inhibition could be shown in Escherichia coli bacteria at the time of formation of amino acid, isoleucine. Formation isoleucine is stopped further when it has been formed in optimum quantity (Figure 4.13).

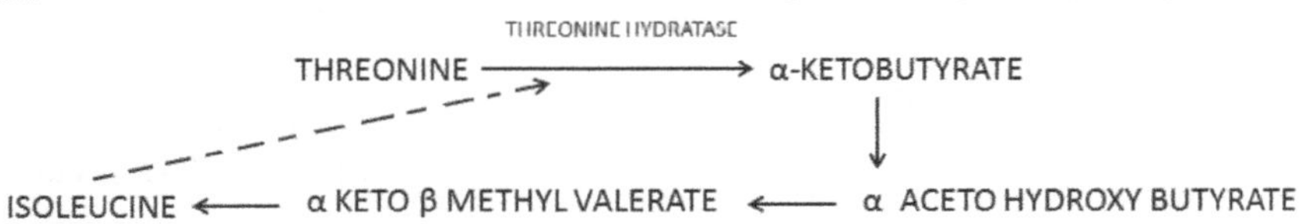

So, the final product isoleucine works as inhibitor or modulator at the first stage. Similarly, hexokinase during glycolysis in respiration acts on glucose to form glucose-6-phosphate. Allosteric inhibition occurs by glucose-6 phosphate thus decreasing the activity of enzyme hexokinase.

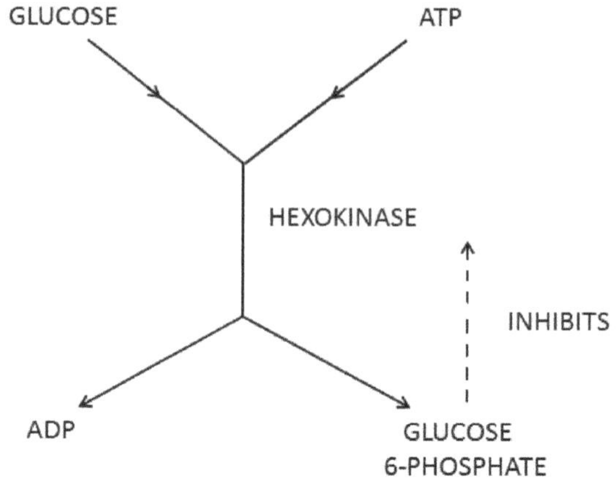

Figure 4.13: Feedback allosteric inhibition

The allosteric enzymes bear two types of sites i.e. active and allosteric. The end product combines with allosteric site of enzyme and checks its catalytic activity. The product which acts as inhibitor brings about change in confirmations in the active sites of enzymes resulting in reduction of catalytic activity. **Jacob, Chengeux** and **Monod** have named this process as allosteric transition. This process is called allostery (Table 4.4).

Table 4.4: Differences between competitive and allosteric Inhibition

S. No.	Character	Competitive Inhibition	Allosteric Inhibition
1.	Binding of inhibitor	At active site	At allosteric site
2.	Feedback regulation	Not exhibited	Exhibits to prevent excess formation of final products.
3.	Structure	Similar to substrate molecules.	Different from substrate molecules.
4.	Mode of action	Occupies the active site to check the binding of substrate molecule at active site.	Occupies allosteric site but induces the change in confirmations in active sites.
5.	Example	Malonic acid showing competitive inhibition to succinic acid.	Formation of isoleucine amino acid from threonine.

4.11 Coenzyme and Cofactor

The term coenzyme might be defined as the substance necessary for the activity the enzyme. But now a days most of the biochemist use the term cofactor in this general way and specify co enzyme for those co factor which are organic molecules and participate in catalytic process, e.g. in metalloprotein enzymes, the metallic ion is not the coenzyme but only organic prosthetic group constitutes the coenzyme. Similarly, other cofactor which are not actually co-enzyme might help in establishing a three dimensional structure necessary for their catalytic property. Every coenzyme is a cofactor but every cofactor is not co-enzyme. Most of the co –enzyme are nucleotide and are composed of vitamins.

Coenzymes are organic biomolecules behaving as prosthetic group of the conjugated enzyme and are involved in some group transfer from one organic compound to another organic compound so are essential for enzymatic activity. On the basis of reaction occurring in the presence of coenzyme, the co-enzymes are divided into three categories. Hydrogen transferring co-enzymes, group transferring co-enzymes isomerase and lyase co-enzyme.

NAD^+ (Figure 4.14) Structurally, NAD^+ is formed of adenosine and N-Ribosyl nicotinamide interlinked by a pyrophosphate bond formed among the 5–hydroxyls of two ribosyl units. It was first discovered by Von Euler (1933) and was called Coenzyme-I or Coenzyme phosphate. The adenine, ribosyl and pyrophosphate components of NAD^+ are being involved in the binding of NAD^+ with the enzyme. The specific binding of NAD^+ to enzyme surface gives a chemically familiar polarity to the planar nicotinamide. its topside is called A-face while its downside is called B-face. NAD^+ takes two electrons and one deuterium from the substrate while other hydrogen is released as a proton. In the case of alcohol dyhydrogenase two electrons and deuterium are added to the A-face. In these dehydrogenation actions, 2H-atoms are received by 4th carbon of the pyridine ring which gets reduced **(Figure 4.15)**

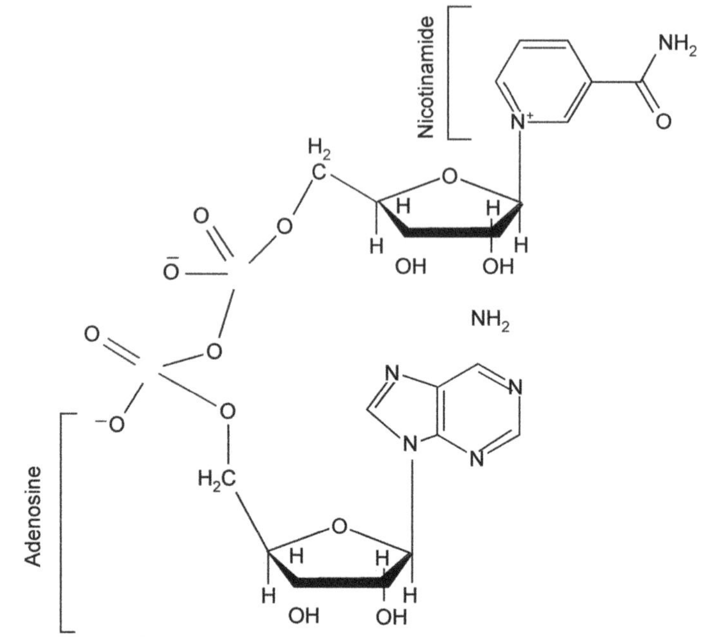

Figure 4.14: Nictotinamine adenine dinucleotide

NAD^+ shows no absorbence at 340nm while NADH absorbs the light at 340 nm showing use in the bioassay of dehydrogenases.

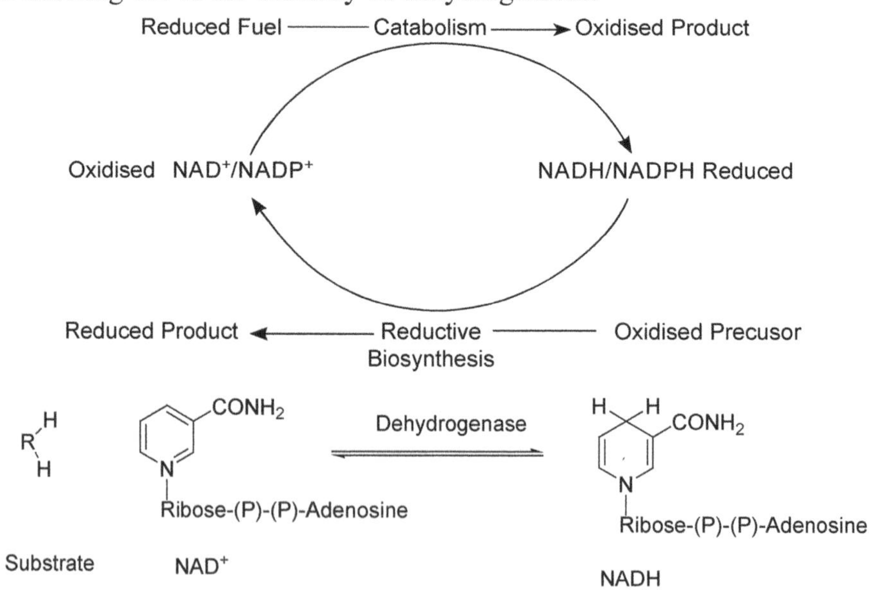

Figure 4.15: General action of NAD^+ linked dehydrdogenase. (Interconversion of NAD^+ and NADH)

Biological Reactions Catalyzed by NAD^+-linked Dehydrogenases (DHases):

(a) In Carbohydrate Metabolism

(1) Isocitric acid + $NAD^+ \xrightleftharpoons{\text{Isocitric Acid DHase}}$ α-Ketoglutaric acid + CO_2 + NADH + H^+

(2) Pyruvic acid +NAD^+ + Co.A $\xrightleftharpoons[\text{Complex}]{\text{Pyruvic Acid DHase}}$ Acetyl Co.A + CO_2 + NADH + H^+

(3) α-Ketoglutaric acid+Co.A+NAD^+ $\xrightleftharpoons[\text{Acid DHase Complex}]{\text{a-Ketoglutaric}}$ Succinyl Co.A + CO_2 + NADH+H^+

(4) L-Malic acid + NAD^+ $\xrightleftharpoons[\text{Acid DHase Complex}]{\text{Malic Acid Dehydrogenase}}$ Oxaloacetic acid + NADH + H^+

(5) Lactic acid + NAD^+ $\xrightleftharpoons{\text{Lactic Acid DHase}}$ Pyruvic acid + NADH + H^+

(6) D-Glyceraldehyde-3-P+NAD^++Pi $\xrightleftharpoons{\text{PGAL-DHase}}$] 1,3-Diphosphoglycerate+NADH+H^+
(3-PGAL)

(b) In Fatty Acid Metabolism

In β-oxidation reaction sequence of fatty acids, fatty acyl Co. A in the presence of enzyme acyl Co.A dehydrogenase and a H-acceptor flavoprotein, undergoes dehydrogenation inside the mitochondria. It produces a trans double bond between carbons 2 and 3 which is hydrated in the presence of enoyl Co.A hydratase to form L-3-hydroxyacyl Co.A.

L-3-hydroxyacyl Co.A is dehydrogenated in the presence of coenzyme NAD^+ and an enzyme L-β-hydroxyacyl Co.A dehydrogenase to form 3-ketoacyl Co.A.

L-3-Hydroxyacyl Co. A + NAD^+ $\xrightleftharpoons[\text{Acyl Co DHase}]{\text{L-Hydroxy}}$ 3-Ketoacyl-Co.A+NADH+H^+

(c) In Amino Acid Metabolism

For excretion, amino groups are transaminated from amino acids to α-ketoglutaric acid to form L-glutamic acid which undergoes dehydrogenation in the presence of enzyme L-Glutamate DHase (present in the mitochondrial matrix) AND NAD^+ behaving as an acceptor of reducing equivalents.

L-Glutamate + NAD^+ + H_2O $\xrightleftharpoons[\text{DHase}]{\text{L-Glutamate}}$ α-Ketoglutarate + NH_4^+ + $NADH^+$ + H^+

NH_3, so formed is changed into harmless urea by ornithine cycle in the liver. Urea is excreted out in urine by the kidneys.

Deficiency of niacin causes pellagra which is characterized by 3D's: dermatitis, diarrhoea and dementia.

Coenzymes of Riboflavin

Riboflavin, also called lactoflavin or vitamin B_2 or vitamin G is a B-complex vitamin commonly found in green leafy vegetables among the vegetable sources and liver, kidney, milk and egg yolk among the animal sources. It was first discovered in milk. It is also synthesized by intestinal bacteria. Daily requirement of vitamin B_2 is about 1.4 mg in adult male and about 2.5 mg for children and pregnant ladies.

The riboflavin consists of the heterocyclic ring, isoalloxazine (flavin) connected to alcohol ribitol through its N_2 (Figure 4.16). Riboflavin is involved in the formation of 2 coenzymes : EMN (Flavin mono nucleotide) and FAD (Flavin adenine dinucleotide).

FMN has a phosphate being the esterified with the 5-hydroxyl of ribitol of riboflavin

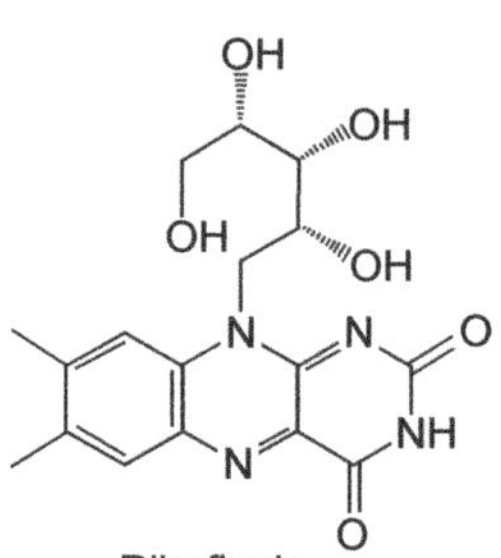

Riboflavin

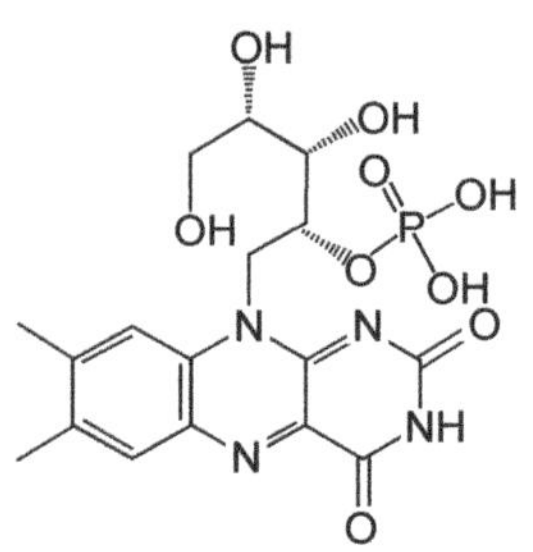

Riboflavin monophosphate (FMN)

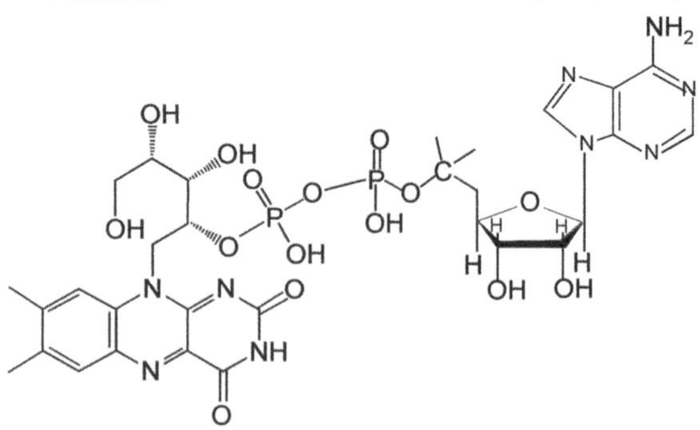

Flavin adenine dinucleotide (FAD)

Figure 4.16: Flavin adenine dinucleotide (FAD)

(Figure 4.16) while FAD is structurally similar to NAD where adenosine is linked via a pyrophosphate linkage to the heterocyclic ring, riboflavin. The flavin coenzymes are bound to their enzymes dehydrogenases) more firmly than the niacin coenzymes, so these are called flavoproteins or flavin dehydrogenases. Deficiency of riboflavin interfere the metabolic activities and is known to cause: cheilosis (the cracking of skin at the corners of mouth), keratitis (inflammation of skin),glossitis (inflammation of the tongue), and eye problems like burning sensation and photophobia.

Mode of action. FMN and FAD act as coenzymes in oxidation-reduction reactions accepting and donating two H-atoms through their isoalloxazine ring.

$$\underset{\text{Substrate Coenzyme}}{S{<}^{H}_{H} \;+\; FAD} \xrightleftharpoons{\text{Dehydrogenase}} \underset{\text{Reduced Coenzyme}}{FADH_2 \;+\; S}$$

Biological actions catalyzed by Flavin-linked dehydrogenases:

(i) FAD

(a) In Carbohydrate Metabolism

(1) Oxidative decarboxylation of Pyruvic acid. During aerobic glycolysis, glucose or other monosaccharides arae converted to pyruvic acid (3C). It is also produced during the degradation of amino acids like alanine, serine and cysteine. Pyruvic acid go through oxidative decarboxylation to form acetyl Co. A (2C) in the presence of pyruvate-dehydrogenase complex.

Pyruvic acid + NAD^+ + NAD^++ Co~ASH $\xrightarrow[\text{COMPLEX}]{\text{PYRUVATE DEHYDROGENASE}}$ Acetyl Co.A + CO_2 NADH + H

$$\Delta G = -8 \text{ kcal/mol}$$

This multienzyme complex is found in the mitochondrial matrix and is formed of 3 enzymes Pyruvate dehydrogenase, Dihydrolipolyl transacetylase and Dihydrolipoyl dehydrogenase) and coenzymes (Thiamine Pyrophosephate, lipoic acid, coenzyme-A, FAD and NAD^+).

(2) Oxidative decarboxylation of α-ketoglutaric acid and the formation of succinyl Co. A in the Krebs cycle is catalyzed by a α-ketoglutarate dehydrogenase complex which is also formed of enzymes and 5 coenzymes as in pyruvic acid dehydrogenase complex.

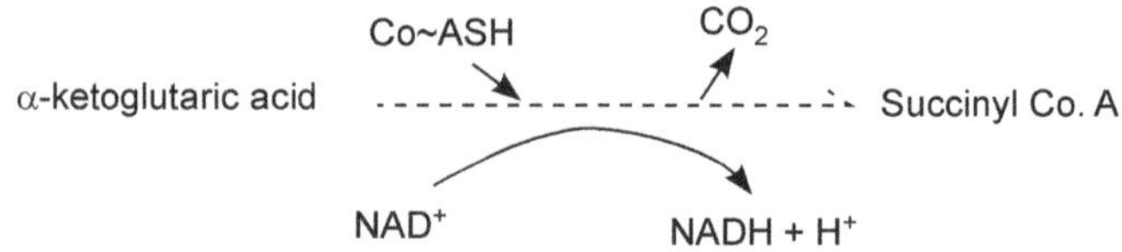

(3) Dehydrdogenation of succinic acid (Figure 4.17) to fumaric acid is catalyzed by a FAD-linked succinic acid dehydrogenase, FAD is covalently bound but in most of flavin dehydrogenases, FMN or FAD is non-covalently bound.

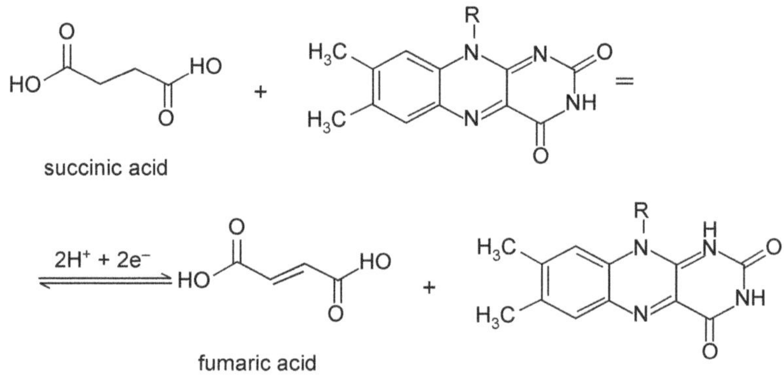

Figure 4.17: Dehydrogenation of succinic acid

(b) In Fatty Acid Oxidation

In mitochondrial matrix, FAD acts as coenzyme of acyl Co.-A hydrogenase enzyme which enzymatically dehydrogenates the saturated fatty acyl Co.A to produce trans- Trans-Δ^2- Enoyl Co.A .e.g.

$$\text{Palmitoyl Co.A. FAD} \xrightleftharpoons{\text{Pamityl Co. A DHase}} \text{Trans-}\Delta^2\text{- Enoyl Co.A+FADH}_2$$

(Δ^2 = trans double bond between carbons 2 and 3)

(ii) FMN role in ETC in The Mitochondria

ETS, also called ETC or cytochrome system is formed seven electron carriers with increasing electron-affinity (Table 4.5). FMN acts as second electron carrier on the route no. I.

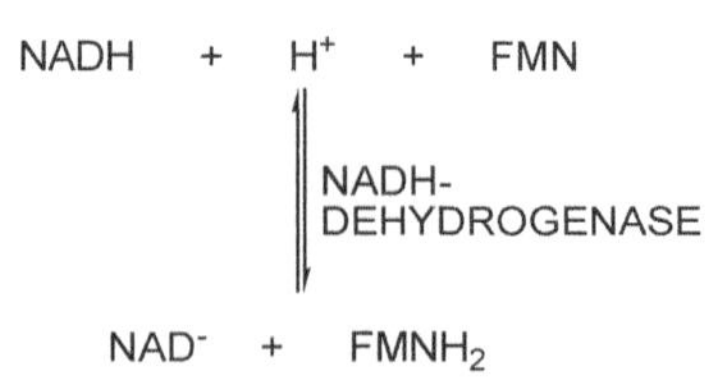

It acts as coenzyme of NADH-dehydrogenase enzyme which transfers the energy equivalents from NADH to FMN so at NADH gets oxidized while FMN is reduced to $FMNH_2$ (Figure 4.18). It passes the energy equivalents to coenzyme Q forming one ATP molecule.

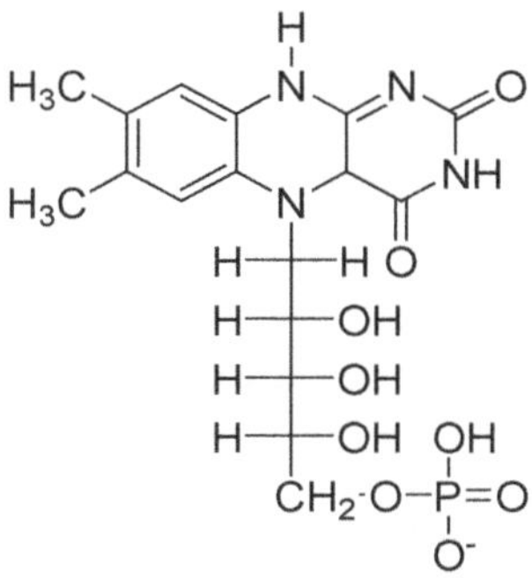

Figure 4.18: $FMNH_2$

Table 4.5: Various Flavin-linked Dehydrogenases

Enzyme	Function	Flavin-nucleotide
1. Succinic dehydrogenase.	TCA-cycle	FAD
2. Dihydrolipyl DHase.	Component of Pyruvate DHase and α-ketoglutarate DHase complex	FAD
3. NADH dehydrogenase.	ETC	FMN
4. Acetyl-Co.A DHase.	Fatty acid oxidation	FAD
5. D-Amino acid oxidase.	Amino acid oxidate ion	FAD
6. Monoamine oxidase.	Oxidation of monoamines	FAD

4.12 Drug–Receptor Interactions

The receptors are the macromolecules that are indulged in chemical signaling within and in between the cells; also they might also be sited within the cytoplasm or on the surface of the cell membrane. Also the drug action site, that is eventually accountable for the pharmaceutical effect, is a receptor, i.e. any biological molecules interacting with the drugs. The activated receptors directly or indirectly controls involved cellular biochemical processes (e.g., protein phosphorylation, ion conductance, enzymatic activity, DNA transcription).

The Molecules (e.g., drugs, neurotransmitters, hormones,) those binding with the receptor are termed **ligands**. The ligand might be activated or inactivated and binding could be specific and reversible. The extent to which a drug acts on any given site relative to the other sites; selectivity narrates broadly to physicochemical binding of the drug with the cellular receptors is called selectivity.

The capability of drug for affecting any given receptor is associated with the drug's affinity (possibility of the drug inhabiting the receptor at any given time) and inherent efficacy (intrinsic activity—the degree to which the ligand is activating the receptors that leading to response in the cell. The affinity or activity of a drug is identified by the chemical structure.

It is observed that the pharmacologic effect is similarly studied by time interval for which the drug-receptor (DR) complex is persisting. The conformation change controlling the rate of drug association and dissociation influences the drug-receptor (DR) lifetime. An extended residence time describes an extended pharmacologic effect. The extended residence time could be a possible drawback when it prolongs the toxicity of a drug. In few receptors, the transient drug occupancy yields the preferred pharmacologic effect, however prolonged tenancy grounds for the toxicity.

Capability of binding with a receptor is influenced by both the intracellular regulatory mechanisms as well as the external factors. The baseline receptor concentration and the efficacy of stimulus-response (SR) mechanisms differ for the tissues. The Drugs, aging and disorders and mutations in gene could enhance or suppress binding capacity of receptors.

In order to escalate the mode of action of drug, it becomes necessary to study the interacting forces binding the drug and the receptors. As because in the biological fluids and the bloodstream, the concentration of drugs and receptors is small, the law of mass action only couldn't justify the ability of small dose of structure specific drugs to prompt a answer by combination with all, or all of the appropriate receptors. To obtain the desired drug-receptor (DR) interaction, the study reveals that something additional than mass action is needed. One mole of a drug consists of 6.02×10^{23} molecules (Avogadro's number). Suppose if 300g /mole is the molecular weight of an average drug and later on 15 mg (is an effective does for various drugs) consisting of 3×10^{19} drug molecules. The human organism is self-possessed with approximately 3×10^{13} cells. And therefore $3 \times 10^{19}/3 \times 10^{13} = 10^{6}$ drug molecules will be acting on every cell. A single erythrocyte cell consists of some 10^{10} molecules and assuming the same number of molecule of the human body, Le Chatelier might have a tough time clarifying the desired drug-receptor interaction, forming a stable complex.

The low energy state of the drug-receptor (DR) complex shown in **scheme 1,** is the driving force for the interaction of drug-receptor (DR), where k_{on} being the rate constant for forming the DR-complex, that totally relies on the concentrations of both the drug as well as the receptor, and k_{off} being the rate constant for breaking the DR-complex, relying on the DR-complex concentration along with the other forces and factors. For a drug the biological activity is correlated with the attraction for the receptor, i.e., formation of a stable drug-receptor (DR) complex. The stability is normally studing how tough is the dissociation, which is denoted by its K_d.

Note that

$$K_a = \frac{\text{[drug] [receptor]}}{\text{[drug-receptor (DR) complex]}}$$

Since K_d is a *dissociation* constant, the smaller the K_d, the higher is the DR-complex concentration, consequently more is the complex stable with more affinity for the receptor of the drug. K_d approximately denotes the amount of the drug needed to attain an equilibrium of amount 50% in the DR-complex. For having an indication of the affinity for its target of a typical drug, it is assessed that the median K_d for enzyme inhibitor drugs is around 20nM (the enzyme is available as fifty percent in the drug-enzyme complex at 20 nM concentration of drug). The DR-complex formation comprises an extravagant equilibrium. The solvated ligands and proteins (like drugs/receptors) usually occur as an equilibrium mixture of numerous conformers (Scheme 4.1). The solvent molecules occupying receptors binding site should be moved by the drug producing a solvated complex for forming a complex,; the interactions in both the drug and the receptor are much robust than drug-receptor interaction with the solvent molecules. The formation of DR-complex is entropically uncomplimentary; as this root for the damaging the confirmation of both the ligand and protein accompanied with the losing three translational and rotational degrees of freedom. Consequently, highly promising enthalpic interactions in the drug and receptor needs to pay for the loss in entropy.

$$\text{Drug + Receptor} \underset{K_{off}}{\overset{K_{on}}{\rightleftharpoons}} \text{Drug-receptor (DR) Complex}$$

Scheme 4.1: Equilibrium between a drug and receptor and DR-complex

4.13 Interactions (Forces) Involved in The Drug-receptor (DR) Complex

The equivalent forces as experienced by different interacting organic molecules including ionic interactions, covalent bonding, dipole-dipole and ion-dipole, the charge-transfer (CT) interactions, H-bonding hydrophobic, cation-π interactions, hydrophobic, halogen bonding, and van der Waals interactions are experienced by the drug-receptor complex. These weaken interactions generally are merely potential when the molecular surfaces are too nearby and correspondingly the strength of the bond bond depends on distance.

Usually, weak non covalent interactions exist in drug and a receptor; and producing reversible result. Therefor the concentration of drug in the extra cellular fluids drops normally by metabolism. For terminating the pharmacological action it is needed that the effect of the drug persists for a limited time. On few occasion, the effect formed by a drug could be carried on or irreversible. For a *chemotherapeutic agent,* it becomes need that a drug is acting selectively acting on a tumor cell's foreign organism forming a complex that is irreversible with the receptor in such a way that the drug could exercise its toxicity for an elongated duration and where the covalent bond is desired.

4.14 The theories for Drug-receptor (DR) Interactions

From the past we already have numerous hypotheses been proposed accounting the ability of a drug-receptor (DR) interaction and prompt a biological response. Numerous of suggestions are talked over here, beginning from the very first the occupancy theory to the latest multistate model.

The Occupancy Theory

Gaddum[49] and Clark[50] *in occupancy theory* stated that the pharmacological effect intensity is directly proportional to the number of receptors that are occupied by the drug.

Once the drug-receptor (DR) complex detaches the response also terminates. Alhough a maximal response is not given by all agonists. Hence, the theory didn't justify partial agonists and inverse agonists.

Ariens[51] and Stephenson[52] revised the occupancy theory accounting for the partial agonists given by Stephenson. They developed the original Langley perception of a receptor, being the DR-complex interactions involving two stages: firstly, termed as the *affinity* is the drug receptor complexation, *and* secondly named the *intrinsic activity* that is the commencement of a biological effect. Affinity is the capability of a drug binding with the receptor totally depending on their molecular complementarity. The maximum response prompted by a compound comparative to that of a given reference compound is the Intrinsic activity (α), whereas the characteristics of a compound giving the extreme response/the capacity of the DR-complex to start a response[53] is the efficacy. For explaining minor variations we will be using the capability of a compound to initiate a biological response as efficacy. This latter property was well-thought-out to be constant in the original theory. Few examples of efficacy and affinity are shown in **Figure 4.19 A** displays the theoretical does-response curves showing same affinity for the five different drugs with the receptor having $pK_d = 8$, but are having variable efficacies with range of 100% to the 20% of the maxima. The drug is a full agonist which is having 100% efficacy; whereas the remaining are partial agonists. The **Figure 4.19 B** displays the does-response curves for the four different drugs having the same efficacy (all being full agonists), howsoever owning variable affinities ranging from a pK_d of 9 to 6.

Antagonists could attach with the receptor i.e. great affinity; but lacking the efficacy. The potent agonists should have a lesser amount of affinity for their receptors compared with the partial agonists/antagonists.

Figure 4.19: Theoretical dose-response curves show (A) drugs with the same affinities and different efficacies (the top compound) is a full agonist and the others are partial agonists and (B) drugs with identical efficacies (all full agonists) but different affinities.

Agonists or Antagonists

Hereafter, these two properties efficacy and affinity gets detached. The expressions agonist, antagonist, partial and inverse agonist not entirely depends on the essentially properties of drugs but also on the biological system. Any compound being an agonist for one receptor might behave as an antagonist/inverse agonist for the other receptor. Any specific receptor is well-thought-out to possess an intrinsic full response; being the major extent of response produce from any ligand by the capability of receptor. Any compound that prompts the maximum response may be a full agonist; also a specific compound could be accomplished of surpassing the maximum response of a tissue, but the witnessed response could only be the maximum response of that specific tissue. The drug being unable for stimulating the maximum response of the tissue which depends on the structure of the drug, may be a partial agonist . The complete agonist or partial agonist is claimed to exhibits affirmative efficacy, an antagonist exhibits zero efficacies, and a full/partial inverse agonist exhibits reduced basal tissue response i.e. negative efficacy

The revised occupancy theory accounts for the presence of partial acceptor activate the acceptor agonists and antagonists but it doesn't explain why two drugs those occupying the similar receptor will act contrarily, i.e., as an agonist/antagonist.

The Rate Theory

The alternate to the occupancy theory, that being recommended by Paton is that the activation of receptors is proportionate to the total number of encounters between the drug and the receptor/unit time. Thus, the *rate theory* recommends that the pharmacological activity is and not the number of occupied receptors, but a function of rate at with the drug and the receptor associates and dissociates. Every association might yield a quantum of stimulus. For agonists, this rate might be faster (the former being slower than the latter). The rate of dissociation might be slow of an antagonist with a receptor but association could be faster. For partial agonists the rate of dissociation might be having intermediate drug-receptor (DR) complex. Both the occupancy theory and rate theories, at equilibrium are mathematically the same. The rate theory also doesn't justify why the several types of compounds displaying the features that are same as in the occupancy theory,

The Induced-Fit Theory

The *induced-fit theory* purposed by *Koshland* was firstly offered for the action of substrates with the enzymes, is applicable to drug-receptor (DR) interactions as well. As per theory, the receptor essentially not exists in the suitable conformation needed for binding with the drug. When the drug is approaching the receptor, a *confirmational change* is encouraged, orienting the binding sites. **(Figure 4.20)**. This conformational variation in the receptor is liable for the movement and interaction of residues with the substrate. This receptor/enzyme was proposed to be flexible, and after releasing the drug/product the original conformation could be attained. The change in confirmation of the drug (substrate) could also experience deformation even if this fetches strain in the drug. As per induced fit theory, an agonist might encourage a change in conformation that produce a response, also an antagonist can also bind without any change in conformation. This theory could be reformed to the rate theory discussed above. An agonist might encourage a conformational change in the receptor, so that the agonist can binds lossely and separate out effortlessly. An antagonist is resulted if drug-receptor (DR) complexation doesn't produce any change in confirmation of the receptor, leading to a stable DR-complex.

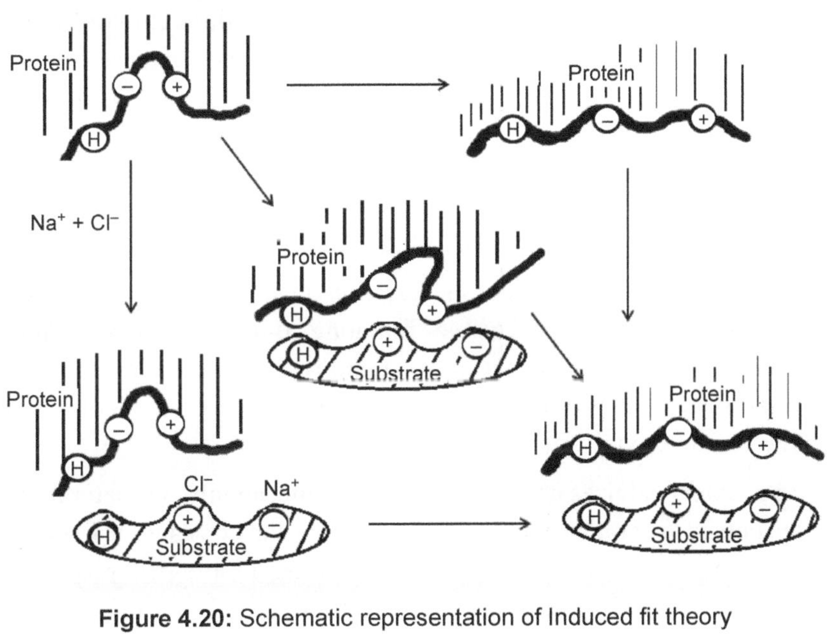

Figure 4.20: Schematic representation of Induced fit theory

Macromolecular Perturbation Theory

Belleau proposed that during drug-receptor interaction, there could result two general types of *macromolecular* perturbations: firstly, *a specific conformational perturbation* making the binding of certain molecules feasible that create a biological response (an agonist) and secondly *a non-specific conformational perturbation* companies other kinds of molecules that don't prompt a response (e.g., an antagonist). The mixture of two complexes will consequence (a partial agonist) if the drug is given to both perturbations. The theory bids a physicochemical root for the rationalizing the molecular process involving receptors, without addressing the inverse agonist concept. .

Activation-Aggregation Theory

The *activation-aggregation theory* of Monad et.al. is an addition lead of the macromolecular perturbation theory (that itself is established on the induced-fit theory) according to which a receptor is in a state of dynamic equilibrium between an activated form (R_0), that is liable for the biological response, and an inactive form (T_0) even when lacking drugs,. On the basis of the theory, agonist is binding with the R_0 form and shifts the equilibrium to the activated form antagonists binding to the inactive form (T_0), and partial agonists binding with both the conformations. As per the model, the agonist binding site in the R_0 conformation could be dissimilar form the antagonist binding site in the T_0 conformation. In case there exists two different conformations and binding sites, accounting for the structural variances in these compounds also easily rationalizing that why an agonist produces a biological response but not so by an antagonist. The theory could elucidate the capacity of partial agonists to have together the properties of agonistic as well as antagonistic.

The Two-State/Multistate Model of Receptor Activation

The conception of a conformation variation in a receptor encouraging an alteration in its action is feasible for long time. The Monod-Wyman-Changeux concept explains that above comprises a two-state model of the receptor activation, that don't go far sufficient. On the base of observations with GPCRs the model was reviewed.

T*wo-state model* advises that in the lack of the agonist or natural ligand, receptors persist in equilibrium (distinct by equilibrium constant *L;* (**Figure 4.21**) in between an active state (R^*), capable of starting a biological response, and a latent state. In lack of the agonist or natural ligand, the receptor's basal activity is explained by the equilibrium between R^* and R. The binding of drug with any of these conformational states as per the equilibrium constants K_d for formation of the drug-receptor (DR) complex with the resting (D.R.) and active (D.R*) states, respectively. The equilibrium is changed by the full agonists entirely to the active state through binding with the active state and affecting maximum response; the partial agonists is favorably binding to the active state however not as much that a full agonist does, hence the maxima response is not achieved; full inverse agonists change the equilibrium completely to the resting state by binding with the resting state, bring about a decline in the basal activity(a negative efficacy); partial inverse agonists favorably binding with the resting state, triggering a decline in the basal activity(a negative efficacy); partial inverse agonists better binding with the resting state, but again not to that level which a full inverse agonist is doing; and antagonists is having the affinities for both being the same. A competitive antagonist is capable of moving both an agonist and the inverse agonist from the receptor.

The two-state receptor model was further extended to a *three-state receptor model* by Leff and coworkers, suggesting that there exist an inactive conformation and two active conformations. This favors experimental results concerning the inconstant behavior (both efficacies and affinities) of agonist and inverse agonist in several systems consisting of the same *receptor promiscuity.* As per the postulate the source for discrepancy agonist efficacies among several agonists is their varying affinities for the different active state.

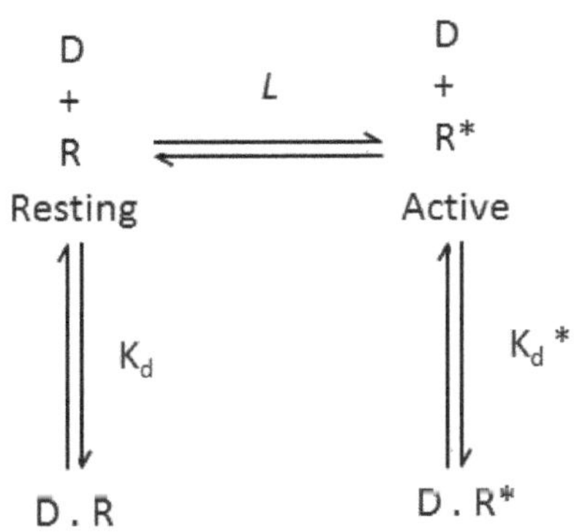

Figure 4.21: Two – state model of receptor activation

Where **D** is the drug

R is the receptor

L is the equilibrium between the resting ® and the active (R*) state of the receptor

4.15 The Structure-Activity Relationships (SAR)

On knowing the structure of a lead compound, the medicinal chemist starts studying its structure-activity relationships (SAR) aiming for identifying the molecular parts those are necessary or not having biological activity.

If it's feasible to crystallize the objective with the available lead compound bound with the binding site, the complex crystal structure might be solved with XRD, then studies with molecular modeling software to find out the necessary binding interactions. However, if the lead structures have not been identified or couldn't be crystallized this becomes impossible. It's necessary for reverting the old-fashioned method for synthesizing a certain number of compounds differing marginally from the original structure, and then reviewing the outcome on the biological activity.

One could imagine the drug as a chemical knight entering the body in order to make battle with an affliction. The drug is armed with a variety of weapons and armour, but it might not be obvious which weapons are necessary to the drug's activity or which armour is essential to its survival. We could only find this out by removing some of the weapons and armour to see if the drug is still effective. The weapons and armour involved are the various structural features in the drug that could either act as binding groups with the target binding site), or assist and protect the drug on its journey through the body . Recognizing functional groups and the kind of bonds (intermolecular) that they could form is necessary in intermolecular bonds that they could form is necessary in understanding the binding of drug and the target.

Let us imagine that we have isolated a natural product with the structure shown in **Figure 4.22.** We shall name it glipine.

There are a range of functional groups existing in the structure and the diagram shows the potential binding interactions that are possible with a target binding site.

It is unlikely that all of these interactions take place, so we have to identify those that do. By synthesizing analogues (such as the examples shown in **Figure 4.23**) where one specific functional group of the molecule is detached or changed, it is likely to find out that which groups are necessary and which are not needed. This comprises testing all the equivalents for comparing them and biological activity with the original compound. In case an analogue displays them with the original compound. If an activity remains similar, then the group is not essential.

The ease with which this task is carried out depends on how easily we could synthesize the necessary analogues. It might be possible to alter few lead compounds straight to the needed analogues, whereas the analogues of other lead compounds might best be prepared by total synthesis. Let us consider the binding interactions that are possible for different functional groups and the analogues that could be synthesized to established whether they are involved in binding or not.

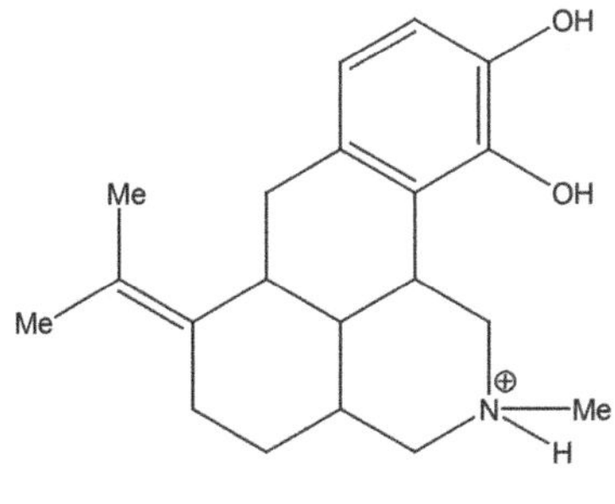

Figure 4.22: Glipine

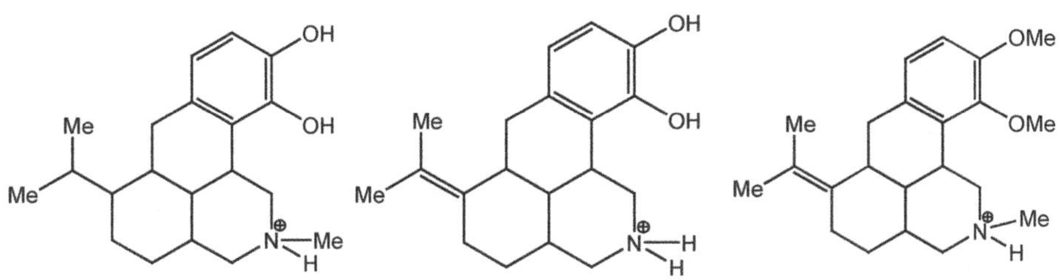

Figure 4.23: Modifications of glipine

The Binding Role of Alcoholic and Phenolic -OH

The alcohol and phenols are the functional groups that are commonly present in drugs and quite often indulge in H-bonding. Here the hydrogen could act as a H-bond donor and oxygen atom could perform as a H-bond acceptor, **(Figure 4.24).** The directional inclination for H-bonding is shown in the Fig. 3.23, but it is necessary to realize that slight deviations are possible. One or all of these interactions might be necessary in binding the drug with the binding site.

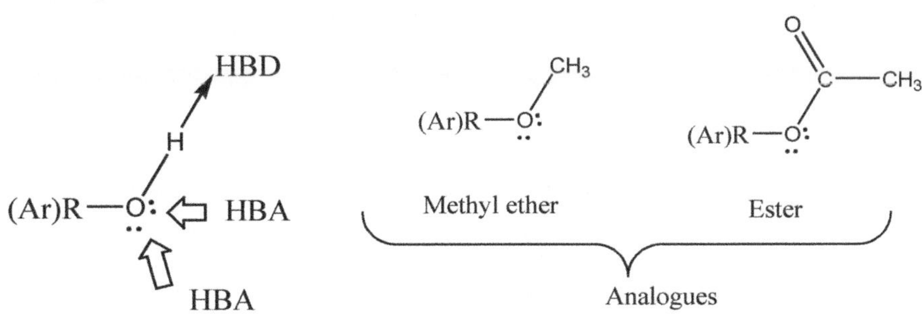

Figure 4.24: Possible hydrozen bonding interactions for an alcohol or phenol

Synthesizing a methylether or an ester analogue would be relevant in testing this, as it is highly likely that the hydrogen bonding might be disturbed in either analogue. Let us consider the methyl ether first.

The hindrance or preventing the H-bonding of phenol/alcohol by the ether can be accounted by two reasons. The obvious elucidation is (i) the proton of the -OH group is indulged as a H-bond donor and, (ii) by removing it; the hydrogen bond is lost (Frames 1 and 2 in **Figure 4.25).** However, supposing the oxygen atom as a hydrogen bond acceptor (Frame 3, **Figure 4.25).** The oxygen still exists in the ether analogue, so could it still take part in hydrogen bonding? Well, it might, but possibly not to the same extent. The extra methyl group should hinder the close approach that was previously attainable and is likely to disrupt hydrogen bonding (Frame 4, **Figure 4.25**). The hydrogen bonding might not be completely prevented, but we could reasonably expect to be weakened.

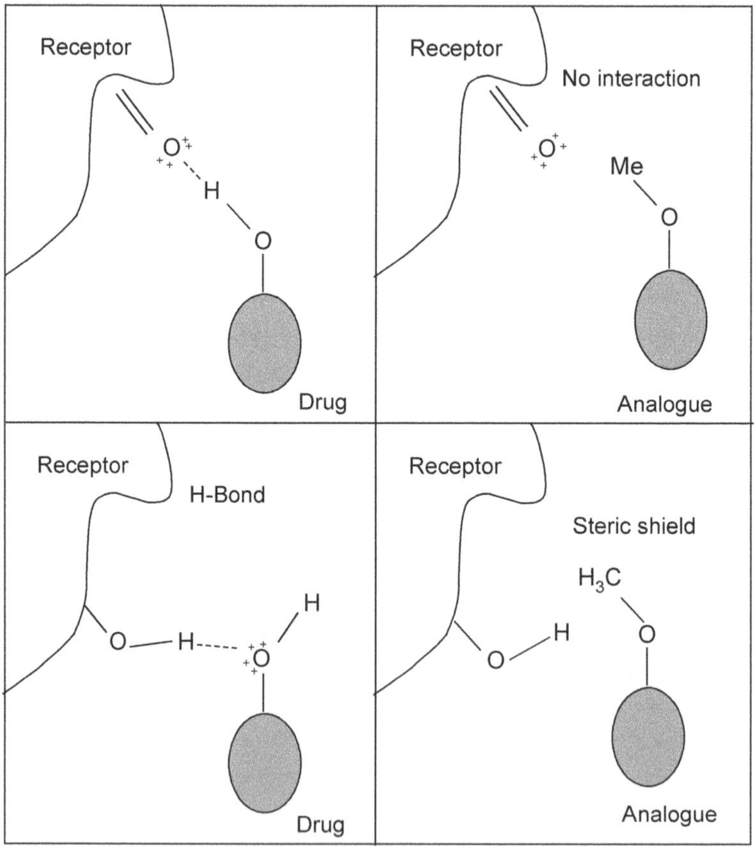

Figure 4.25: Hydrogen bond interactions for an alcohol/phenol in contrast with an ether analogue.

An ester analogue couldn't perform as a hydrogen bond donor either. There is still the possibility of it performing as a H-bond acceptor, but the additional acyl group is even bigger as compared to the $-CH_3$ group of the ether, which must hinder the H-bonding interaction, and also there's exists a dissimilarity between the electronic properties of an ester and an alcohol. The carboxyl group, from the neighboring oxygen is having a fragile pull on the electrons, which gives the resonating structure shown in **Figure 4.26.** Since the lone pair is participating in these type of interaction, this might be not as effective as a hydrogen bond acceptor. Of course, one could then argue that the carbonyl oxygen is potentially a more effective hydrogen bond acceptor; however, it is in a different position relative to the rest of the molecule and might be poorly positioned to form an effective hydrogen bond interaction with the target binding region.

It is relatively easy to acetylate alcohols and phenols to their corresponding esters, and this was one of the early reactions that was carried out on natural products such as morphine . Alcohols and phenols could also be converted easily to ethers.

In this section, we considered the OH group of alcohols and phenols. It should be remembered that the OH group of a phenol is linked to an aromatic ring, which could also bc involved in intermolecular interactions.

The Binding Role of Aromatic Rings

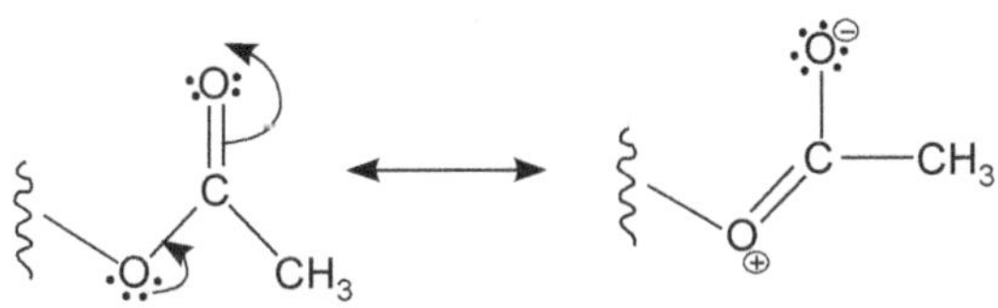

The aromatic rings being planar, hydrophobic structures, are commonly indulge in van der Waals interactions with the smooth hydrophobic area of the binding site. The analogue comprise of a cyclohexane ring instead of the aromatic ring will not binding so well, because the no longer flat ring. Interaction of axial protons could be weak but they too working as a buffer keeping the cyclohexane ring at a distance. **(Figure 4.27)** For the aromatic ring the binding region might be narrow also instead of having a planar surface. In case, the cyclohexane ring might be incompetent to fit into it, due to the bulky structure.

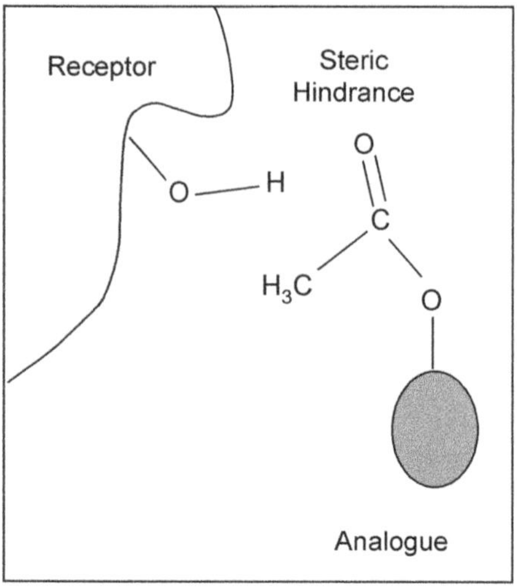

Figure 4.26: Factors by which an ester group could disrupt the H-bonding of the original hydroxyl group.

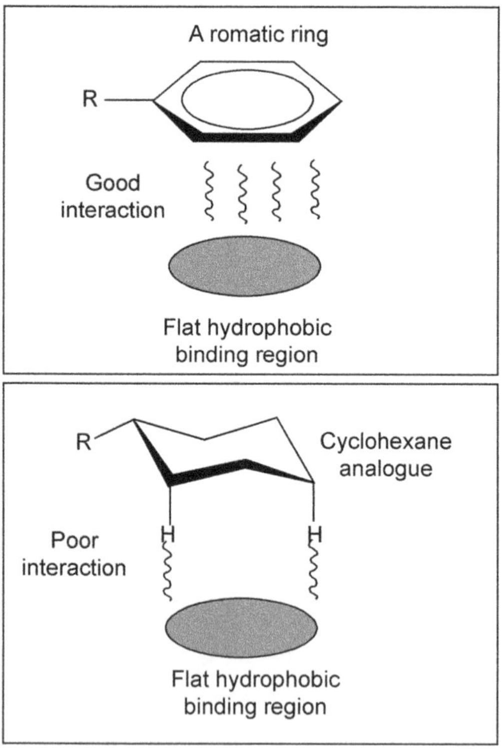

Figure 4.27: Binding comparison of an aromatic ring with a cyclohexyl ring

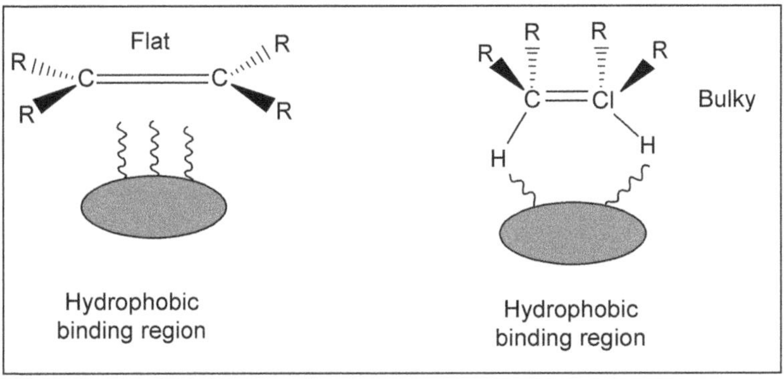

Figure 4.28: Binding comparison of an alkene with an alkane

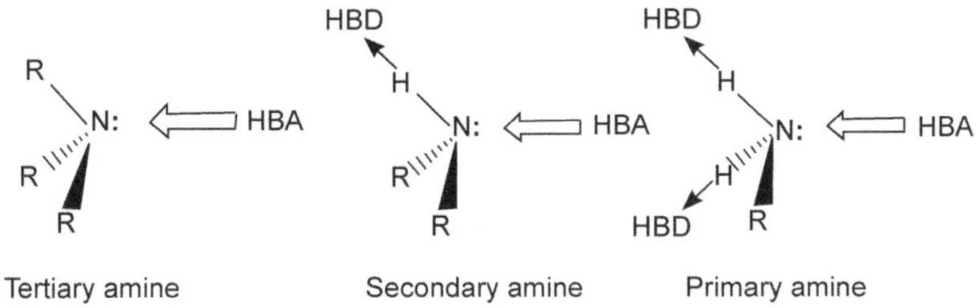

Figure 4.29: Possible hydrogen bonding interactions for amines

Binding Role of Alkenes

Alkenes can act together with the hydrophobic part of the binding site by means of van der waals interactions because they are planer and hydrophobic like aromatic ring. Activity of the equal saturated correspondents might be wealth testing, as the saturated alkyl areas is massive and cannot line the appropriate section of the binding place so diligently **(Figure 4.28)**.

Binding role of Amines

One of the most important functional group present in medicinal chemistry and in many drugs are Amines. Either as a H-bond donor/acceptor amines may well be tangled in H-bonding **(Figure 4.29)** Nitrogen atom is having one lone pair of electrons and work as H-bond acceptor for one hydrogen bond. The primary and secondary amines perform as H- bond donors as because they have -NH groups.

In most of the cases, the amino group might be protonated when interacting with the aimed binding site, meaning that it is ionized and could not perform as a H-bond acceptor. Though, it could still behave as a H-bond donor and will form a firm hydrogen bonds than when it was not ionize **(Figure 4.30)**.

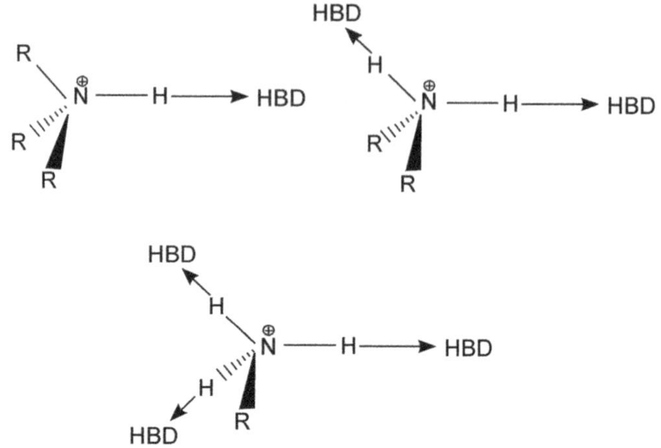

Figure 4.30: Possible hydrogen bonding interactions for ionized amines

Whereas, a strong ionic interaction might arise with a carboxylate ion in the binding site. **(Figure 4.31).**

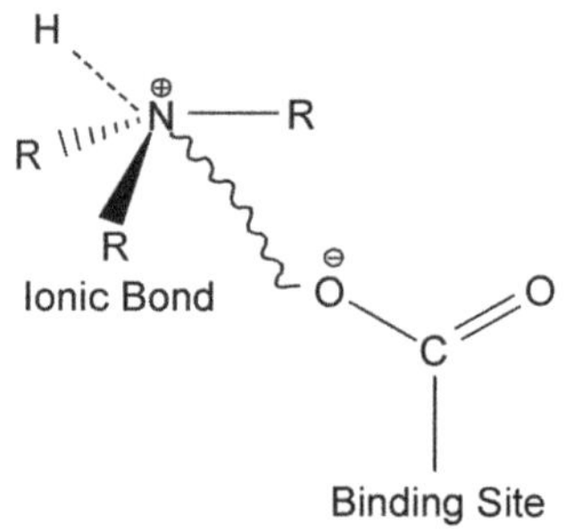

Fig. 4.31: Ionic interaction between an ionized amine and a carboxylate ion (R^{1-4}=H, alkyl)

Questions

1. Differentiate apoenzymes, cofactor, coenzyme and holoenzyme.
2. Give the importance of enzymes.
3. Define turnover number of enzymes.
4. Explain prosthetic group of conjugated enzymes.
5. Differentiate enzyme activator and inhibitor.
6. Differentiate lyases and ligases.

7. What do you mean by E.C number of enzyme.
8. Define Michaelis-Menton constant.
9. Why the enzymes are called biocatalyst.
10. Differentiate allosteric site and catalytic site.
11. Differentiate competitive and non-competitive inhibition.
12. What is feedback inhibition? Give one example.
13. What do you means by denaturation of enzymes?
14. Explain chemical nature of enzyme.
15. Describe lock and key hypothesis of mechanism of enzyme action.
16. Explain induced-fit. Hypothesis of enzyme action.
17. Describe the effect of temperature on enzyme activity.
18. What is structure activity relationship of drug molecules? Explain the binding role of $-NH_2$ group.
19. Explain the binding role of –OH group in structure activity relationship of drug molecules.
20. Explain various Drug action – receptor theories.

❑❑❑

5

Lipids

5.0 Introduction

Lipids are complex and heterogeneous in nature. At room temperature lipids may exist either as liquids or non-crystalline solids. Molecules belonging to the category of lipids are mostly soluble in organic or non-polar solvents such as chloroform, hexane, diethyl ether *etc*. Thus, lipids can be defined as *water insoluble biomolecules that are highly soluble in organic solvents*. Lipids mostly possess long-chain hydrocarbon groups which explain their being readily soluble in non-polar solvents. They either are derived from or naturally occur in most plants, animals, and microorganisms. They have varied biological roles such as they serve as fuel molecules, are used as cell membrane components, energy storage molecules, signal molecules, for insulation, and hormones. Fatty acids, waxes, sterols, vitamins A, D, E, and K that are fat-soluble, triglycerides, and phospholipids belong to the broad category of lipids. Lipids are basically soluble in organic solvents, insoluble in water and are derived or present in living organisms.

Despite being commonly used, definition of lipids based on solubility has some associated issues. Very short-chain fatty acids with one or four carbon chain are still considered lipids, though they are completely miscible in water and insoluble in nonpolar solvents. Additionally, some compounds such as some *trans* fatty acids are not derived directly from living organisms but still belong to the broad class of lipids (Figure 5.1).

DOI: 10.1201/9781003727453-5

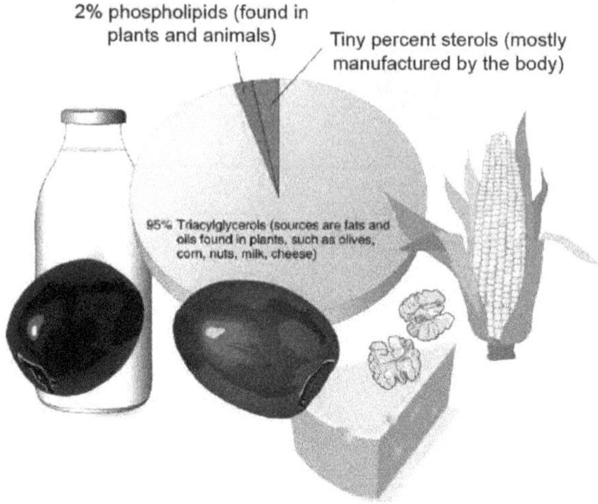

Figure 5.1: Sources of lipids

5.1 Classification of Lipids

Classification of lipid structures can be centred on physical properties at room temperature such as oils and fats. The oils are liquid and fats are solid at room temperature.

On the basis of polarity, lipids are grouped as polar and neutral lipids. Fatty acids, glycerides, waxes, and sterols belong to neutral lipids. Polar lipids consist of glycerophospholipids, glyceroglycolipid, sphingoglycolipid and sphingophospholipid.

In phospholipids two of the hydroxyl groups of glycerol are condensed with fatty acids while the third hydroxyl group is attached to a highly polar molecule and thus it is a polar lipid with two nonpolar hydrophobic tails and a charged hydrophilic head (Figure 5.2).

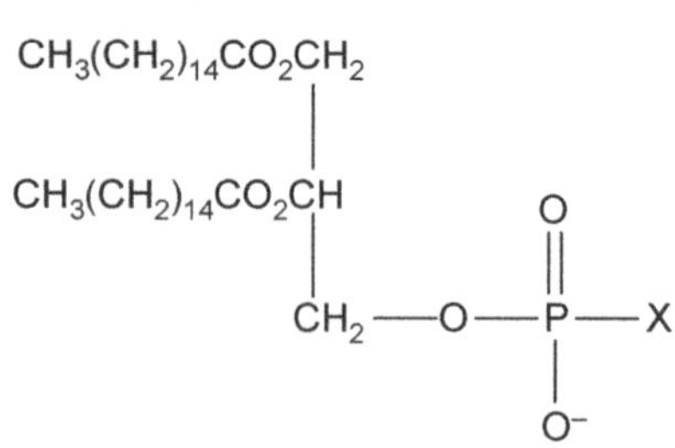

$X = OCH_2CH_2N(CH_3)_3^+$ phosphatidyl choline

$X = OCH_2CH_2NH_3^+$ phosphatidyl ethanolamine

Figure 5.2: Polar lipids

Based on their essentiality for humans, lipids and in particular fatty acids can be classed as essential and nonessential fatty acids.

Hydrolyzable/Saponifiable Lipids are those lipids that can be hydrolysed under basic conditions. The functional group is usually an ester and neutral fats such as waxes, phospholipids, and glycolipids are included in this class. **Non-hydrolyzable/Non-saponifiable Lipids** lack hydrolysable functional groups and include fat-soluble vitamins and steroids.

The most preferred classification is the one that is based on structure. Lipids can also be broadly classified as follows:

1. **Simple Lipids:** This class consists of fatty acid esters with various alcohols. It can be further separated into two subclasses.
   - **(a) Triacylglycerol/ Triglycerides:** They are composed of esters derived from glycerol and three fatty acids.
   - **(b) Waxes:** They are made up of fatty acid esters bearing monohydric alcohols with a higher molecular weight.
2. **Complex or Compound Lipids:** They are esters of fatty acids containing other groups also apart from the alcohol and fatty acids.
   - **(a) Phospholipids:** They contain a phosphoric acid residue in addition to fatty acids and alcohol. They also bear nitrogen containing bases and other substituents. There are further two sub-classes that differ in the alcohol substituent.
     - **(i) Glycerophospholipids** *or* **Phosphoglycerides** have glycerol
     - **(ii) Sphingophospholipids** possess the alcohol, sphingosine.
   - **(b) Glycolipids (Sphingoglycolipid or Glycosphingolipids):** They contain a fatty acid, sphingosine, and carbohydrate. Sphingoglycolipid include cerebrosides, gangliosides and globosides.
   - **(c) Other Complex Lipids:** Lipids such as sulfolipids, aminolipids, lipopolysaccharides and lipoproteins belong to this category.
3. **Precursor and Derived Lipids:** They are building blocks for simple and complex lipids. Fatty acids, glycerol, steroids, fatty aldehydes, and ketone bodies, hydrocarbons, lipid soluble vitamins, and hormones belong to this class.

Lipids are also be categorized based on their function:

1. Energy-storage Lipids

Carbohydrates and fats are two such biomolecules which provide us energy. While carbohydrates are primarily used for short-term energy storage, lipids on the other hand are used for long-term energy storage. This is because each gram of carbohydrates stores 4 calories of energy. However, each gram of lipid stores 9 calories. Thus, lipids help in compact energy storage, since it contains more energy per gram. The triglycerides are energy storage lipids. Each triglyceride contains three fatty acids component bound to a glycerol backbone. Adipose tissue that is distributed throughout our body is composed of fat cells, known as adipocytes, and they store the trigylcerides. Small amounts of tryiglycerides is also present in our blood.

2. Membrane Lipids

The membrane lipids are the phospholipids, glycolipids, and cholesterol. Lipids are known to be amphiphilic that means that it has one end which is soluble in water, thus it is 'polar' and other nonpolar end is soluble in fat. Thus a lipid bilayer is formed with polar ends pointing outwards the cell which is hydrophilic and the nonpolar ends pointing inwards which is hydrophobic. In this way interior of the cell is separated from the exterior.

3. Emulsification Lipids

Bile acids are mainly steroidal acids. Bile acids in the intestine regulate the digestion and absorption of cholesterol, triglycerides, and fat-soluble vitamins. Oxidation of cholesterol in the liver leads to the synthesis of primary bile acids. They are conjugated to hydrophilic amino acids such as glycine or taurine to increase their solubility. Bile acid is secreted into bile, stored in the gallbladder, and released in the intestine in response to dietary fat. Bile salts are also amphiphiles since they have a hydrophobic and a hydrophilic side. The hydrophobic portion faces the lipids forcing it to disperse, while the hydrophilic portion faces the water, increasing solubility. Thus a micelle is formed and the large lipid droplets get converted into smaller ones. This increases the surface area and effectively facilitates the action of digestive enzymes.

4. Chemical Messenger Lipids

Steroid Hormones are fat-soluble molecules that are derived from cholesterol. Steroid hormones include the sex hormones produced by male and female gonads and hormones produced in the adrenal glands. They are broadly classified as glucocorticoids, mineralcorticoids, androgens, estrogens and progestogens. Eicosanoids also act as signalling molecules and are derivatives of polyunsaturated C_{20} polyunsaturated arachidonic fatty acid. Prostaglandins, thromboxanes, prostacyclins, and leukotrienes belong to eicosanoids.

5. Protective-coating Lipids

Esters of medium chain fatty acids with long-chain alcohols are known as waxes. They act as protective coatings in animal, insect, and plants because of their hydrophobic nature. Waxes are of importance in foods as additives, adhesives, cosmetics, lubricants etc. Most commonly used wax includes beeswax, carnauba wax, and candelilla wax.

6. Fat-soluble Vitamins

Vitamins A, D, E, and K are known as fat soluble vitamins because they are soluble in organic solvents. They are transported in the body and absorbed in a similar manner to that of fats. Vitamin A is essential for normal vision and maintenance of epithelial tissue (Figure 5.3). Vitamin A is found in sufficient quantity in foods of animal origin. Body can also convert certain carotenoids, such as β-carotene, to vitamin A. Carotenoids are present in green leafy vegetables, yellow and orange fruits and vegetables. Ergocalciferol (Vitamin D2) and cholecalciferol (Vitamin D3) are biologically important and are formed in human skin from 7-dehydrocholesterol, *via* light irradiation. Vitamin D helps in the absorption of calcium and phosphorus. Vitamin D can be procured from animal foods such as butter, fish, and egg yolks. Vitamin E has antioxidant activity and protects cells from the damaging effects of free radicals. Vitamin E molecules include tocopherols and tocotrienols. Sources of vitamin E are whole grains, nuts, seeds, and some vegetable oils. Out of the numerous vitamin K forms, phylloquinone (vitamin K1) that is found in green leaves, and menaquinone (vitamin K2), synthesized by intestinal bacteria are the most important. Vitamin K play a significant role in preventing excessive bleeding and helps in blood clotting.

Lipid Soluble Vitamins

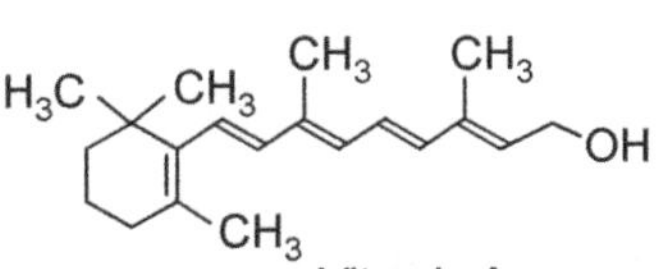

$C_{20}H_{30}O$ Vitamin A
Part of the visual pigment

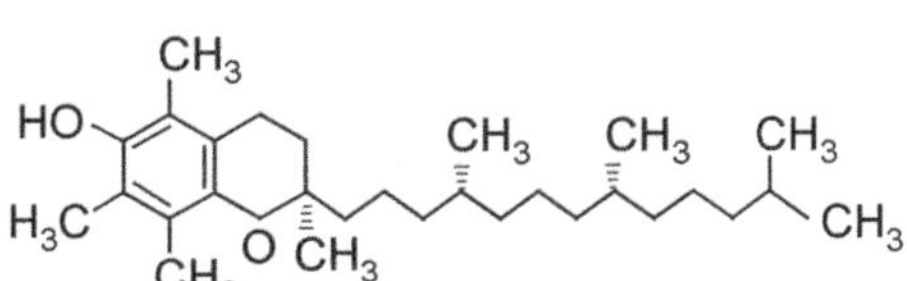

$C_{29}H_{50}O_2$ Vitamin E an antioxidant

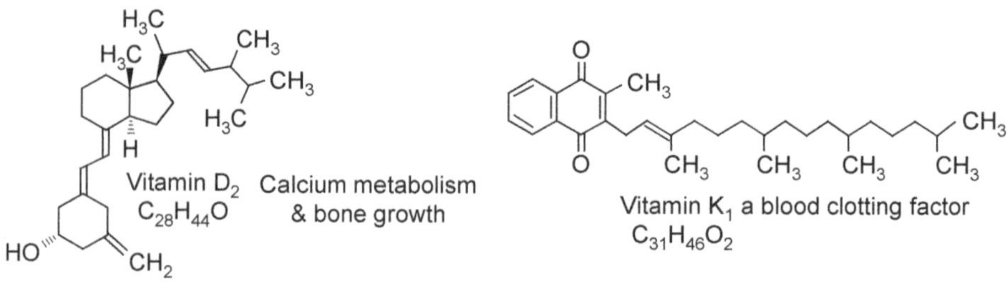

Figure 5.3: Lipid soluble vitamins

Lipids Functions in the Body

1. Triglycerides are stored in the body and is a major reservoir of energy.
2. They form important constituents of cell membrane in the form of phospholipids, glycolipids and sterols.
3. Absorption of vitamins in the body such as A, D, E and K depend on the presence lipids in the diet.
4. The basic unit of lipids, i.e. acetyl CoA is used for the synthesis of cholesterol and hence steroid hormones.
5. The lipids maintain the membrane structure and integrity.
6. Lipids act as insulators and protects internal organs from shock.
7. They play a vital role in blood coagulation.
8. Dipalmitoyl lecithin, a phospholipid act as a surfactant and is needed for the normal working of the lung alveoli.

5.2 Fatty Acids

In order to understand the lipid structures, it is essential to first learn about the fatty acids. Fatty acids are carboxylic acids having hydrocarbon chains which range from 4 to 36 carbons (C4 to C36), with C18 being the most common. At physiological pH, the terminal carboxyl group (–COOH) ionizes, becoming $-COO^-$. Thus it has amphipathic nature which means it has both hydrophilic and hydrophobic character. However, for long-chain fatty acids (LCFAs), the hydrophobic portion is predominant. More than 90% of the fatty acids found in plasma are in the form of fatty acid esters (primarily triacylglycerol, cholesteryl esters, and phospholipids). Biosynthesized fatty acids almost always possess an even number of carbon atoms and an unbranched carbon chain.

Four Major Physiological Roles of Fatty Acids

Phospholipids and glycolipids have fatty acids as their building blocks. These amphipathic molecules are vital components of biological membranes. Proteins through *covalent attachment to fatty acids reach to membrane locations.* Third, *fatty acids are fuel molecules. Triacylglycerols* or *triglycerides* stored in adipose tissue to meet the energy needs of a cell or organism, are esters of fatty acids with glycerol. Fourth, *fatty acid derivatives serve as hormones and intracellular messengers.*

Fatty acids can be classified into four classes:

1. Straight chain fatty acids
2. Branched chain fatty acids
3. Substituted fatty acids
4. Cyclic fatty acids.

Straight chain fatty acids can be further classified as short-chain fatty acids (C4 and C6), medium chain fatty acids (C8 and C10) and long-chain fatty acids with carbon chain, C12 to C26.

Branched Chain Fatty Acids are less abundant than straight chain acids in animals and plants, such as isovaleric acid and isobutyric acid (Figure 5.4).

Figure 5.4: Branched chain fatty acids

Cyclic Fatty Acids are produced by some bacteria and plants *e.g.* hydnocarpic acid (16:1), chaulmoogric acid (18:1), and gorlic acid (18:2) (Figure 5.5).

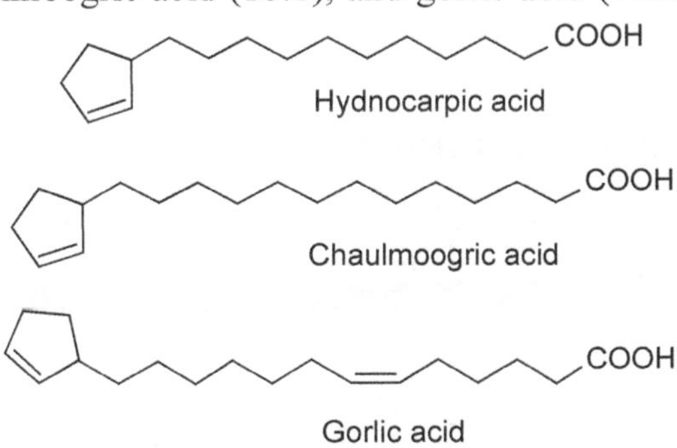

Figure 5.5: Cyclic fatty acids

Cyclopropane fatty acids (CPFA) are cyclic fatty acids containing three membered ring. Lactobacillic acid, sterculic acid, dihydrosterculic acid and malvalic acid are common examples of CPFA (Figure 5.6).

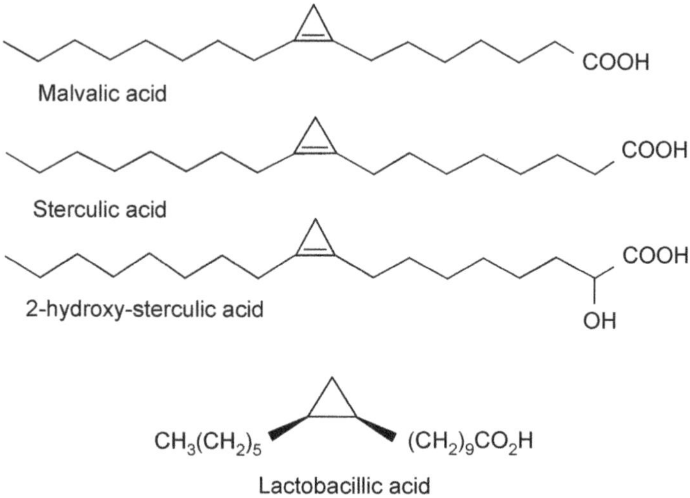

Figure 5.6: Cyclopropane fatty acids

Substituted Fatty Acids are the ones in which one or more hydrogen is replaced by other groups such as hydroxyl. **Cerebronic acid** and **hydroxynervonic acid** are two prominent examples of this class that also have key role in cerebroside synthesis (Figure 5.7).

$$CH_3-(CH_2)_7-CH=CH-(CH_2)_{12}-\underset{\displaystyle OH}{\underset{|}{CH}}-COOH$$

hydroxynervonic acid (24 : 1 (n-9) OH)

$$CH_3-(CH_2)_{21}-\underset{\displaystyle OH}{\underset{|}{CH}}-COOH$$

cerebronic acid (24 : 0 OH)

Figure 5.7: Substituted fatty acids with hydroxyl groups

5.3 Nomenclature of Fatty Acids

According to IUPAC terminology, the fatty acids are named after the parent hydrocarbon. For example, the 16-carbon carboxylic acid is called hexadecanoic acid. More often, common names are more used than standard IUPAC terminology. For example, palmitic acid is more commonly used for hexadecanoic acid. Double bonds are designated using the Greek letter delta Δ which indicates the distance of the double bond from the carboxyl carbon. By starting the numbering of carbons from the carboxylic acid group, a double bond between ninth and tenth carbons will be termed as Δ^9 bond. Thus, fatty acid with eighteen carbon skeleton and a double bond between the ninth and tenth carbons can be named as Δ^9-octadecenoic acid or oleic acid. Similarly, palmitoleic acid can be termed as Δ^9-hexadecenoic acid.

Geometry of the double-bond can be explained using *cis–trans* or *E/Z* nomenclatures. *Cis-trans* can also indicate relative positions of groups in ring structures. Atoms or groups are *cis* if they lie on same side or *trans* if they are on the opposite sides of a reference plane in the molecule. Also in *E/Z* nomenclature if the preferred atoms or groups are on the same side of the plane, then it is termed as *Z* configuration. If they lie on the opposite sides of the plane, it is the *E* configuration. Thus, oleic acid can also be written as *c*-9-octadecenoic acid or 9*Z*- octadecanoic acid, since it has a *cis* double bond or *Z*-configuration between the ninth and tenth carbon.

There are some other shorthand alternatives to name the fatty acids. One such terminology includes writing the chain length and number of double bonds separated by a colon (Table 5.1). For example, octadecenoic acid with 18 carbons and 1 double bond can be written as 18:1. In fatty acids the carbon atom that is adjacent to the carboxyl group is referred to as α, and carbon atom 3 is denoted as β. The methyl carbon atom at the terminal end of the chain is called the *ω*-carbon atom. Thus, the position of a double bond can be represented by counting from the methyl end and using ω or n denotation. It can be expressed as n-x or ω-x, where x is the number of carbons and is counted from the methyl end. Thus, oleic acid can also be written as 18:1 9c; 18:1 n-6 or 18:1 ω-6.

Table 5.1: Shorthand alternatives for naming the fatty acids.

Structural Notation		Common Name	Structure
Saturated Fatty Acids			
12:0		Lauric acid	
14:0		Myristic acid	
16:0		Palmitic acid	
18:0		Stearic acid	
20:0		Arachidic acid	
Monounsaturated Fatty Acids			
16:1 Δ^{9}	ω-7	Palmitoleic acid	
18:1 Δ^{9}	ω-9	Oleic acid	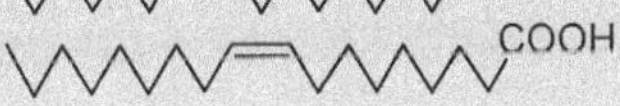
Polyunsaturated Fatty Acids			
18:2 $\Delta^{9,12}$	ω-6	Linoleic acid	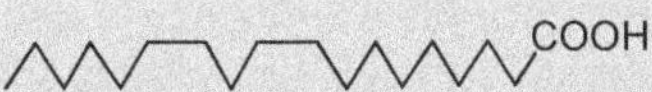
18:3 $\Delta^{9,12,15}$	ω-3	Linolenic acid	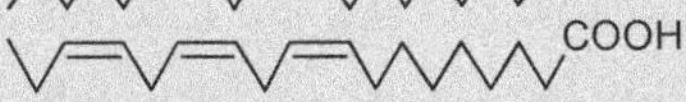
20:4 $\Delta^{5,8,11,14}$	ω-6	Arachidonic acid	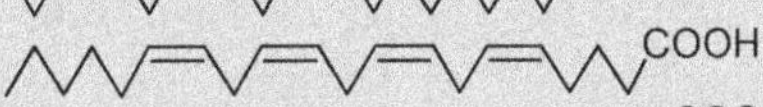
20:5 $\Delta^{5,8,11,14,17}$	ω-3	EPA (eicosapentaenoic acid)	
22:6 $\Delta^{4,7,10,13,16,19}$	ω-3	DHA (docosahexaenoic acid	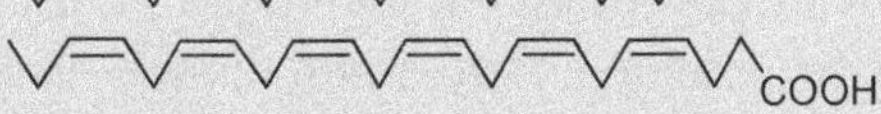

5.4 Saturated and Unsaturated Fatty Acids

Saturated fatty acids (SFAs) does not bear any double bond. A **saturated fatty acid** *is a fatty acid with a carbon chain bearing carbon–carbon single bonds.* On the basis of unsaturation, fatty acids can be classified as monounsaturated fatty acids (MUFAs), or polyunsaturated fatty acids (PUFAs). A **monounsaturated fatty acids** *bear a carbon chain with only one carbon–carbon double bond* with mostly the cis configuration. *Fatty acid with a carbon chain having two or more carbon–carbon double bonds are termed as* **polyunsaturated fatty acid**. Up to six double bonds can be found in biochemically important PUFAs.

Unsaturated fatty acids are synthesized from its saturated counterpart by desaturases enzymes, found in the endoplasmic reticulum and it introduces the *cis* double bonds. For this, hydrogen is removed and combined with molecular oxygen to form water. Human body cannot catalyse such reactions beyond carbons 9 and 10. Thus, humans can make oleic acid which has double bond between carbon 9 and 10, but cannot synthesize linoleic acid which has double bond between carbon 9 and 10 as well as between carbon 12 and 13.

5.5 Essential Fatty Acids

An essential fatty acid is fatty acid that cannot be synthesized within the human body and is essentially needed for the optimal health (Figure 5.8); therefore it is needed from dietary sources. The polyunsaturated fatty acids, namely linoleic acid and linolenic acid (Figure 5.9) are two essential fatty acids. Linoleic acid (LA) is a key member of the omega-6 acid family, and linolenic acid (ALA) is the major member of the omega-3 acid family. Scientific abbreviation for α-linolenic acid is 18:3n-3, where 18:3 means that ALA is an 18-carbon fatty acid with three double bonds and n-3 explains that the first double bond is in the n-3 position, which means it is an omega-3 fatty acid. Similarly, linoleic acid can be written as 18:2n-6 meaning 18-carbon fatty acid with two double bonds and first double bond at n-6 position. Longer omega-6 and omega-3 fatty acids can be obtained from linoleic acid (LA) and α-linolenic acid (ALA), *via* a series of desaturation and elongation reactions. Arachidonic acid (AA; 20:4n-6) can be biosynthesized from linoleic acid and thus inadequacy of linoleic acid makes arachidonic acid an essential fatty acid. Likewise, linolenic acid is the starting material for the biosynthesis of two omega-3 fatty acids namely eicosapentaenoic acid (EPA; 20:5n-3) and docosahexaenoic acid (DHA; 22:6n-3).

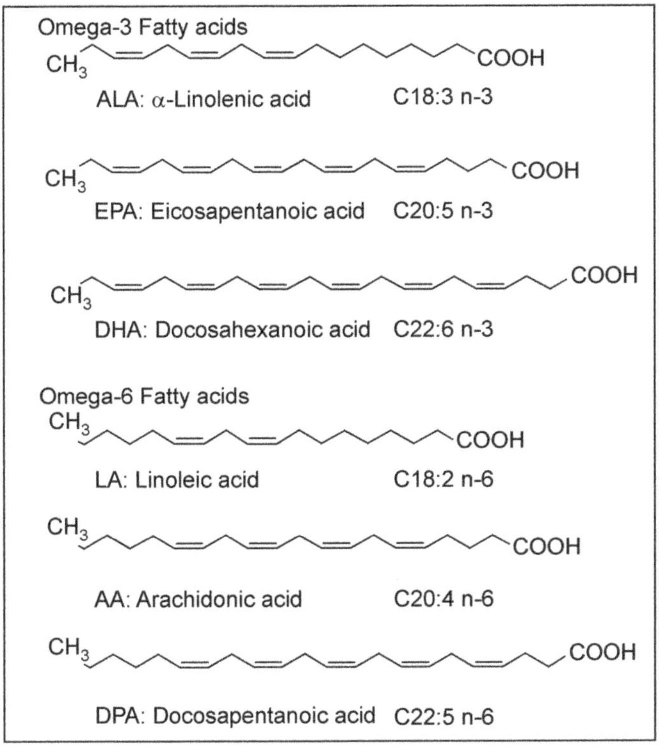

Figure 5.8: Class of essential fatty acids

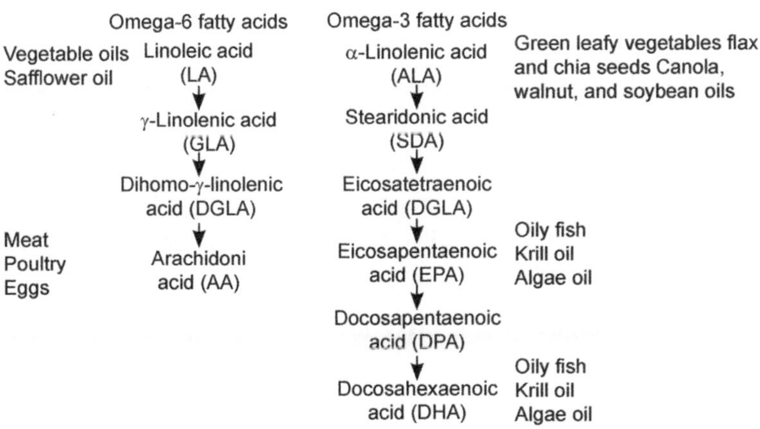

Omega-6 (n-6) and omega-3 (n-3) fany acids comprise the two classes of essemial fatty acids (EFA). The parent compounds of each class, linoleic acid (LA) and a-linolenic acid (ALA) (bold font), give rise to longer chain derivatives inside the body. Due to low efficiency of conversion of ALA to the long-chain omega-3 PUFA, eicosapentaenoic acid (EPA) and docosahexaenoic acid (DHA), it is recommended to obtain EPA and DHA from additiorial sources. Dietary sources of the different long-chain PUFA are listed in the colored boxes (25).

Figure 5.9: Dietary sources of polyunsaturated fatty acids (PUFA)

Omega-6 and omega-3 fatty acids are key structural components of cell membranes. They affect properties of cell membrane and membranes of cell organelle, such as permeability, fluidity, membrane-bound enzymes and cell-signalling. They protect the liver from accumulation of fat and therefore, prevent fatty liver. DHA plays predominant role in normal function and development of vision and nervous system function. DHA intake is important for foetus and infants for the growth of their brain and development of brain functions. Various scientific studies have revealed that intake of EPA and DHA may benefit people with type 2 diabetes mellitus and cardiovascular issues. Since, essential fatty acids are involved in the esterification, transport and metabolism of cholesterol; it lowers the serum cholesterol level, increases HDL and reduces the risk of heart diseases. Arachidonic acid is a crucial precursor of eicosanoids that regulates blood pressure, clotting, and some other important body functions. If essential fatty acids are not in sufficient quantity in the diet then it can lead to dry, scaly and red rashes, proneness to infection, poor wound healing, liver abnormalities and decreased growth in infants and children.

5.6 Physical Properties of Fatty Acids

The physical properties and structure of fatty acids are mostly dependent on the length and degree of unsaturation in the fatty acids. Double bonds introduce kinks in the hydrocarbon chain that can influence some features including the melting point. Water solubility of fatty acids depends upon the carbon chain length. Short chain fatty acids are relatively soluble in water and the solubility decreases as the chain length increases. The slight solubility of short-chain fatty acids is attributed to the polarity of the carboxyl group. In longer chain fatty acids, the nonpolar nature of the hydrocarbon chain dominates the polarity of carboxyl group. Therefore, long-chain fatty acids are fundamentally insoluble in water.

As the molecular weight of the fatty acid increases, their melting point increases. For example lauric acid, a C_{12} fatty acid melts at 44°C, whereas palmitic acid (C_{16}) has a melting point of 63°C. As the chain length increases, surface area becomes greater which facilitates better intermolecular interactions between fatty acid molecules, leading to increase in the melting point.

On the other hand, unsaturated fatty acids have lower melting points than saturated fatty acids of the same length. This can be explained by decreased molecular attractions between carbon chains. The "bends" in unsaturated fatty acids due to the *cis* bond prevents the chain to closely pack together as tightly as the saturated fatty acids. Thus saturated fatty acid chain tends to be solids at room temperature, whereas long-chain unsaturated fatty acids are liquids at room temperature. The introduction of more than one double bond in the hydrocarbon chain in unsaturated fatty acids results in more "bends" in the molecule. Thus, the melting points of polyunsaturated fatty acids are much lower than mono unsaturated fatty acids. For instance, the melting point of stearic acid is 69.6°C and exists as a waxy solid (Figure 5.10), whereas oleic acid with one *cis* double bond has a melting point of 13.4°C. The melting points of polyunsaturated fatty acids such as linoleic acid are even lower and thus they exist as liquids at room temperature. The shape of a *trans* fatty acid is comparable to that of its saturated counterpart. Thus, trans fatty acids are known to have a higher melting point than its cis isomer. The *trans* isomer of oleic acid, known as elaidic acid, exists in a linear shape and a melting point of 45 °C.

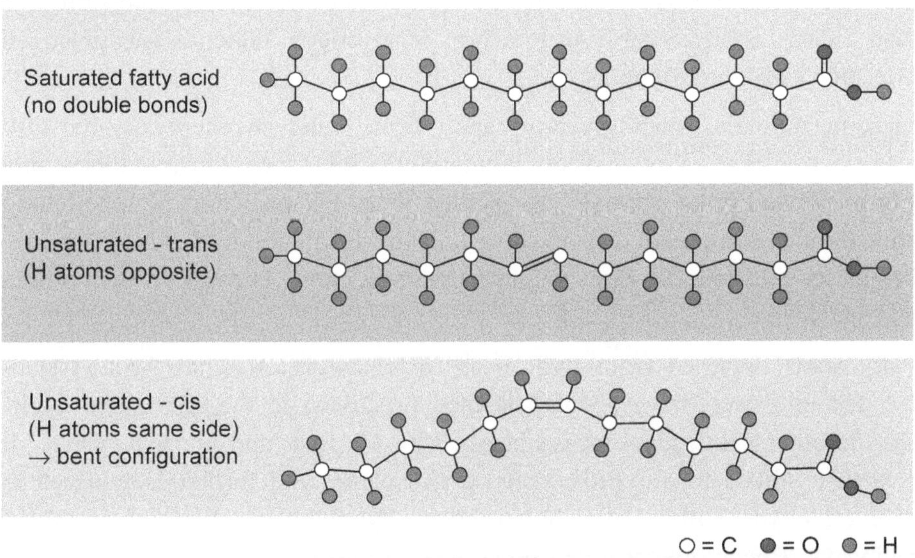

Figure 5.10: Spatial orientation of saturated and unsaturated fatty acids

5.7 Triacylglycerols

Triacylglycerols or triglycerides exist in nearly all commercially important fats and oils of animal and plant origin. They are primary components of **adipose tissue** and **sebum.** They serve as efficient energy-storage molecules and provide caloric content much higher than both **carbohydrates** and **proteins**.

In most eukaryotic cells, triacylglycerols form microscopic, oily droplets serving as metabolic fuel. In vertebrates, adipocytes, or fat cells, store large amounts of triacylglycerols as fat droplets. Triacylglycerols provide energy during seed germination. Triacylglycerols in adipocytes and germinating seeds is hydrolyzed by lipases, which releases fatty acids that goes to sites where fuel is required.

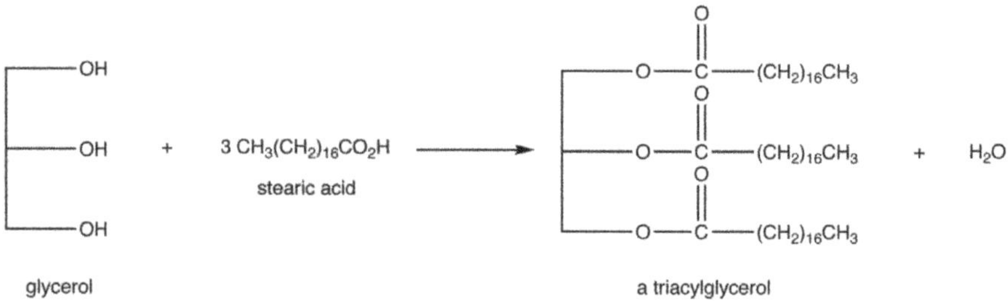

Humans have fat tissue under the skin, in the abdominal cavity, and mammary glands in the form of adipocytes. Triacylglycerols deposited in adipocytes can meet the energy needs for months. In contrast, the human body can store less than a day's energy supply in the form of glycogen. Glucose and glycogen offer quick source of energy.

In some animals, triacylglycerols apart from being an energy source also acts as an insulation against low temperature. Seals, penguins, walruses, and warm-blooded polar animals are padded with this insulator. In hibernating animals, the accumulated fat reserve helps in insulation and energy storage. Migratory birds can fly huge distances without eating because of their stored triacylglycerol.

Fatty acids obtained from hydrolysis of triacylglycerols are broken step-by step into acetyl CoA, which is then processed in the citric acid cycle. On the other hand, glycerol is absorbed by the liver and phosphorylated. It is then oxidized to dihydroxyacetone phosphate, and further isomerized to glyceraldehyde 3-phosphate. It is used in glycolytic and the gluconeogenic pathways. Hence, glycerol can be converted into pyruvate or glucose in the liver.

As we know that, in triacylglycerol, three fatty acids are esterified through their carboxyl groups to the glycerol backbone. If the same fatty acid is esterified at all the three positions, then triacylglycerol is termed as simple triacylglycerols. Mixed triacylglycerols means two or more different fatty acids (Figure 5.11).

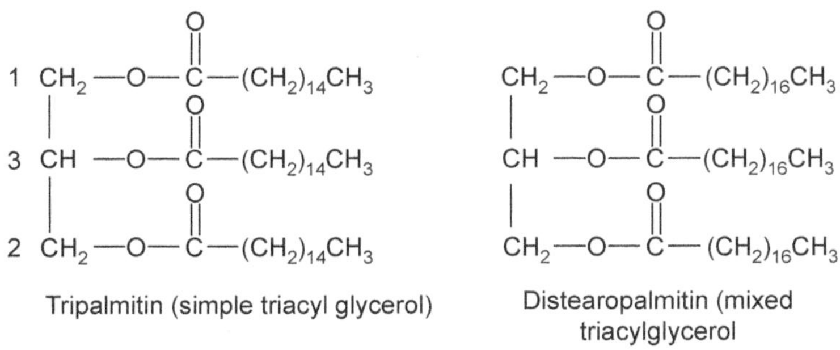

Figure 5.11: Examples of simple and mixed triacylglycerol

At room temperature, triacylglycerols with unsaturated fatty acids are liquids, while those with saturated fatty acids are solids. When the primary hydroxyls of glycerol are esterified by different fatty acid groups, the resulting molecule is chiral and exists as two enantiomers. The prefix *sn*- which means stereospecific numbering is used for a particular enantiomer, an equal mixture of enantiomers can be denoted by *rac*-, and *x*- is used for an unknown stereochemistry, for e.g., 1-palmitoyl-2-linoleoyl-3-oleoyl-*sn*-glycerol (Figure 5.12).

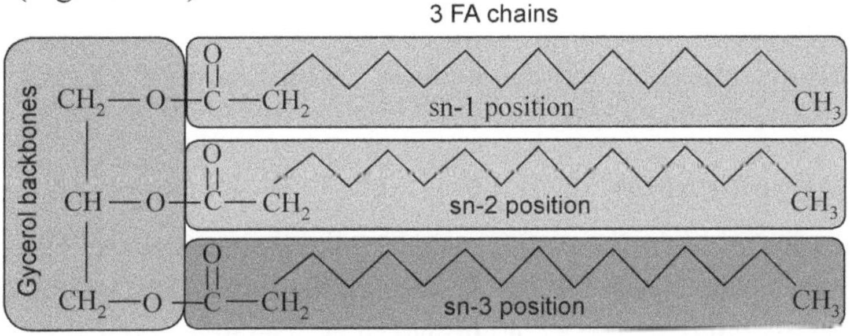

Figure 5.12: Structure of 1-palmitoyl-2-linoleoyl-3-oleoyl-sn-glycerol

Different kinds of triacylglycerol molecules naturally occur in fats and oils. The difference between the two is their physical state at room temperature. At room temperature, fat exists as a solid or a semi-solid. Generally, they are obtained from animal sources; hence termed as *animal fat*. Oil is a liquid at room temperature and is obtained from plant sources. As oils and fats contain mixtures of triglycerides, no fat or oil can be represented by a single definite chemical formula.

Fats chiefly contain triacylglycerols predominated by saturated fatty acids, although some unsaturated fatty acids may be present in small quantities. Oils are composed of triacylglycerols with larger amounts of mono- and polyunsaturated fatty acids. A notable exception to this is coconut oil, which contains saturated fatty acid, particularly lauric acid (12:0) and is rich in *shorter-chain* fatty acids. High concentration of mono- and polyunsaturated fatty acids in oils, results in lower melting points. At room temperature fats are solids, but due to the warmer body temperature, fat exist as liquid or semi-solid for its movement in the body of living animals.

5.8 Chemical Reactions pf Triacylglycerols

5.8.1 Hydrolysis

Chemical properties of triacylglycerols are similar to that of ester and alkene functional groups. Triacylglyercol hydrolysis can be catalysed by an acid, a base or a lipolytic enzyme. Hydrolysis of triacylglycerides can result in products such as free fatty acids (FFAs), diacylglycerides (DGs), and monoacylglycerides (MGs). When all three fatty acids are removed from the triglyceride leading to glycerol and fatty acids, it is referred as complete hydrolysis. Under acidic conditions, fatty acids are formed and under basic conditions, fatty acid salts are obtained. The fatty acid salts can be converted to fatty acids by the action of mild acids. Intestinal lipase carries out stepwise hydrolysis of triacylglycerols, first to diacylglycerides, then monoacylglycerides and lastly liberating all three fatty acids and glycerol. Hydrolysis is important for metabolism, mobilization of fats and for the growth of germinating seeds.

5.8.2 Saponification

Saponification is hydrolysis of fats under alkaline conditions. The products obtained are glycerol and Na^+ or K^+ salts of the fatty acids, which are called soaps. The overall reaction occurs in two steps. The first step is the hydrolysis of fatty esters to produce glycerol and three fatty acid molecules. In the second step, the fatty acids and the base (NaOH) reacts, resulting in neutralization of acid and formation of soap. The industrial production of soaps now uses the direct reaction between water and fats under high temperature and pressure conditions.

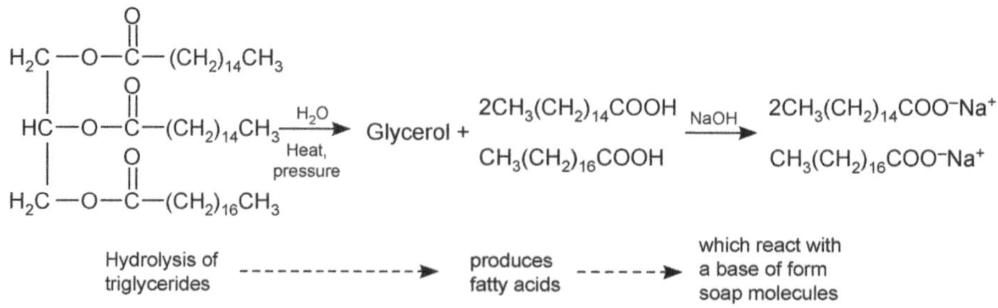

The cleansing action of soap is related to micelle formation where the carboxylate ions act as the polar group. A spherical cluster is formed in which the polar portions are on the water surface, and the nonpolar hydrocarbon portions are located in the interior. The non-polar ends of soap are repelled by water but it interacts with the nonpolar grease or dirt. The polar end of soap are hydrophilic and thus attracted towards the water molecules. Thus, the dirt gets trapped and removed during washing (Figure 5.13).

Soaps are converted into their insoluble calcium or magnesium salts, when used in "hard" water having high concentrations of Ca^{2+} and Mg^{2+}. Synthetic detergents such as sodium dodecylsulfate (SDS) are less prone to precipitation in hard water.

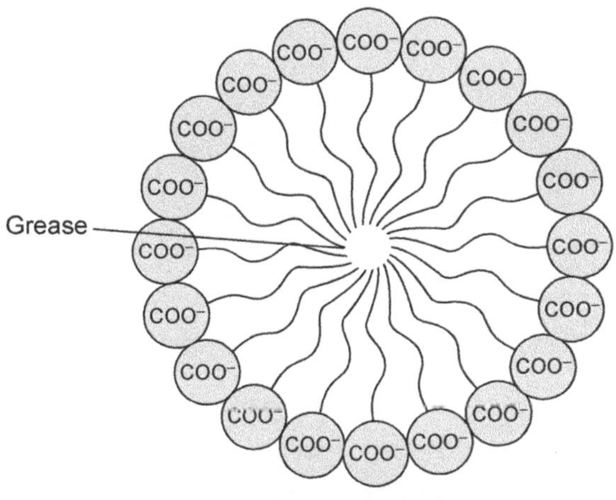

Figure 5.13: Micelle formation during the cleansing action of soap

5.8.3 Ester Exchange Reactions

Exchanging the fatty acids or alcohol groups present in the triglyceride modifies the composition and properties of oils and fats. It can be done in a number of ways such as by reaction with an excess of other fatty acids known as acidolysis catalysed by an acid or enzyme. Thus, a triglyceride rich in long chain fatty acids when treated with a medium fatty acid in the presence of acid or enzyme leads to triglyceride rich in medium chain fatty acids.

Methanolysis or alcoholysis of triacylglycerols is used to prepare methyl esters for fatty acid analysis, a process frequently referred to as transesterification. Biodiesel can be produced in large scale by methanolysis of vegetable oils or waste fat.

Glycerolysis is a type of transesterification reaction of triacylglycerols with glycerol. Using a basic catalyst such as sodium hydroxide or sodium methoxide, it produces mono- and diacylglycerols.

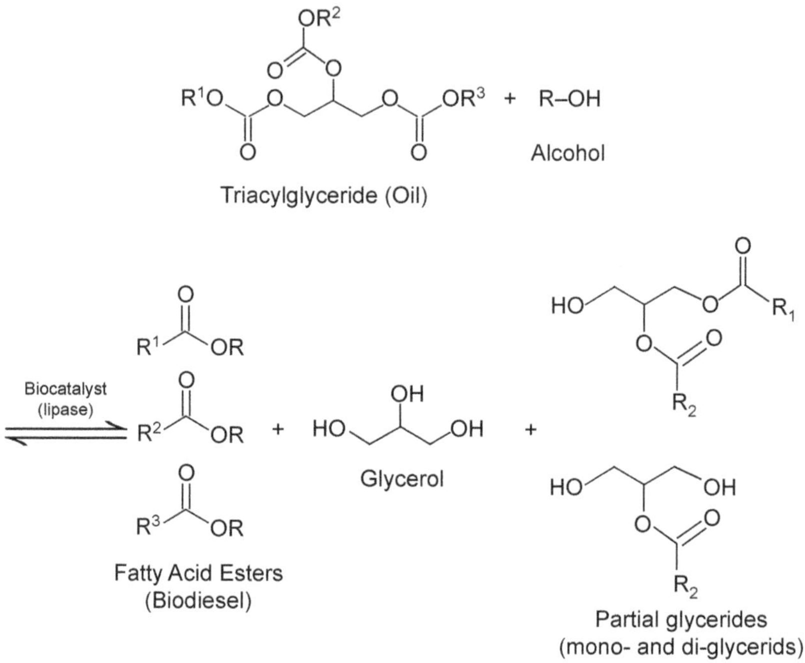

Interesterification is a rearrangement reaction as it leads to intra- and intermolecular exchange of fatty acids on the glycerol backbone of triacylglycerols using base catalyst or lipase. It alters the melting point and improves the functional properties of the fat and oils. Triglyceride bearing three saturated fatty acids is mostly solid at room temperature. Switching the saturated fatty acids with unsaturated fatty acids in triglycerides lowers the melting point. Chemical interesterification is carried out using low concentration of a base catalyst at moderate temperatures. However, chemical process leads to random product formation and side products and there is no selectivity for position of fatty acid on glycerol. The product composition can be controlled by doing interesterification at lower temperatures. Fats produced by partial hydrogenation of oils contain trans isomers which have undesirable health concerns. Liquid oils can be hardened by interesterification with fully saturated fats without isomerizing any of the fatty acids. Enzyme catalyzed reactions are more selective as compared to a chemical reaction. Lipase catalyzed interesterification takes place at a lower temperature and with fewer side reactions. They can be selective for fatty acids at different positions on the glycerol backbone (sn-1 and sn-3 rather than sn-2) or for a particular fatty acids.

Interesterification

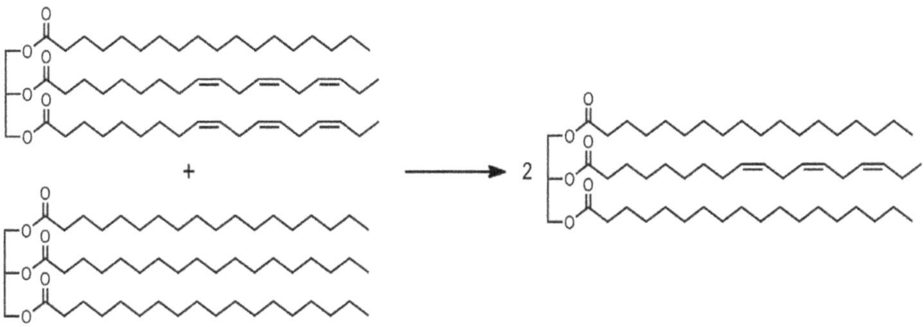

5.9 Hydrogenation

Hydrogenation involves addition of hydrogen across carbon–carbon multiple bonds, which increases the degree of saturation. Transition metals such as palladium on charcoal or platinum oxide also known as Adam's catalyst can be used for saturation of fatty acids at ambient temperature and pressure. Hydrogenation is accompanied by exchange and movement of hydrogen atoms along the chain in the region of the double bonds.

When some, but not all, double bonds are converted into single bonds it is known as partial hydrogenation. Partial hydrogenation reduces the polyunsaturated content of oils while maintaining or increasing the monoene content. As some double bonds are converted into single bonds, there is an increase in the melting point of triglyceride. In this way, plant oils are converted into semi-solid fats such as peanut butter and margarine. However, such a kind of hydrogenation converts some *cis* double bonds within fatty acid residues into *trans* double bonds producing trans unsaturated fatty acids.

Partial Hydrogenation

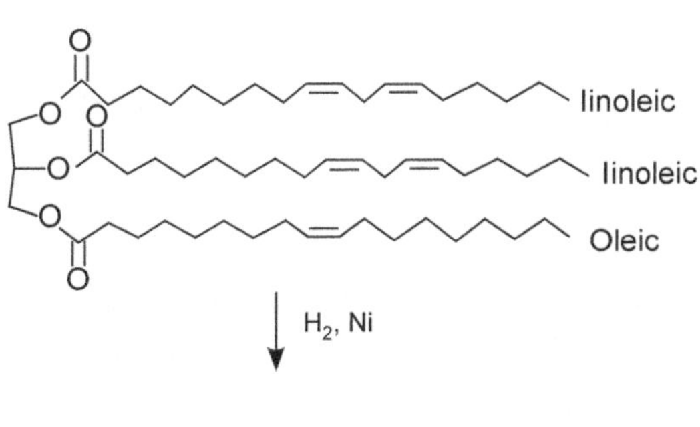

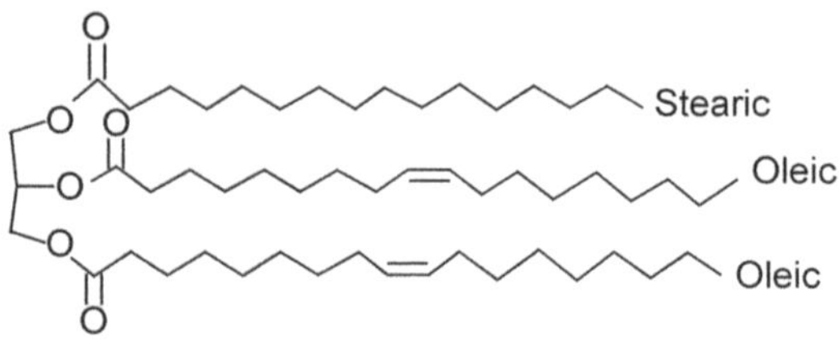

Trans Fats

Our body mostly have cis fatty acids and thus the enzymes have active sites that preferentially metabolize cis fatty acids. Trans fats can accumulate in the body as the cell's enzymes can't either break them or break them inefficiently. When cis fatty acids are incorporated into the cell membranes, they do not pack very compact and thus it provides fluidity to the cell membrane. When membranes will have trans fats, the membrane fluidity and its receptor function will be affected.

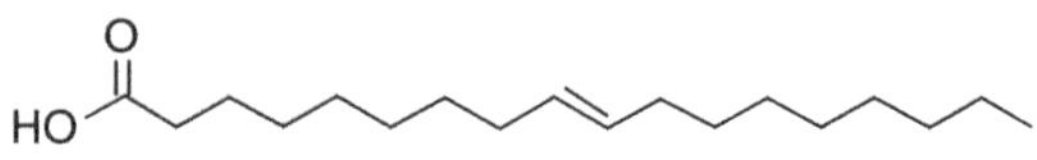

trans-Oleic acid

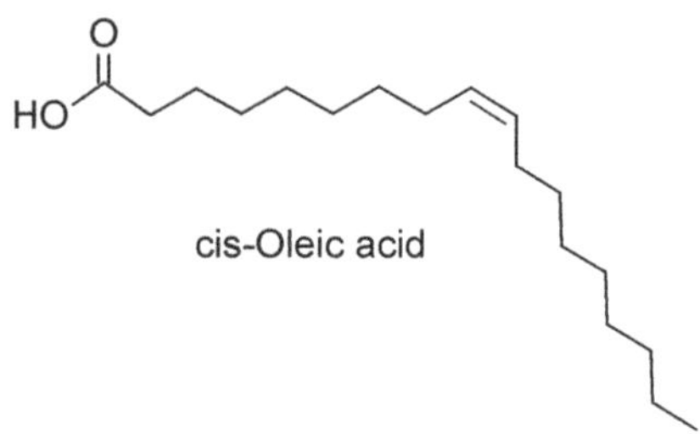

cis-Oleic acid

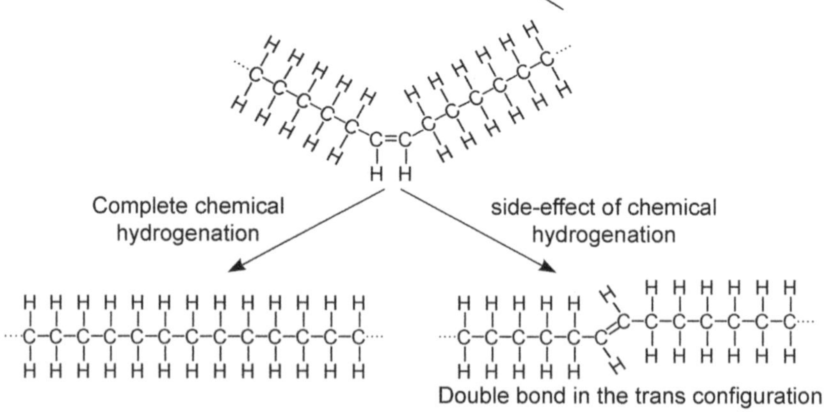

Hydrogenation imparts desirable features such as increased shelf life, spreadability and better texture. However, it also introduces some serious health problems because of the production of semisolid products. At times, not all cis-double bonds in the fatty acid chains are completely saturated in the hydrogenation process, some isomerize to their trans configuration. These unnatural trans-fats leads to heart diseases, cancer, diabetes and obesity, as well as problems related to immune response and reproductive problems. Trans fatty acids also compete with essential fatty acids leading to essential fatty acid deficiency. Their structures are similar to saturated fatty acids and they have a tendency to increase cholesterol level leading to formation of atherosclerosis.

5.10 Rancidification

Rancidification is known as the process which causes a substance to become rancid which means having an unpleasant smell or taste. It can be of three types:

1. **Hydrolytic Rancidity:** It refers to the odor that develops when triglycerides are hydrolyzed and free fatty acids are released.
2. **Oxidative Rancidity:** It is associated with the degradation of fats and oils by oxygen in the air. Through a free radical process, the double bonds in an unsaturated fatty acid undergo cleavage, and can form peroxides, which then decompose to release volatile aldehydes and ketones. Oxidation in unsaturated fats gets accelerated by exposure to light and heat. It can be curbed by addition of antioxidants. Thus, airtight packaging will slow rancidity development due to exclusion of oxygen.
3. **Microbial Rancidity:** It refers to a process in which microorganisms, such as bacteria or molds, partially hydrolyse glycerides using their enzymes such as lipases.

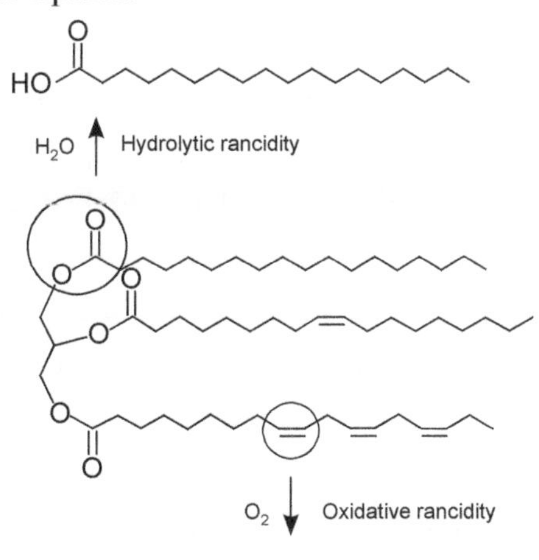

Degradation products

↓

Hydroperoxides, alcohols, aldehydes, ketones, carboxylic acid etc.

Natural vegetable fats and oils containing antioxidants like vitamin E prevents rancidity. Therefore, vegetable fats can be preserved for a longer time as compared to animal fats. Antioxidants like vitamin E, vitamin C, butylatedhydroxy toluene (BHT) and butylatedhydroxy anisole (BHA) can prevent oxidation of fats and thus the development of rancidity.

5.11 Oxidation

The fatty acid alkyl chain is vulnerable to oxidation due to the presence of double bonds and adjacent allylic carbons. Auto-oxidation, photooxidation and enzymatic oxidation are responsible for deterioration of unsaturated oils and fats and can result in undesirable odour and degradation of nutritional quality. The external factors that influence lipids oxidation include oxygen, light, heat, sensitizers, enzymes and microorganisms *etc.*

The allylic hydrogen can be easily removed as it requires activation energy of approximately 50 kcal/mol. Also hydrogen abstraction can result in rearrangements where cis double bond can change to trans double bond. In general, lipid oxidation includes three phases: initiation where unsaturated fatty acid reacts with initiator to form a radical. Oxygen can react with radicals of fatty acid and produce peroxyl radical. The peroxyl radicals abstract hydrogen from other fatty acids and become hydroperoxide and generate free alkyl radicals, which is called Propagation step. Termination includes interaction of high concentrated radical with each other forming non-radical compounds. The primary oxidation product in the first stage can further decompose producing short-chain aldehydes and carboxylic acid products that create objectionable odors. Photooxidation can happen in fatty acids when they are exposed to ultraviolet (UV), incandescent, and fluorescent light due to the presence of photosensitizers, such as chlorophyll, riboflavin, and methylene blue. To avoid this, commercial foods containing fats and oils contain antioxidants- substances that are more easily oxidized than the food. Two naturally occurring antioxidants are vitamin C and vitamin E. Synthetic oxidation inhibitors are butylated hydroxyanisole (BHA) and butylated hydroxytoluene (BHT).

Oxidation of Fats and Oils: Decomposition of Allylic Peroxides

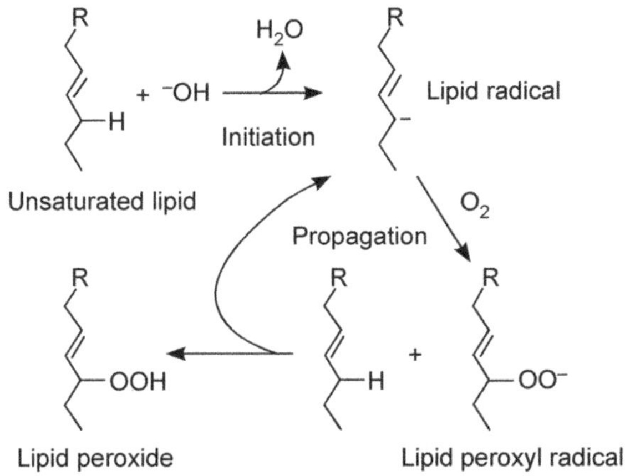

In animals and some bacteria, the oxidation of long-chain fatty acids to acetyl-CoA is a central energy-yielding pathway. The electrons removed during fatty acid oxidation goes through the mitochondrial respiratory chain, and this drives the ATP synthesis, and acetyl-CoA produced from the fatty acids is completely oxidized to CO_2 *via* the citric acid cycle, resulting in further energy conservation.

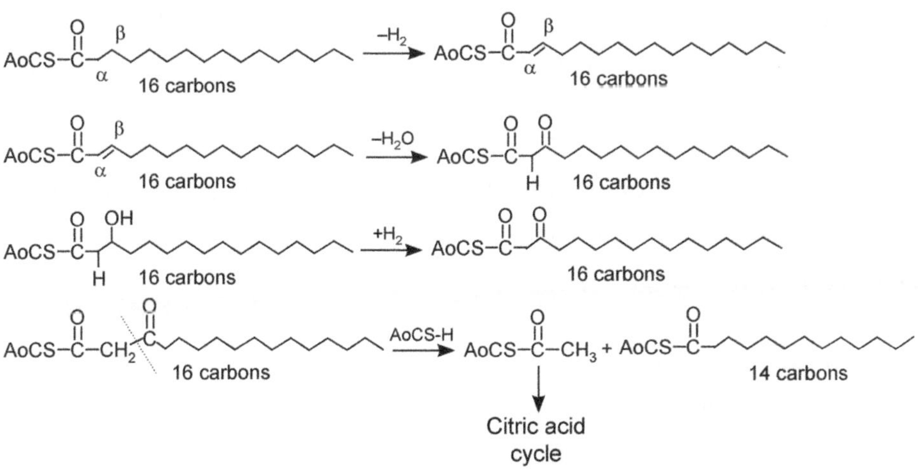

Saponification Number is expressed as grams of fat saponified by one mole of potassium hydroxide. It indicates average molecular weight and hence the chain length. Fats with long chain fatty acids have low saponification value because there are relatively less number of carboxylic functional groups present per unit mass of the fat as compared to short chain fatty acids.

Iodine Number is used to measure the degree of unsaturation of a fat. It is defined as the number of grams of iodine that combines with 100 g of a fat. High iodine numbers indicate higher degree of unsaturation. Iodine is incorporated into the double bonds present in the fatty acids. Determination of iodine number will help to know the degree of adulteration of given oil. Coconut oil is very saturated and used in making soaps. Linseed oil is highly unsaturated, and thus is a drying oil used for making oil paints.

Peroxidation (auto-oxidation) of lipids when exposed to oxygen is responsible not only for deterioration food but also for damage to tissues *in vivo*. Peroxide is the most widely used method for the prediction of rancidity. It measures the extent of primary oxidation. Secondary oxidation can be determined from *p*-anisidine test. The **peroxide value** is termed as the amount of peroxide oxygen consumed per 1 kilogram of fat or oil.

5.12 Reichert-Meissl Number

The number of ml of alkali required neutralizing the volatile fatty acids from 5 g of fat, e.g. Reichert-Meissl value for butter is 26 and coconut oil is 7.

Acid Value is also known as neutralization number or acid number or acidity. It is the mass of potassium hydroxide in milligrams needed to neutralize one gram of chemical substance having carboxylic acid groups. In a typical procedure, a known amount of sample is dissolved in an organic solvent and titrated with a solution of potassium hydroxide of known concentration using phenolphthalein as an indicator.

Hydroxyl Value is known as the amount of potassium hydroxide in milligrams required to neutralize the acetic acid that is left after acetylation of one gram of a chemical substance that contains free hydroxyl groups. Thus, hydroxyl value measures the free hydroxyl groups in a chemical substance. This analytical method involves acetylation of the free hydroxyl groups with acetic anhydride in pyridine solvent. On completion, water is added, and the remaining unreacted acetic anhydride is converted to acetic acid, which is then measured by titration with potassium hydroxide.

5.13 Waxes

Biological waxes are monoesters of long-chain fatty acids with long-chain alcohols (Figure 5.14). The fatty acids are generally saturated with 14 to 36 carbon atoms. The alcohols may be saturated or unsaturated with 16 to 30 carbon atoms. Melting points of waxes are higher than those of triacylglycerols. They are known for their insolubility and water-repellent nature because of the dominant hydrophobic components.

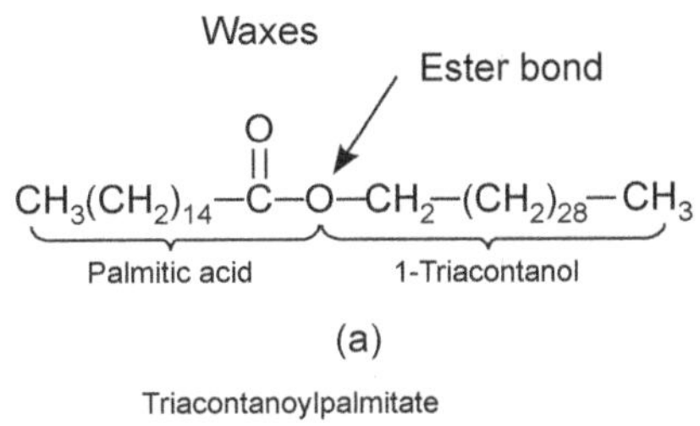

Figure 5.14: Waxes

Skin glands of humans and animals secrete waxes to protect the hair and skin and to keep them supple, lubricated, and waterproof. Certain birds such as the aquatic birds secrete waxes from their preen glands to make their feathers water-repellent and also minimize heat loss in cold waters. The shiny leaves of many plants in arid regions are coated with a layer of waxe, and this prevents excessive evaporation of water and also protects them against parasites.

Biological waxes have several applications in the pharmaceutical, cosmetic, polishing and lubricant industries. Lanolin from lamb's wool, beeswax, carnauba wax from a Brazilian palm tree, and spermaceti oil from whales are widely used waxes.

5.14 Phospholipids

Membranes give cells their individuality by separating them from their outside environment. Phospholipids is the main component of cell membranes. A **phospholipid** is a lipid that contains one or more fatty acids, a phosphate group, and an alcohol that is attached to the phosphate group. The platform molecule on which a phospholipid is built may be the three-carbon alcohol such as glycerol or a more complex C18 aminodialcohol called sphingosine. Glycerol-based phospholipids are called glycerophospholipids, and those based on sphingosine are called sphingophospholipids (Figure 5.15).

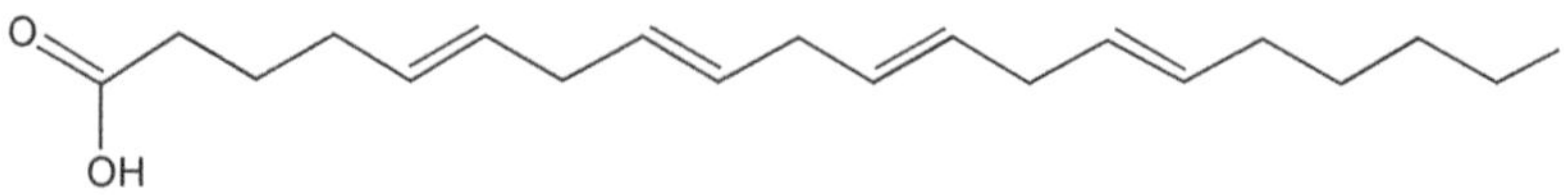

(a) Free fatty acid (arachidonic acid)

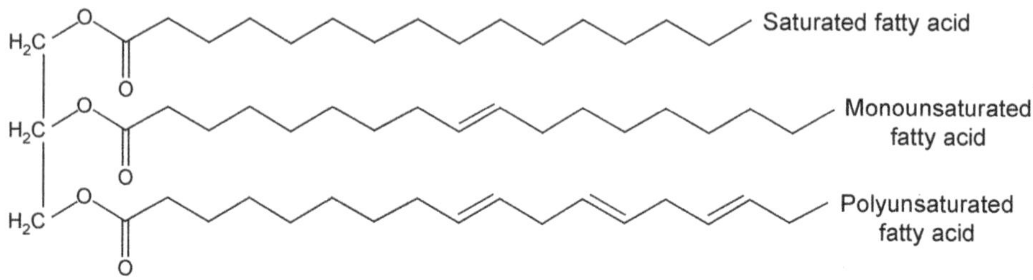

(b) Triacylglycerol (1-palmitic-2-oleic-3-linolenic-triacylglycerol)

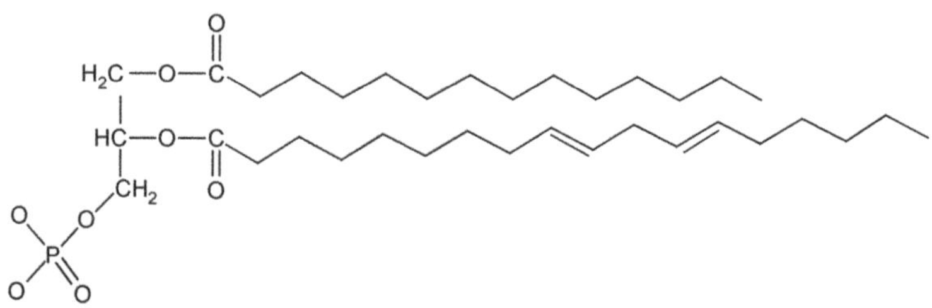

(c) Phospholipid (L-α-1-myristyol-2-linoleoyl-phosphatidic acid)

Figure 5.15: Example of free fatty acid, triacylglycerol and phospholipids

5.14.1 Glycerophospholipids

Glycerophospholipids function almost exclusively as components of cell membranes. A glycerophospholipid is a lipid that contains two fatty acids and a phosphate group esterified to a glycerol molecule and an alcohol esterified to the phosphate group. Glycerophospholipids have four ester linkages as compared to three ester linkages in triacylglycerols (Figure 5.16).

When the alcohol attached to the phosphate group in a glycophospholipid is choline, ethanolamine, or serine, glycerophospholipids are called phosphatidylcholines, phosphatidylethanolamines, or phosphatidylserines, respectively. The fatty acid, glycerol, and phosphate portions of a glycerophospholipid structure constitute a phosphatidyl group. Phosphatidyl inositol, cardiolipins and plasmalogens are other classes of glycerophospholipids

Phosphatidylcholines

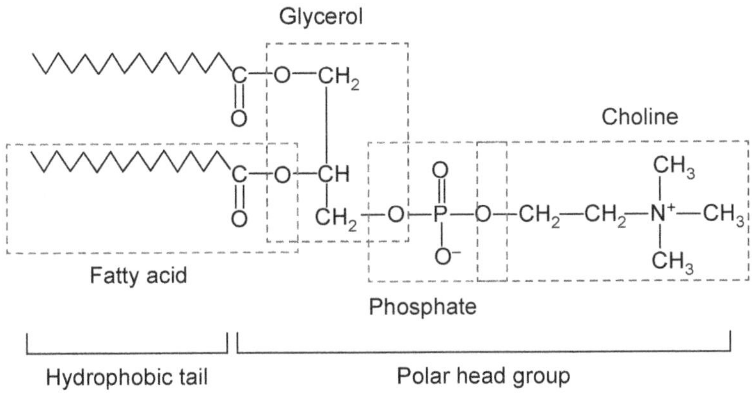

Figure 5.16: Phosphatidylcholines

All glycerophospholipids have a "head" and two "tails." The head constitutes phosphate and the associated alcohol and is polar in nature. Two tails are the two fatty acid carbon chains that constitute the nonpolar part. The polar head group of a glycerophospholipid is soluble in water. The nonpolar tail chains are insoluble in water but soluble in nonpolar substances.

Made by the liver, phosphatidylcholines have choline attached to the phosphate group and are also known as *lecithins.* There are a number of different phosphatidylcholines because different fatty acids may be bonded to the glycerol portion of the structure. Lecithin is present in brain, nervous tissue, and sperm and egg yolk. Lecithins helps in fat-emulsification. Dipalmitoyl lecithin, a lung surfactant is needed to prevent lung alveoli to collapse. It's absence in premature infants may produce respiratory distress syndrome or hyaline membrane disease.

Phosphatidylethanolamines and Phosphatidylserines

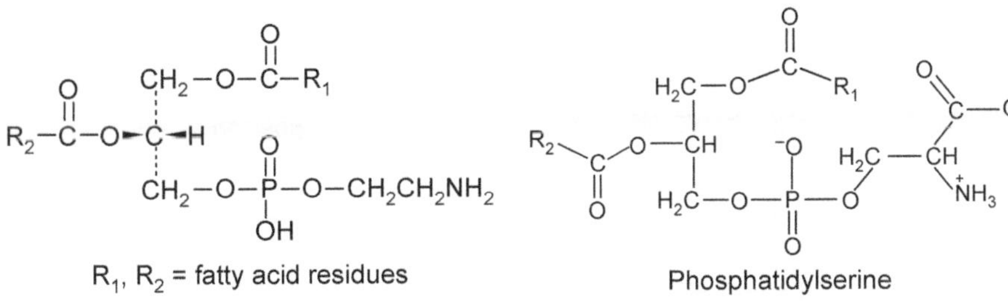

Figure 5.17: Phosphatidylethanolamines and phosphatidylserines

Phosphatidylethanolamines and phosphatidylserines (Figure 5.17) are also known as *cephalins.* These compounds are found in heart, liver tissue and brain. They play important role in blood clotting. Phosphatidyl inositol contains inositol as the nitrogenous base. The action of certain hormones such as oxytocin, vasopressin is facilitated through phosphatidyl inositol (Figure 5.18). In response to hormonal action, it is cleaved to diacyl glycerol and inositol triphosphate, both of which act as secondary messengers. Cardiolipin is diphosphatidyl glycerol and contains two molecules of phosphatidic acid held by glycerol. It has antigenic properties and is present in the inner mitochondrial membrane.

Phosphatidyl inositiol

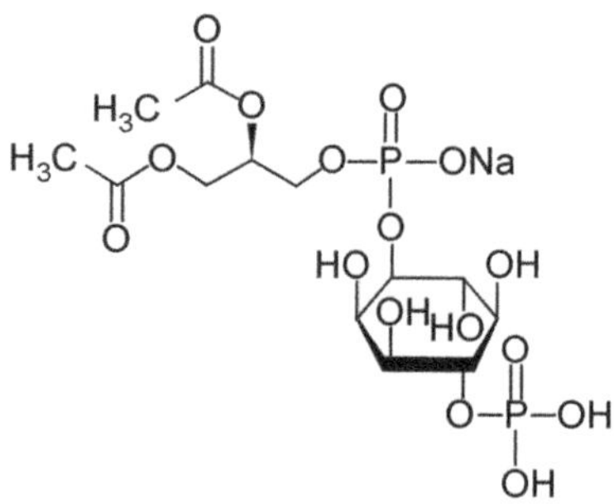

Cardiolipin

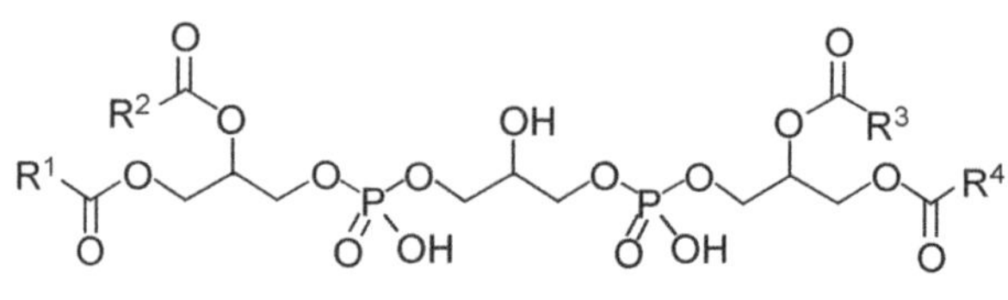

Figure 5.18: Phosphatidyl inositiol and cardiolipin

In Plasmalogen, α-1 position of glycerol has a vinyl-ether linkage. Plasmalogens are present in the skeletal muscle, cardiac muscle and in semen (Figure 5.19).

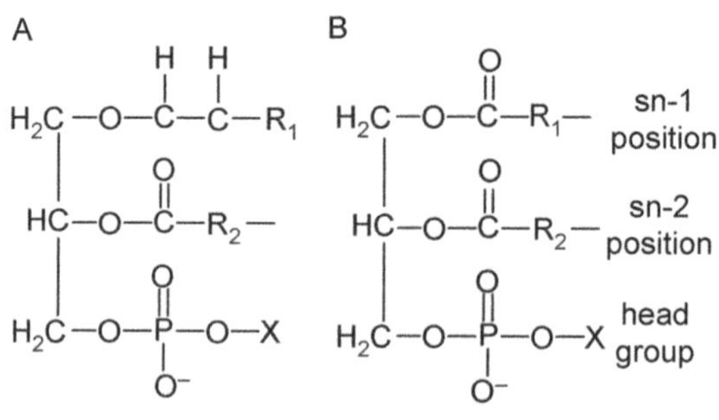

Figure 5.19: Plasmalogen versus other glycerophospholipids

5.14.2 Sphingophospholipids

Sphingophospholipids have structures based on the 18-carbon monounsaturated aminodialcohol, sphingosine. It contains one fatty acid and one phosphate group attached to a sphingosine molecule and an alcohol attached to the phosphate group. The fatty acids are attached to the sphingosine *via* an amide linkage and the phosphate group is attached *via* an ester linkage. Additionally, there is an alcohol esterified to the phosphate group.

For sphingophospholipids, the fatty acid is one of the tails, and the long carbon chain of sphingosine itself is the other tail. The polar head is the phosphate group with its esterified alcohol. Sphingophospholipids in which the alcohol esterified to the phosphate group is choline are called sphingomyelins (Figure 5.20). Sphingomyelins are found in all cell membranes and are important structural components of the myelin sheath, the protective and insulating coating that surrounds nerves. Their presence is also seen in brain, lungs, nerve and other tissues. Deposition of sphingomyelin and deficiency of sphingomyelinase enzyme in liver, lymph nodes, bone marrow and central nervous system results in Niemann-Pick disease. Ceramide is formed by the esterification of sphingosine which is an amino alcohol with a fatty acid of high molecular weight. Principally found in white matter of brain in myelin sheath and medullated nerves. Ceramide is common for all glycolipids and sphingomyelin.

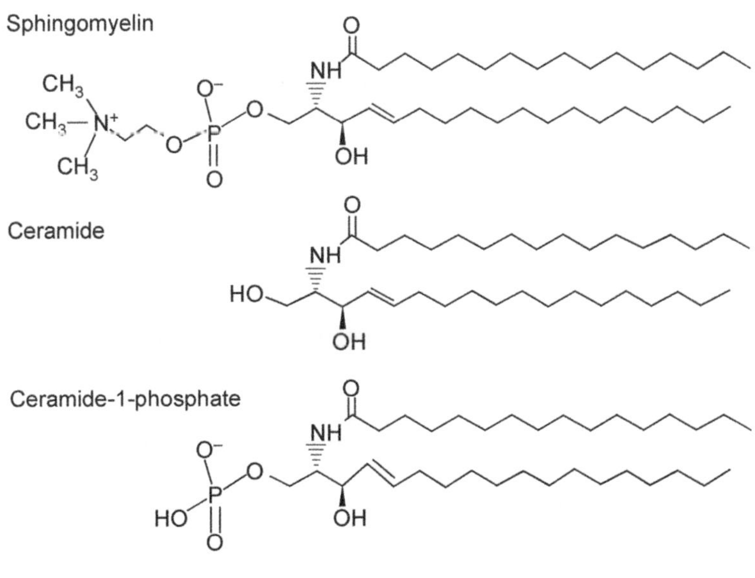

Figure 5.20: Examples of sphingophospholipids

5.15 Sphingoglycolipids

A sphingoglycolipid contains both a fatty acid and a carbohydrate component attached to a sphingosine molecule. A fatty acid is attached to the sphingosine through an amide linkage, and a monosaccharide or oligosaccharide is attached to the sphingosine at the terminal -OH carbon atom through a glycosidic linkage.

The simplest sphingoglycolipids, which are called cerebrosides (Figure 5.21), contain a fatty acid, an amino alcohol sphingosine and a single monosaccharide unit—either glucose or galactose. They are present in the myelin sheath of nerves and primarily in the brain white matter. In Gaucher's disease their level increases in tissues like reticuloendothelial cells of spleen, liver, lymph node and bone.

More complex sphingoglycolipids, called gangliosides, contain *N*-acetylneuraminic acid (sialic acid), fatty acid, alcohol sphingosine and a branched chain of up to seven monosaccharide residues. They occur in the gray matter of the brain as well as in the myelin sheath. In Tay-Sachs disease, ganglioside level increases.

Cerebroside

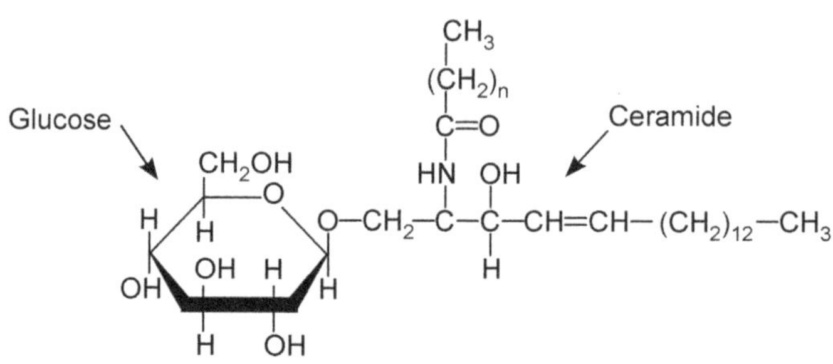

Glucocerebroside

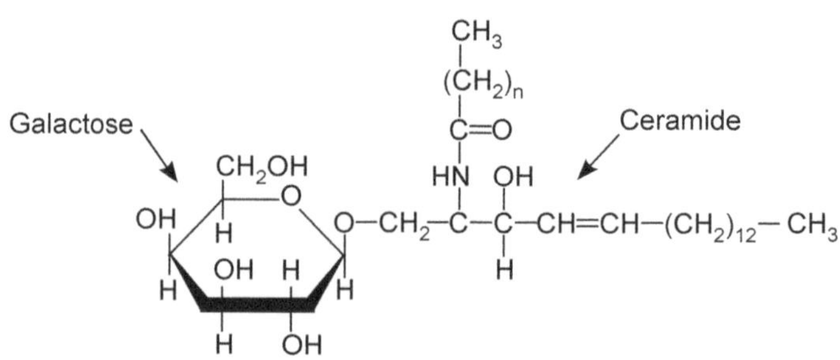

Galactocerebroside

Figure 5.21: Examples of sphingoglycolipids

5.16 Cholesterol

Cholesterol has fetched thirteen Nobel Prizes to scientists who devoted major parts of their careers to understand this biomolecule. It was isolated from gallstones in 1784, since then has been a fascinating topic for scientists from the most diverse areas of science and medicine.

Cholesterol is a lipid having four fused hydrocarbon rings and thus forms a bulky steroid structure (Figure 5.22). Cholesterol is called a "sterol" because it is made out of a alcohol and steroid. There is a hydrocarbon tail at one end of the steroid and a hydroxyl group at the other. The hydroxyl group forms hydrogen bonds with nearby carbonyl oxygen of phospholipid and sphingolipid. Cholesterol is is absent in prokaryotes and intracellular membranes.

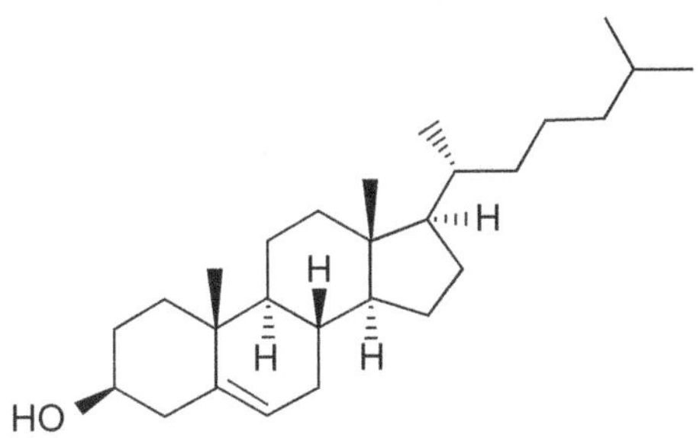

Figure 5.22: Cholesterol

Cholesterol serves many functions in human body. Some of them are listed below:

1. It helps in the manufacture of steroids, or cortisone-like hormones such as vitamin D, testosterone, estrogen and cortisone. Thus controls various bodily functions.
2. It assists the liver in the synthesis of bile acids, needed for digestion and absorption of fat-soluble vitamins like vitamin A, D, E and K.
3. It helps in formation of the myelin sheath, which insulates nerve cells. This ensures proper function of our brains by enhancing electrical impulses. The absence of cholesterol might lead to memory problems and inflammation of central nervous system.
4. It can also be used as a source of energy in the body.
5. It maintains body temperature.
6. It protects internal organs.
7. It helps in modulation of the fluidity of cell membranes.

Cholesterol and triglycerides are insoluble in the blood and therefore need to be transported by carriers called lipoproteins. *Low-density Lipoprotein (LDL) and very low-density lipoprotein (VLDL) are known as the "bad" cholesterols. High-density Lipoprotein (HDL) is known as "good" cholesterol.*

HDL is called "good" cholesterol because it prevents clogging of the arteries by removing excess cholesterol in the arteries. HDL transports the excess cholesterol to the liver for excretion or re-utilization. Hypercholesterolemia is a condition where a person has extremely high concentration of LDL and low concentration of HDL. When too much LDL circulates the blood cell, it can built up the inner walls of arteries that feed the heart and brain, therefore, cause the clogging of the arteries. Hypocholesterolemia is a condition when there is an extremely low concentration of cholesterol in the body. This condition is usually rare, but if they do occur, it might be because of other illness that has caused the body to generate low or no cholesterols.

Atherosclerosis refers to the fatty deposits block the arteries. Healthy arteries are smooth inner lining and blood can flow though them easily. However, if arteries are damaged, the infection can cause inflammation. White blood cells go to the damaged arteries and take up lipids, including cholesterol. Fatty acids start to grow in the affected area and makes the artery become stiff and obstructs the blood flow.

5.17 Cell Membrane (Lipid Bilayer)

Illustration of a Cell Membrane

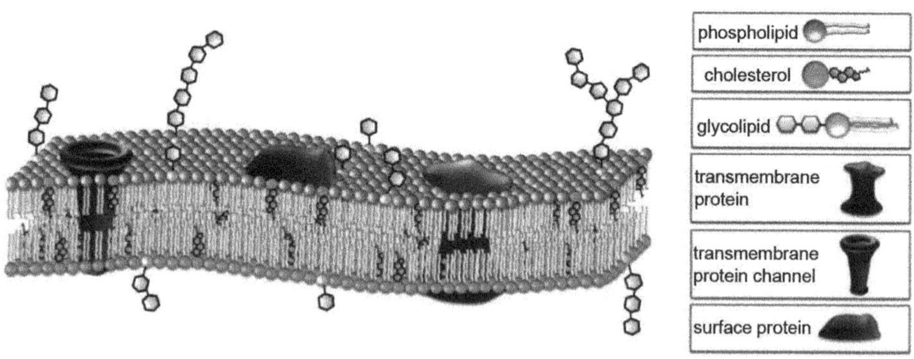

Figure 5.23: Lipid bilayer in biological membranes

The boundaries of cells are formed by *biological membranes*, the barriers that *define the inside and the outside of a cell*. These barriers prevent molecules generated inside the cell from leaking out and unwanted molecules from diffusing in; yet they also contain transport systems that allow specific molecules to be taken up and also unwanted compounds can be removed from the cell (Figure 5.23). Such transport systems confer on membranes the important property of *selective permeability.* In addition to an external cell membrane (called the plasma membrane), eukaryotic cells also contain internal membranes that form the boundaries of organelles such as mitochondria, chloroplasts, peroxisomes, and lysosomes.

The compositions of the inner and outer membrane in many naturally occurring bilayers is different. In human red blood cells, the inner (cytoplasmic) leaflet is mainly composed of phosphatidylethanolamine, phosphatidylserine and phosphatidylinositol. The outer (extracellular) leaflet is made of *p hosphatidylcholine, sphingomyelin and a variety of glycolipids.* Thus, the lipid components of the membrane form the permeability barrier, and protein components act as a transport system of pumps and channels that endow the membrane with selective permeability.

Biological membranes serve several additional important functions indispensable for life, such as energy storage and information transduction, that are dictated by the proteins associated with them.

5.18 Steroid Hormones

Hormone a chemical substance produced by a specialized organ or gland. It enters the bloodstream and regulates the actions of other cells or organs.

Types of Steroid Hormones

- Glucocorticoids; cortisol is the major representative in most mammals
- Mineralocorticoids; aldosterone being most prominent
- Androgens such as testosterone
- Estrogens, including estradiol and estrone
- Progestogens (also known a progestins) such as progesterone

In addition to bile salts, which facilitate the digestion of lipids, five major classes of steroid hormones are derived from cholesterol: **progestagens, glucocorticoids, mineralocorticoids, androgens, and estrogens**.

Pregnenolone (C21) is an essential intermediate in the synthesis of steroids. This steroid is formed by scission of the side chain of cholesterol.

Progesterone (C21), synthesized from pregnenolone, is the precursor of **cortisol and aldosterone**. Hydroxylation of progesterone and cleavage of its side chain yields androstenedione, an **androgen** (C19). **Estrogens** (C18) are synthesized from androgens by the loss of an angular methyl group and the formation of an aromatic A ring. Vitamin D, which is important in the control of calcium and phosphorus metabolism, is formed from a derivative of cholesterol by the action of light.

Progesterone, a *progestagen,* prepares the lining of the uterus for implantation of an ovum and is also essential for the maintenance of pregnancy. *Androgens* (such as testosterone) are needed for male secondary sex characteristics. *Estrogens* (such as *estrone*) are required for the development of female secondary sex characteristics. Estrogens and progesterone, also participate in the ovarian cycle. *Glucocorticoids* (such as *cortisol*) promote gluconeogenesis, formation of glycogen, enhance the degradation of fat and protein, and inhibit the inflammatory response. *Mineralocorticoids* (primarily *aldosterone*) increase the reabsorption of Na^+ and the excretion of K^+ and H^+, in kidneys which leads to an increase in blood volume and blood pressure. The major sites of synthesis of these classes of hormones are the corpus luteum, for progestagens; the ovaries, for estrogens; the testes, for androgens; and the adrenal cortex, for glucocorticoids and mineralocorticoids (Table 5.2).

Table 5.2: Steroid Hormones and Their Functions

Hormone	Functions
Progesterone	Helps in uterus lining for the implantation of ovum
Glucocorticoids (cortisol)	It is produced in adrenal cortex and is a catabolic steroid It helps in gluconeogenesis, in breakdown of fat and protein, and in anti-inflammatory reactions in the body
Mineralocorticoids (aldosterone)	It is produced in adrenal glands. Helps in maintaining blood pressure by increasing sodium reabsorption by kidney
Androgens	Testosterone is produced in testes but other androgens in adrenal cortex It is a anabolic steroid Helps in development of male secondary sex characteristics It prevents bone resorption

Contd.../(Table 5.2)

Hormone	Functions
Estrogen	It produced in ovaries primarily but also in adipose cells of males and females It helps in development of female secondary sex characteristics It prevents bone resorption
Vitamin D (not a steroid hormone)	It produced in the skin in response to UV light and processed to active form in kidney Helps in intestinal calcium absorption It promotes bone formation It prevents phosphate loss by kidneys

5.19 Lipoprotein

Lipids by definition are insoluble in water. In order to transport lipids such as fatty acids, triacylglycerols, steroids and fat soluble vitamins in the blood plasma, a carrier protein is required. Cholesterol, phospholipids and triacylglycerols, are transported in the plasma by large complexes called lipoproteins (LP). It's external layer comprises of phospholipids, unesterified cholesterol, and specialized proteins, called apolipoproteins or apoproteins. Hydrophilic portions of phospholipids, unesterified cholesterol is oriented outward toward the surrounding water and lipophilic portions of each molecule oriented inwards.

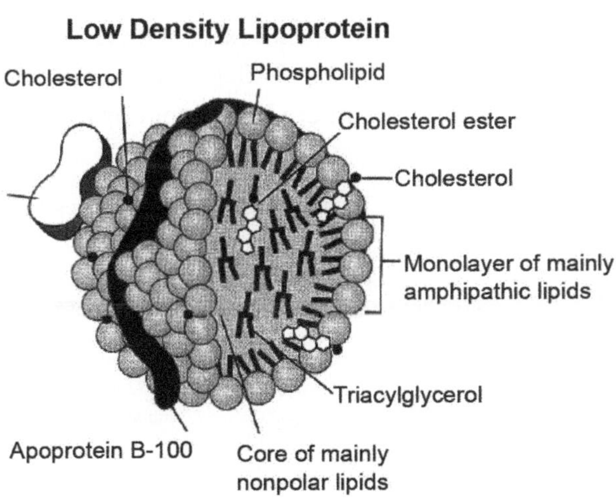

Figure 5.24: Lipoprotein

Apoprotein is needed to solubilize lipid and it helps to maintain the structural integrity of LPs. They also help to regulate the activity of LP metabolic enzymes. All of the apoproteins have amphipathic α-helixes with the hydrophobic side chains facing the lipid interior of the lipoprotein and the hydrophilic residues interacting with the polar head groups of the phospholipids or interacting with the aqueous solvent.

Lipoproteins are classified according to their density. The densities of these lipoproteins are related to the relative amounts of lipids to proteins in the complex (Figure 5.24). Higher protein content means higher density of the lipoprotein. The lowest density lipoproteins are the **chylomicrons** followed by the **chylomicron remnants**, very low density lipoproteins **VLDL**s, intermediate density lipoproteins **IDL**s, low density lipoproteins **LDL**s, and high density lipoproteins **HDL**s.

Chylomicrons are synthesized in the epithelial cells that line the small intestine. Lipids from the intestine get transported using *chylomicrons* to adipose, cardiac, and skeletal muscle tissue. The triglyceride component of the lipid and not cholesterol get hydrolyzed by lipoprotein lipase, and free fatty acids is released and absorbed by the tissues. The remaining chylomicron structure is called a *chylomicron remnant*, which bears cholesterol and it again goes to the liver and is degraded.

When diet contains more fatty acids than needed immediately as fuel, they are converted in the liver to triacylglycerols and exported with *very-low-density lipoprotein* (VLDL) (Figure 5.25). Excess carbohydrate is also converted to triacylglycerols and packed as VLDLs. VLDLs can be transported from the liver to muscle and adipose tissue, where lipoprotein lipase releases free fatty

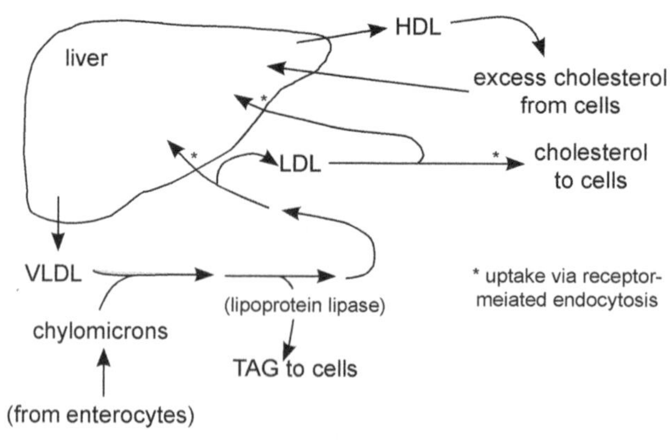

Figure 5.25: The role of HDL, LDL and VLDL in liver

acids. The resulting remnants, which are rich in cholesteryl esters, are called *intermediate-density lipoproteins*. Half of these are taken up by the liver for processing, and the other half get converted to *low-density lipoprotein*. Low-density lipoprotein is known to be the major carrier of cholesterol in blood. The role of LDL is to transport cholesterol to peripheral tissues. HDL binds and esterifies cholesterol released from the peripheral tissues and then transfers cholesteryl esters to the liver or to tissues that use cholesterol to synthesize steroid hormones.

High levels of cholesterol leads to the formation of plaques in arteries which can be fatal. Excess cholesterol is present as low density lipoprotein particle and is thus called "bad cholesterol. "The ratio of cholesterol in the form of high density lipoprotein, sometimes referred to as "good cholesterol," to that in the form of LDL can be used to evaluate susceptibility to the development of heart disease. For a healthy person, the LDL/HDL ratio is 3.5.

5.20 Liposomes

Liposomes are small artificial vesicles made up of the same material as a cell membrane. They are of spherical shape and can be created from cholesterol and phospholipids. Due to their small size and amphiphilic character, they are promising drug delivery systems. Properties of liposome can differ by changing lipid composition, size, surface charge, and method of preparation.

Multilamellar Vesicles (MLVs)–These liposomes have more than one lamella, and can vary in size between 100 to 1000 nm.

Small Unilamellar Vesicles (SUVs)–These liposomes are smaller than 0.1 μm with just a single lamella. The composition of the membrane, as well as the aqueous medium has an influence on the minimum size that can be attained.

Large Unilamellar Vesicles (LUVs)–These liposomes sizes start from 0.1 μm and can reach sizes of up to 1000nm, which is close to the size of living cells. These liposomes have just a single lamella.

Lipids self assemble in aqueous solution into liposomes (Figure 5.26). Liposomes thus have an aqueous hydrophilic core surrounded by hydrophobic part like a lipid bilayer though in a spherical structure. They can transport hydrophobic as well as hydrolphilic drugs due to their amphiphatic nature. Hydrophilic compounds inside the aqueous core of liposome cannot pass through the hydrophobic part and is trapped. Hydrophobic drugs place themselves inside the bilayer of the liposome. In order to deliver the contents of the liposome, it moves through the cell membranes effectively due to its hydrophobic outer part (Figure 5.27). Toxicity of the drugs, its oxidation and degradation and its interaction with other non-target cells can be minimized using liposomes. Cancer drug Doxorubicin is delivered to tumor cells by liposome.

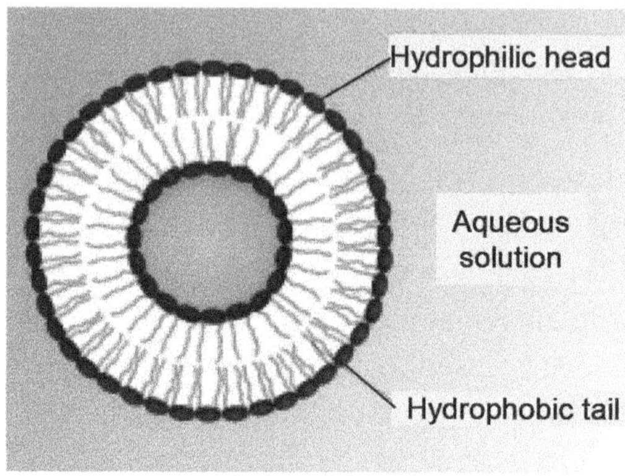

Figure 5.26: Liposomes

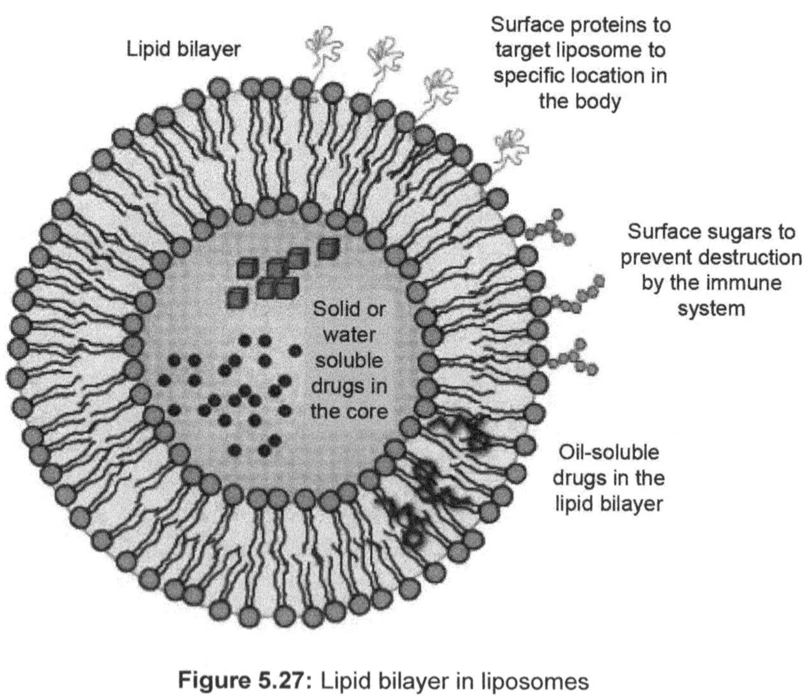

Figure 5.27: Lipid bilayer in liposomes

Eicosanoids

Eicosanoids ("eicosa" is Greek for 20) are derived from 20-carbon polyunsaturated fatty acids, most frequently arachidonic acid. They include the prostaglandins, thromboxanes, leukotrienes, and lipoxins. They exists widely in the cells and tissues of the body and have various biological functions at extremely low concentrations. They do not get transported in the bloodstream to their site of action as hormones do but instead exert their effects in the tissues where they are synthesized (Figure 5.28). Eicosanoids usually have a very short "life," and gets broken and eliminated in urine within seconds of their synthesis. They regulate inflammatory response, pain, fever, blood pressure, blood clotting, reproductive functions, sleep/wake cycle etc.

Prostaglandins are messenger lipids that contain a cyclopentane ring and oxygen-containing functional groups. From the twenty-carbon fatty acids, the eighth and twelfth carbon connects to form a five-membered ring. They are named after the prostate gland, which was first thought to be their only source. Today, more than 20 prostaglandins have been discovered in a variety of tissues in both males and females. They are involved in many functions, including regulation of body temperature, pain, inflammation and fever. They are also involved in secretion of gastric juices, protective mucus formation in the stomach, and contraction of uterus. Aspirin and corticosteroid inactivates the enzyme needed for prostaglandin synthesis.

Thromboxanes contains a cyclic ether ring and oxygen-containing functional groups. The cyclic structure involves a bond between carbons 8 and 12. An important function of thromboxanes is to promote the formation of blood clots. Thromboxanes are produced by blood platelets and promote platelet aggregation.

Leukotrienes contains three conjugated double bonds and hydroxy groups. Leukotrienes are found in leukocytes (white blood cells). Various inflammatory and hypersensitivity (allergy) responses are associated with elevated levels of leukotrienes.

Lipoxygenase interactions results in the formation of **Lipoxins**. They are linear eicosanoids and are similar to leukotrienes. Lipoxins can cause anti-inflammatory activity and its production is enhanced in response to the administration of aspirin.

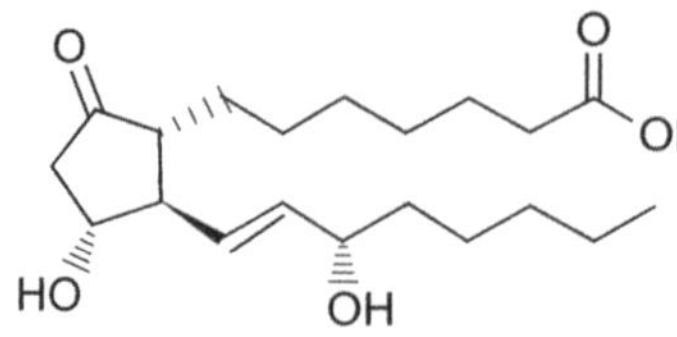

Prostaglandin E_1, The 5-member ring is characteristic of the class.

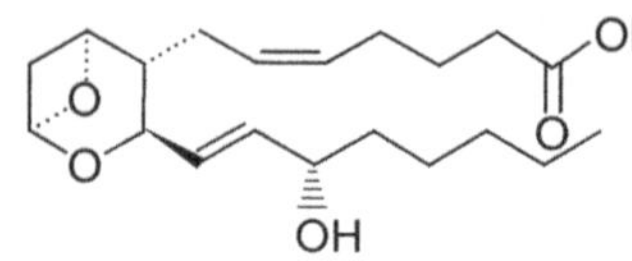

Thromboxane A_2, oxygens have moved into the ring.

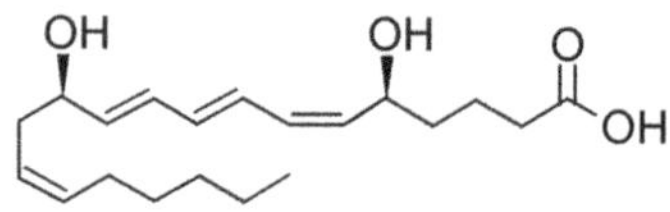

Leukotriene B_4, note the 3 conjugated double bonds.

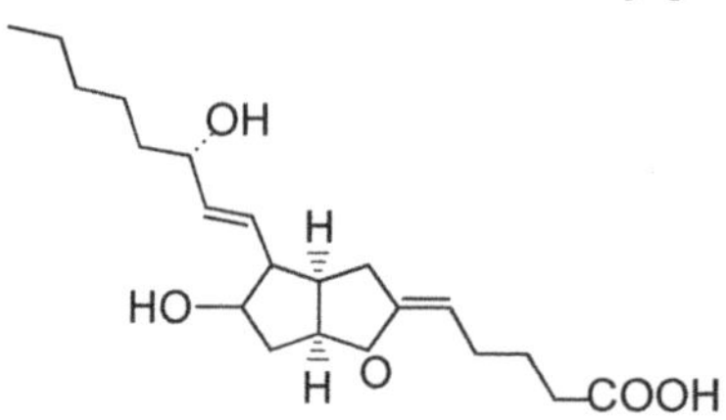

Prostacyclin I_2. The second ring distingushes it from the prostaglandins.

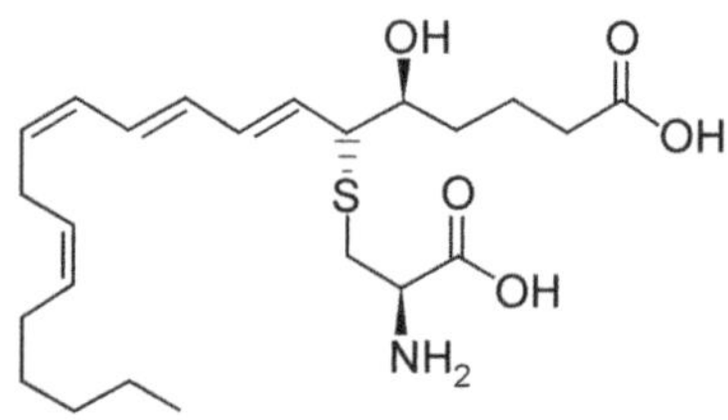

Leukotriene E_4, an example of a cylsteinly leukotriene.

Figure 5.28: Examples of Eicosanoids

Lipotropic Factors are substances, which facilitate mobilization of fat from liver. Various lipotropic factors (agents) are choline, betaine, methionine and inositol. They help in the conversion of triglyceride to phospholipid and thus help in normal transport and utilization of lipids especially in liver. The deficiency of lipotropic factors leads to a condition known as fatty liver, i.e. increased accumulation of fat content in liver.

Questions

1. What are different ways to classify lipids? Give suitable examples.
2. Write one example, each of ω-3, ω-6 and ω-9 fatty acids.
3. What are the two characteristic features of unsaturated bonds found in fatty acid chains

4. Identify the following fatty acids with trivial and systematic names. Also write shorthand notation and indicate the type of ω fatty acid.

   (i) $COOH(CH_2)_7HC{=}CH(CH_2)_5CH_3$

   (ii) $COOH(CH_2)_{14}CH_3$

   (iii) $COOH(CH_2)_7HC{=}CHCH_2CH{=}CHCH_2CH{=}CHCH_2CH_3$

5. Define the following terms and what type of information do we get from these?

   (a) Saponification value

   (b) Acid number

   (c) Iodine number

6. What are the major biological functions of lipids?
7. Explain different physical and chemical properties of fatty acids.
8. Give one example each of ω-6 and ω-9 fatty acid.
9. What are soaps and explain their cleaning action.
10. Which different parameters are used to determine the quality of lipids?
11. What are waxes and how are they different from TAGs?
12. What is the basis of lipid classification? Name the various types.
13. List the physical and the chemical differences between fats and oils?
14. What are emulsions? Give example of a naturally occurring emulsion.
15. What does a high saponification number and a high iodine value of a fat indicate?
16. What are the advantages and disadvantages of hydrogenation of oils?
17. Define rancidity and name the type of chemical reactions involved.
18. What are essential and nonessential fatty acids?
19. Write short note on the nutritional applications of lipids.
20. What are phospholipids?
21. Name three fat soluble vitamins.
22. Explain MUFAs and PUFAs.
23. What are unnatural trans-fats and how do they lead to health issues?

24. What is the difference between low-density lipoprotein (LDL) and high-density lipoprotein (HDL)?
25. Write a short note on lipid bilayers.
26. What are the different types of steroid hormones

❑❑❑

6

Oxidation of Foodstuff (Organic Molecules) as A Source of Energy for Cell

6.0 Introduction

Hans Krebs has defined three stages (Figure 6.1) in the production of energy by the oxidation of foodstuff.

First Stage

In the very first stage termed as digestion, the big molecules existing in diet are transformed to the smaller unit. The Protein is hydrolyzed into the twenty different amino acids, the polysaccharide are hydrolyzed into simple sugar like glucose, and the fats gets hydrolyzed to glycerol and fatty acids. The cell exist in the intestine absorbs the degraded products which are then dispersed across the body. This is a preparation stage in with no energy capturing occurring here.

Second Stage

Here, the various small molecules gets hydrolyzed to limited simple units playing a crucial role in metabolism, rather most of being like- sugar, fatty acids, glycerol, and many amino acids those get changed into the acetyl unit of acetyl Co. A. A very small amount of ATP is produced here which is very small than that achieved in final III stage.

DOI: 10.1201/9781003727453-6

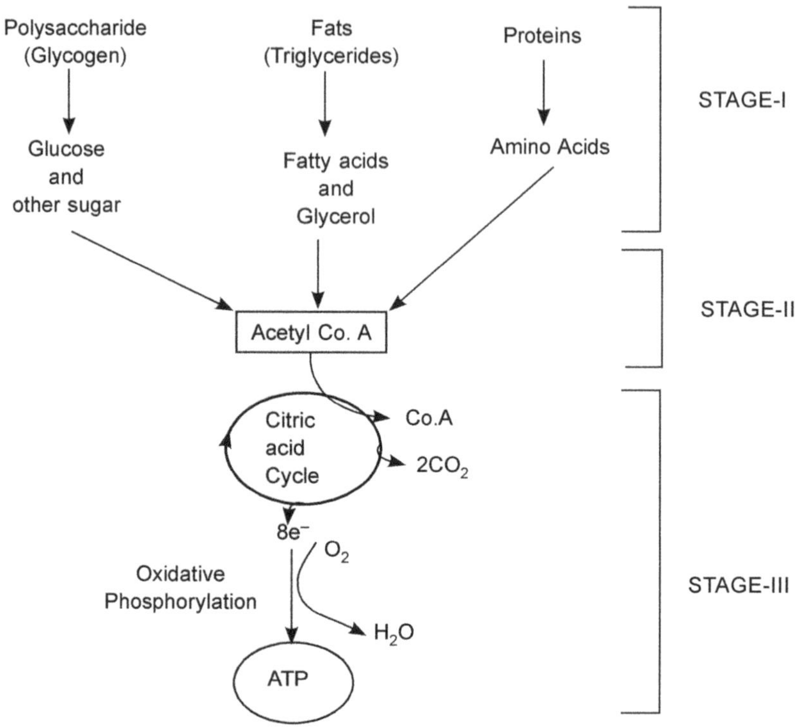

Figure 6.1: Production of energy by the oxidation of foodstuff

Third Stage

Herein, by the complete oxidation of the acetyl Co. A, ATP is produced. This stage comprises of the citric acid cycle (CAC) and that oxidative phosphorylation, that's the final common pathways in the oxidation.

Acetyl CO.A. bring acetyl units in the citric acid cycle (TCA, Krebs Cycle) where these are entirely oxidized to CO_2 .The four pair of e- are transferred (one to FAD and three to NAD^+) for every acetyl group being the oxidized. After this a proton is utilized for synthesizing ATP.

6.1 Metabolism

Metabolism (Greek–change) involves all the chemical reactions/ Chemical change / transformation occur inside the living organism these chemical transformations are catalyzed by enzymes and allow organism to grow and reproduce.

These chemical transformations are categorized in two types

- Catabolism
- Anabolism

Catabolism or destructive activities: - (Figure 6.2) Involving the breakdown of complex

Compounds / molecules by cellular respiration into simpler compounds. Energy is released for the duration of this process.

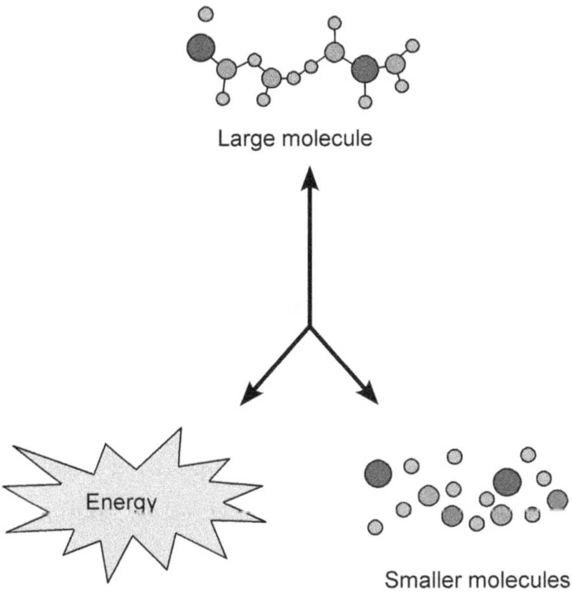

Figure 6.2: Catabolism

$$C_6H_{12}O_6 + 6O_2 \longrightarrow 6CO_2 + 6H_2O + \text{Energy}$$

Anabolism or constructive activities: (Figure 6.3) Involving biosynthesis of complex compounds / molecules like protein, nucleic acids. Energy is consumed and stored in the course of this procedure.

Energy out in catabolism couldn't be used directly and need to store it and release it gradually.

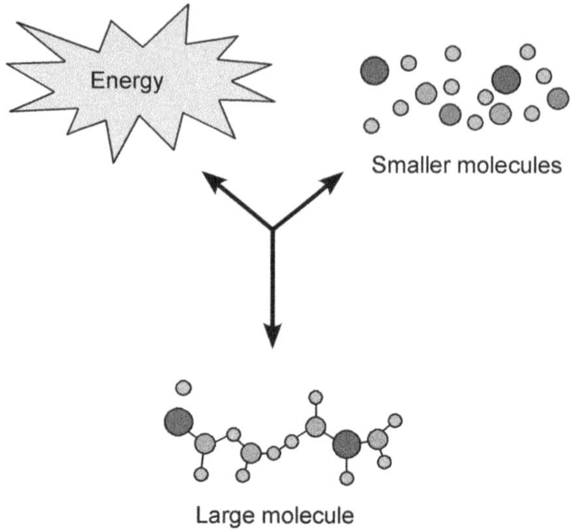

Figure 6.3: Anabolism

6.2 ATP: Universal Currency of Cellular Energy

All living organism utilize ATP as the energy source to derive enormous number of endergonic reactions. It was proposed by Fritz Lipmann in 1940 that adenosine triphosphate /ATP is the agent that links all energy requiring and energy yielding processes of the cell (Figure 6.4). When fuel molecules are degraded, bond energy is out and recovered by coupling phosphate with ADP to yield

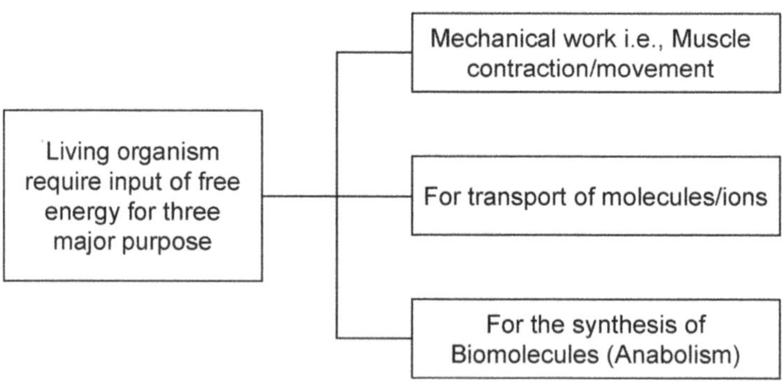

Figure 6.4: Input of free energy

ATP through phosphorylation. Living organism are isothermal in nature and all chemical reactions proceed without appreciable rise in temperature. ATP is an energy rich compound. It yield a large negative free energy when hydrolyzed at physiological pH.

The environment is the driving force for the free energy needed. The ATP/ adenosine triphosphate is special carrier of free energy, better known as the universal currency of cellular energy in biosystem.

ATP (Figure 6.5) is an energy rich compound, which is highly charged and gets entirely ionized at pH 7.0. It consist of sugar (5-carbon D–ribose), base (adenine) and three phosphate groups.

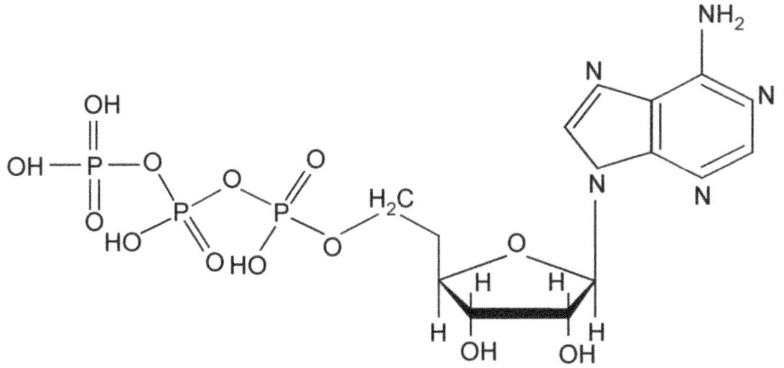

Figure 6.5: Structure of ATP

6.3 ATP Hydrolysis and Free Energy

The free energy of hydrolysis of ATP, varies from cell to cell means it is not constant in a cell, and depending upon pH and Mg^{2+} ion concentration. The active form of ATP is generally a pair of ATP with Mg^{2+}. This makes us focus on its phosphate moiety.

During hydrolysis of ATP/adenosine triphosphate to ADP/adenosine diphosphate and orthophosphate (Pi) or to AMP and pyrophosphate (PPi) a huge amount of free energy is produced. Hydrolysis of ATP and the corresponding standard free energy changes are given in the Table 6.1.

Table 6.1: Hydrolysis of ATP and standard free energy changes with corresponding equilibrium constant at 25°C and pH 7.0

Reaction	G° joules / mole	K
ATP + H_2O ⟶ ADP + Pi	36,800	2.8×10^6 M
ADP +H_2O ⟶ AMP + Pi	36,000	2.0×10^6 M
ATP + H_2O ⟶ AMP + PPi	40,000	1.3×10^7 M
AMP + H_2O ⟶ A + Pi	12,600	1.6×10^2 M
PPi + H_2O ⟶ 2Pi	31,800	$3,7 \times 10^5$ M

The change in Standard free energy associated with the hydrolysis of terminal phosphate is -7.3 kcal /mole, when the second phosphate is hydrolyzed another 7.3 kcal/mole of energy is obtain. The phosphate groups were linked through anhydride bonds which trap a great quantity of energy. When the last phosphate/ P group is hydrolyzed, the energy free has a smaller value, i.e., about -2.5 k cal. The smaller values are linked with ester bonds. Hence, the terminal and second phosphate bonds in ATP are high energy bonds.

The free energy produced when ATP is hydrolyzed. This free energy is utilized to derive the reactions that need an input of free energy like contraction of muscle.

6.4 Conversion of Food Into Energy

The maximum of the food we consumed contains polysaccharides, lipids and proteins. Before they are utilized by our cell are changed into smaller molecules - either as constructing blocks for other molecules or as a energy source. This processes of breakdown acting on foodstuff consumed from outside, however not on the macromolecules present inside our own cells. As discussed earlier that the conversion of food into energy occurs in three stages.

Stage 1

Digestion is the enzymatic breaking down of the foodstuff, take place in a specific organelle, the lysosome within the cells, or in our intestine outside the cells. In this case, the bulky polymeric molecules present in food are fragmented in the course of digestion into the monomer subunits— like fats gets converted to glycerol and fatty acids , proteins is changed to amino acids, and polysaccharides into sugars—by enzymatic action. After the digestion process, the smaller organic molecules go in the liquid found inside the cells i.e. cytosol, for the gradual oxidation. The oxidation take place in two stages of cellular catabolism (Figure 6.6).

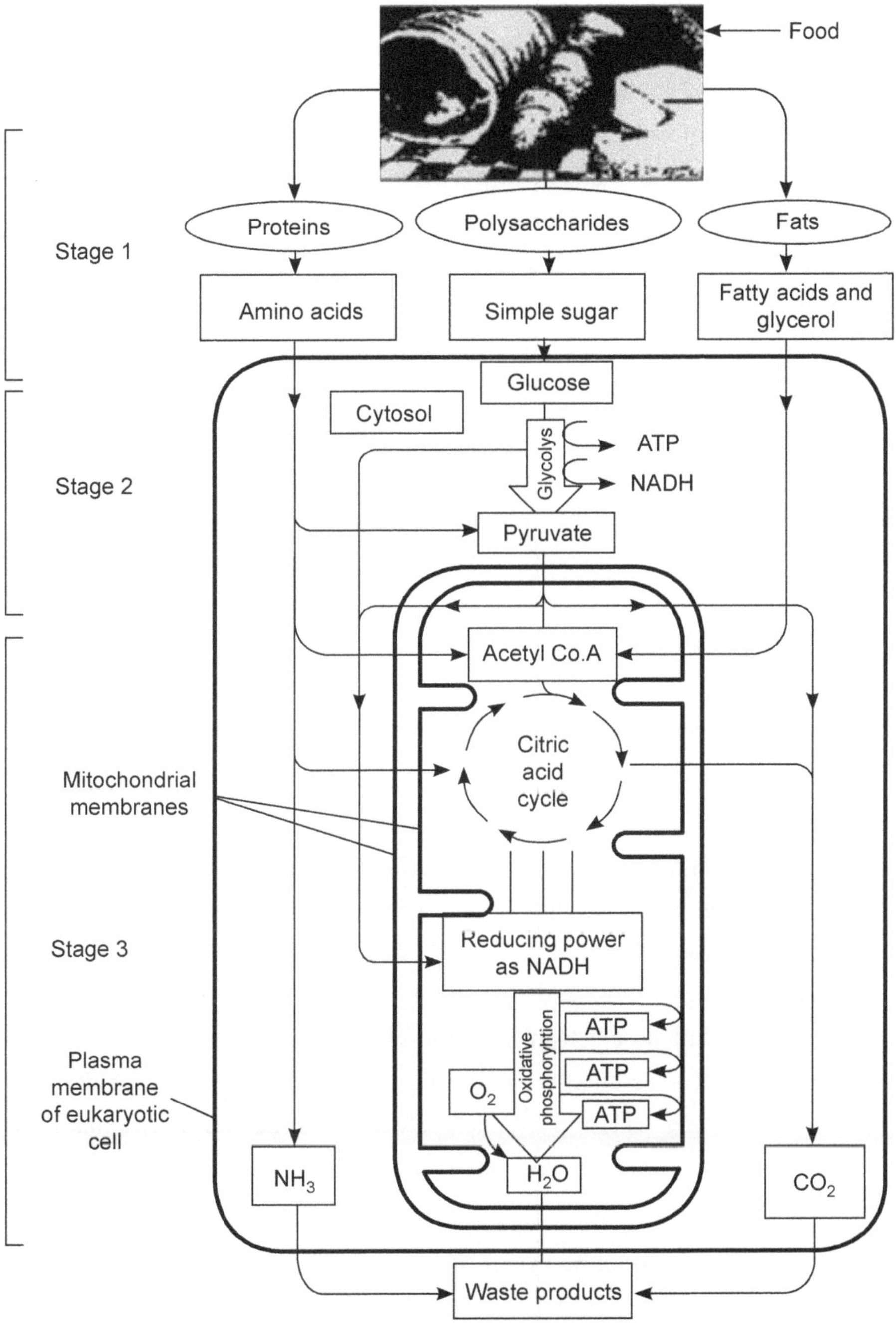

Figure 6.6: Oxidation of foodstuff

Stage 2

Stage 2 Starts from the cytosol and finished in the mitochondria (major energy-converting cell-organelle). On this stage, the glycolysis process changes glucose's each molecule into the two small pyruvate molecules. In glycolysis all sugars are likewise changed to pyruvate after their transformation to one of the sugar intermediates. During the pyruvate formation, ATP and NADH the two activated carrier molecules gets generated. The pyruvate travels from cytosol to the mitochondria. Here, every pyruvate molecule gets converted to CO_2 /carbon dioxide along with two-carbon acetyl group—attaching with coenzyme A (CoA), with formation of another activated carrier molecule i.e. acetyl CoA. Huge amounts of acetyl CoA are generated through breakdown and oxidation of fatty acids those obtained from fats, being carried into the bloodstream, and later for the production of acetyl CoA in the mitochondria.

Stage 3

Stage III is wholly limited to the mitochondria. Here the oxidative break down of food molecules occurs completely. In this stage the acetyl group of the acetyl CoA is allied to coenzyme-A by a high-energy link, and consequently can be transmitted to the other molecules very smoothly.

After the transferal to the 4-carbon molecule i.e. the oxaloacetate molecule, the acetyl group move in a series of reactions termed as the citric acid cycle (CAC). In these reactions the acetyl group gets oxidized to CO_2, with generation of a huge amount of NADH (the electron carrier). To end, the high-energy electrons produced from NADH. These high energy electrons are delivered along with an electron-transport chain inside the inner membrane of mitochondria. Here the energy out by the transferal, is utilized for driving a procedure producing ATP and eats up molecular oxygen (O_2). The maximum energy produced by oxidation is joined to generate maximum ATP in the final steps.

Hence the oxidative breaking down of food molecules is the energy deriving the ATP synthesis in the mitochondria.

6.5 Metabolic Pathway of Carbohydrates

The major pathway of carbohydrates metabolism either begin or end with glucose. Major metabolic pathways of carbohydrates are summarized in the Figure 6.7.

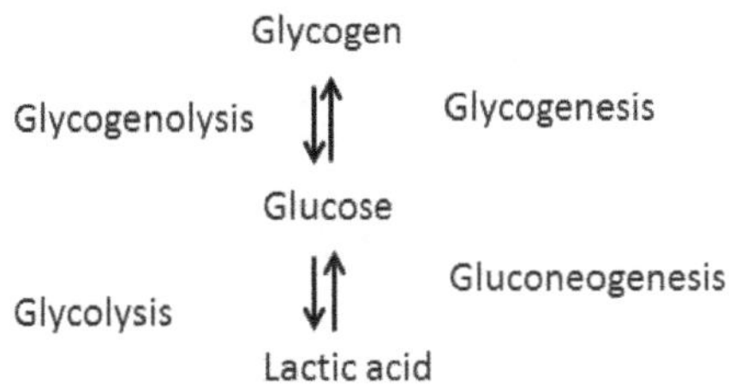

Figure 6.7: Various metabolic pathways of carbohydrates

Main aim of these metabolic pathways is to maintain a persistent blood sugar level within certain physiological limits, called homeostasis. It is controlled by certain hormonal and nervous factors.

6.6 Glycolysis

It is also called EMP-pathway because its details were given by **Embden, Meyerhof** and **Paranas**.

It is an enzymatically-controlled catabolic pathway where one single molecule of glucose (6C) breaks into two pyruvic acid (having 3C atoms) molecules. It happens in cytoplasm exterior the mitochondria. It is an anaerobic process and is similar to both the aerobic and anaerobic (fermentation of glucose in yeasts) respiration. It is suppressed by the presence of oxygen (Pasteur Effect).One molecule of glucose breaks into two molecules of pyruvic acid. In anaerobic conditions, the pyruvic acid gets converted to produce lactic acid in mammalian tissues (e.g. muscle cells) while in fermentation, gets changed to ethanol. But under aerobic condition, pyruvic acid produces acetyl Co A which enters in the Krebs cycle.

Mechanism (Figure 6.8) Glycolysis is an enzymatically controlled stepped process (controlled by 11 enzymes) and is completed in following nine steps most of which are reversible:

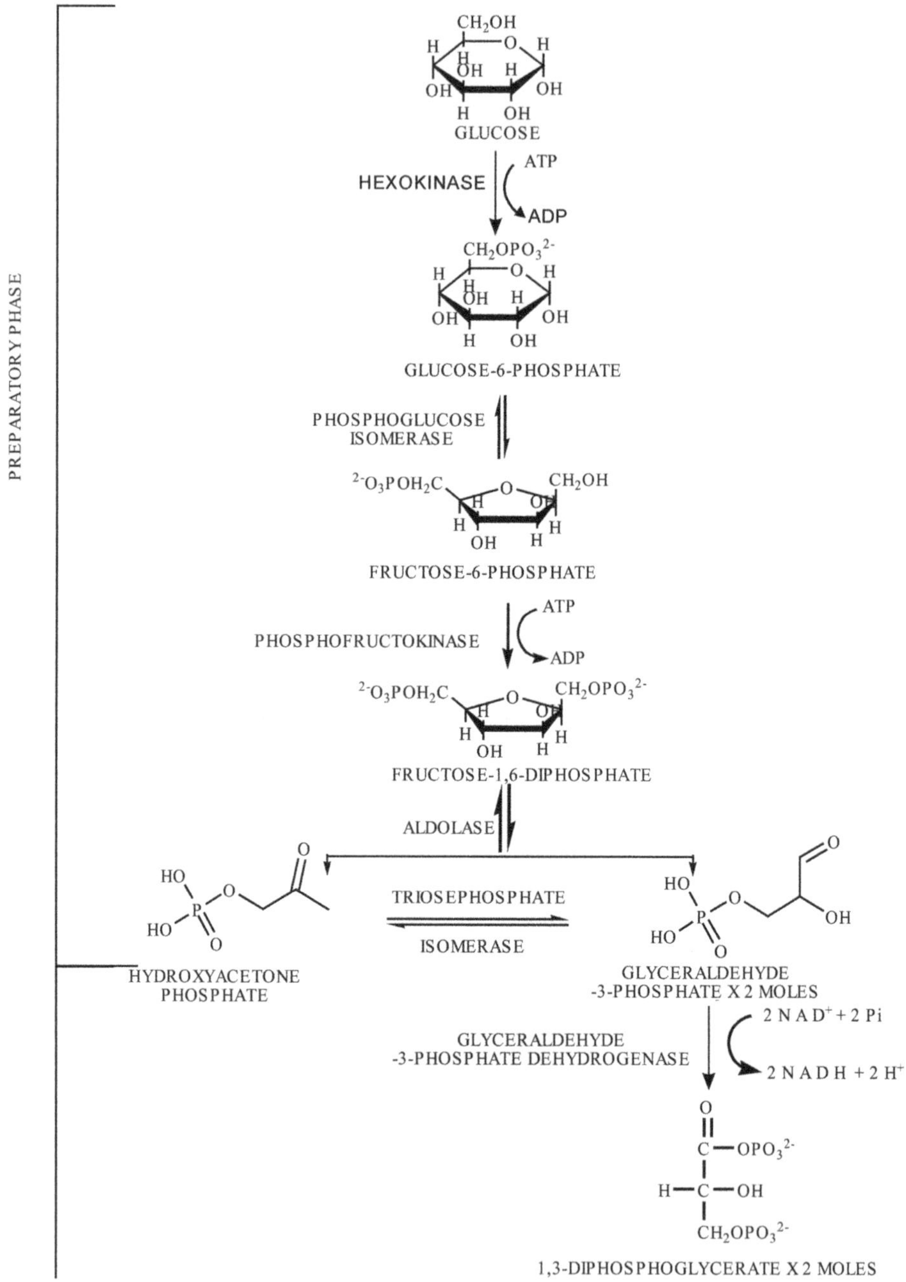

Figure 6.8: Glycolysis detailed phosphorylation (Continued)

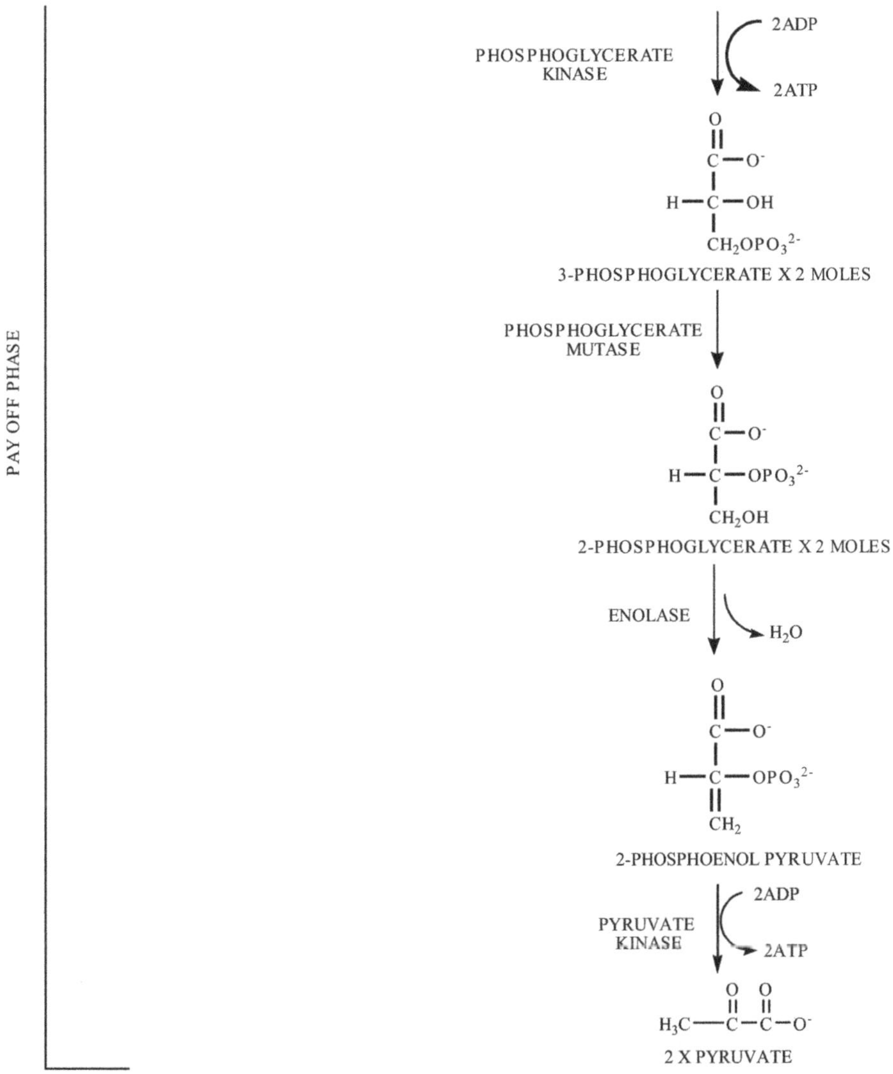

Figure 6.9: Glycolysis detailed phosphorylation

1. Phosphorylation

It involves the addition of a phosphate molecule to glucose. In this, phosphate donor is not phosphoric acid but is energy currency ATP molecule. Phosphate of ATP is moved to carbon-6 of glucose in the existence of enzyme hexokinase or glucokinase and cofactor Mg. A phosphoester bond is formed producing Glucose 6-P. It is an irreversible and highly exergonic ($\Delta G° = -4$ kcal/mole) step. Thus this step consumes one ATP molecule. Hexokinase catalyses the phosphorylation of a number of hexoses like D-glucose, D-fructose, D-mannose and D-galactose; and is inhibited by high concentration of glucose-6-P. On the other hand, glucokinase enzyme phosphorylates only D-glucose and is also not inhibited by glucose-6- P. Glucokinase is predominately present in liver cells while it is absent in muscles. It is a "locking device" by which glucose is trapped in the form of glucose-6- P within the cytosol of the cell because cell membrane is impermeable to it while is permeable to glucose. Glucose-6 P occupies a central position in numerous metabolic pathways **(Figure 6.10).**

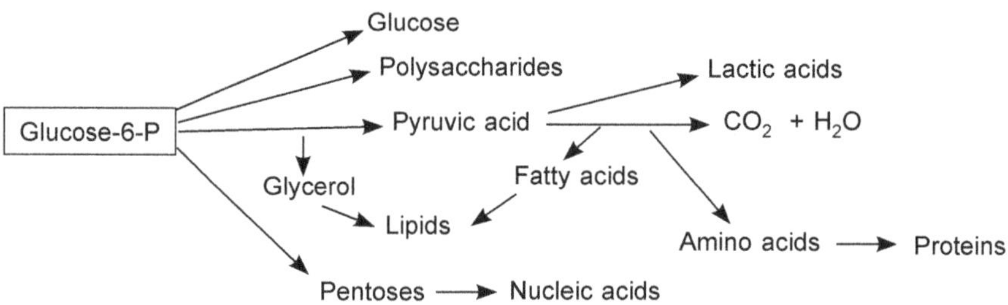

Figure 6.10: Various metabolic pathways starting from Glucose-6(P)

2. Isomerization

In this step, glucose-6-P changes into its isomer fructose-6-P in the existence of enzyme phosphoglucoisomerase and cofactor magnesium. It is a reversible and endergonic (ΔG° = + 0.4 kcal/mol.) step of glycolysis. Fructose-6-P might also be formed directly by the phosphorylation of fructose.

3. Phosphorylation

It comprises the irreversible conversion of Fructose-6-P to Fructose-1, 6-bi-phosphate in the existence of enzyme phosphofructokinase (activator Mg") and a phosphate-donor ATP molecule. In this, a phosphoanhydride bond of ATP is broken and a phosphoester bond is formed with –OH group of carbon-1' of Fructose-6-P. It is an exergonic step (ΔG° = –3.40 Kcal./mołe). This enzyme is inhibited by glucagon, excess of ATP and citrate but is stimulated by ADP or AMP.

These three steps of glycolysis collectively form the priming stage and involves the utilizing of 2ATP molecules:

$$\text{D-Glucose} + 2\text{ATP} \longrightarrow \text{D-Fructose-1, 6-biphosphate} + 2\text{ADP}$$

4. Splitting or Cleavage

It involves the splitting of fructose-1, 6-biphosphate between carbon atoms 3 and 4 in the existence of an enzyme fructose biphosphate aldolase. This produces two isomeric triose phosphate molecules- a molecule of dihydroxyacetone-P (DHAP) and glyceraldehyde-3- P (3-PGAL). It is an endergonic reaction ($\Delta G°$= + 5.73 kcal/mol). These triose phosphate molecules are reversibly interconvertible ($\Delta G°$ = + 1.83 kcal/mol) in the presence of another enzyme triose phosphate isomerase. The equilibrium of this isomerization reaction is initially in favour of DHAP (96.5%) but it moves towards the 3-PGAL which is used in subsequent glycolytic reactions. So we consider that in this way two molecules of 3-PGAL are produced. The fate of every 3-PGAL molecule is same so subsequent reactions must be considered as operating twice.

These four steps transform one single molecule of 6-carbon glucose into two molecules of 3-carbon 3-PGAL molecules and collectively form "preparatory phase" of glycolysis

Every molecule of 3-PGAL undergoes :

5. Phosphorylation and Oxidative Dehydrogenation

It is a 2-stepped reaction in which 3-PGAL first undergoes dehydrogenation and then phosphorylation in the presence of enzyme glyceraldehyde-3-phosphate dehydrogenase. In oxidative dehydrogenation, one of four –SH (sulphahydryl) groups of enzyme combines with aldehydic group of 3-PGAL to form a thiohemiacetal. This enzyme -substrate complex then loses 2H-atom to a universal H-acceptor called NAD (Nicotinamide Adenine Dinucleotide-oxidised form) which is reduced to NADH + H^+. 3-PGAL changes into 3-PGA (3-Phosphoglycerate) as aldehyde group is oxidised to carboxylic group. It is an exergonic reaction ($\Delta G°$=-10.3 Kcal/mol). Dehydrogenation is followed by phosphorylation in which the phosphate group of phosphoric acid is added to the carbonyl group at the position-1 to produce 1,3-biphosphoglycerate. It is an endergonic reaction ($\Delta G°$ = + 11.8 kcal/mol).

So overall reaction is endergonic and involves free energy change of +1.5 kcal/mol. Overall reactions involved in oxidative dehydrogenation and phosphorylation:

$$\text{(a) } R-CHO + H_2O + NAD^+ \underset{\text{Exergonic}}{\rightleftharpoons} RCOOH + NADH + H^+ \quad \Delta G^o = -10.3 Kcal/mol$$

$$\text{(b) } R\ COOH + H_3PO_4 \underset{\text{Endergonic}}{\rightleftharpoons} R\ COO{-}PO_3H_2 + H_2O \quad \Delta G^\circ = +\ 11.8\ Kcal/mol$$

Sum:

$$RCHO + H_3PO_4 + NAD^+ \rightleftharpoons R\ COOPO_3H_2 + NADH + H^+ \quad \Delta G^\circ = +\ 1.5\ Kcal/mol$$

6. Dephosphorylation

It involves the trans phosphorylation of high energy phosphate group from C-1 (acyl phosphate group) of 1,3-biphosphoglycerate to ADP to synthesize ATP molecule in the presence of enzyme phosphoglycerate kinase (cofactor Mg"). 1,3-PGA changes into 3-PGA. This is the first step of ATP synthesis in the glycolytic pathway so this step recovers 2 molecules of ATP invested during the priming stage of glycolysis. It is highly exergonic reaction ΔG = -4 50 Kcal/mol. It is a substrate-level phosphorylation in which substrate participates in ATP or GTP formation.

7. Isomerization

In this, 3-PGA is reversibly changed to 2-PGA by the shifting of PO_4^{3-} group from the C-3 to C-2 in the existence of an enzyme-phosphoglyceromutase (cofactor Mg^{2+}). In this isomerization, 2, 3-biphosphoglycerate functions as an obligatory intermediate at the active site of the enzyme as shown in following equation:

$$2,3-\text{ Biphosphoglycerate} + 3-PGA \rightleftharpoons \text{2-PGA+2,3 Biphosphoglycerate} \quad \Delta G^\circ = +\ 7.06\ Kcal/mol$$

8. Dehydration

It involves the loss of water molecule from 2-PGA in the presence of enzyme enolase (activator Mg") so producing phospho-enol-pyruvic acid. The latter undergoes molecular rearrangement producing a high energy enolic phosphate bond. It is a reversible and endergonic reaction (ΔG° = +0.44 Kcal /mol).

9. Dephosphorylation

It contains the transfer of energy rich phosphate of phospho-enol-Pyruvic acid to ADP to form enol-pyruvic acid changes into enolpyruvate which spontaneously and non-enzymatically changes to pyruvate (Keto tautomeric form). It is an irreversible and exergonic reaction ($\Delta G°$-7.5kcal/mol).

10. Reduction

If the conditions are aerobic, then the pyruvic acid is either changed to Acetyl Co.A to enter Krebs cycle inside the mitochondrial matrix or used in lipogenesis. But if the conditions are anaerobic, in the presence of enzyme lactic acid dehydrogenase (LDHase), the pyruvic acid is changed to lactic acid and H-donor $NADH_2$ which gets oxidised to NAD^+ (explained in fermentation).

The steps of glycolysis involving the conversion of 3-PGAL molecules into 2 molecules of pyruvate is called "pay-off phase" and produces 4 ATP molecules. So the overall reaction of glycolysis is:

$$\text{Glucose} + 2\text{ Pi} + 2\text{ ADP} + 2\text{ NAD}^+ \longrightarrow 2\text{ Pyruvate} + 2\text{ ATP} + 2\text{ NADH} + 2H^+ + 2H_2O$$

(iv) Energy Balance of Glycolysis

Glycolysis represents an emergency energy-yielding pathway in which net gain from one glucose molecule is 2 ATP molecules, 2 Pyruvate and $2NADH_2$, molecule (Table 6.2). So glycolysis is a low-energy producing process as most of energy is still trapped in the pyruvate and NADH, molecules. In aerobic conditions, every NADH, is reoxidised and produces 3 ATP molecules on electron transport chain. So in aerobiosis, there is a gain of 8 molecules of ATP.

Table 6.2: Energy balance of glycolysis

Stage of Reaction	ATP used	ATP formed
Glucose ⟶ Glucose-6-P	1	
Glucose-6-P ⟶ Fructose 1,6- bi-P	1	
2 x (1,3-di-P-glyceric acid ⟶ 3 -P- Glyceric acid)		2
2 x (phosphoenol pyruvic acid ⟶ pyruvic acid)		2
2 x (reoxidation of NADH by electron carrier system		6
Total	2	10

(v) Importance of the Glycolysis

(a) Glycolysis is a necessary source of energy at the period of birth. During delivery, the blood circulation declines to most of the parts of the body of the neonate except the brain so conserving oxygen supply to the brain while other body parts show glycolytic pathway for their energy supply. This continues until circulation returns to normal.

6.7 Fermentation

Fermentation (L. fermentum = to boil), also called anaerobic respiration, comprises the generation of energy from the different food nutrients in lack of oxygen. This was first reported by **Kostychev**. It is found in some bacteria (e.g Clostridium tetani), yeasts (e.g. Saccharomyces) and intestinal endoparasitic helminthes (e.g Taenia, Ascaris, etc.). Out of these, tetanus-bacelli, denitrifying and methanogenic bacteria are obligatory or strict anaerobe (exclusively in the absence of oxygen) while yeasts and Ascaris are facultative anaerobes (could use oxygen when available). Pasteur studied anaerobic respiration for first in yeasts. Chemical pathway of fermentation was reported by **Neuberg** (1912). Anaerobic respiration comprises glycolysis where one single molecule of 6-C glucose is changed to two molecules of pyruvic acid (3C). In fermentation, pyruvic acid acts as H-acceptor that depends on the metabolic pathways and end products, fermentation is of 2 types:

(a) Alcohol Fermentation

This process is used in brewing-industry producing beverages like beer, rum, whiskey, etc. It is done by brewing yeast– Saccharomyces cerevisiae Buchner of (1897) reported the alcoholic fermentation in the cell-free yeast extracts (Figure 6.11). The following steps are involved here.

(1) The decarboxylation of pyruvic acid by an enzyme pyruvate decarboxylase occurs in the presence of Zn^{2+} as cofactor and co-enzyme thiamine pyrophosphate (TPP). This yields acetaldehyde. This decarboxylation of pyruvate continues through a series of intermediates that are covalently bound to TPP.

(2) In the presence of enzyme ethanol dehydrogenase, acetaldehyde receives 2H-atoms form NADH, and is reduces into ethyl alcohol.

So overall reaction of alcoholic fermentation is:

$$\text{Glucose} + 2\text{ Pi} + 2\text{ ADP} \longrightarrow 2\text{ Ethanol} + 2\text{ CO}_2 + 2\text{ ATP} + 2\text{ H}_2\text{O}$$

So both glycolysis and alcoholic fermentation yield only two molecules of ATP.

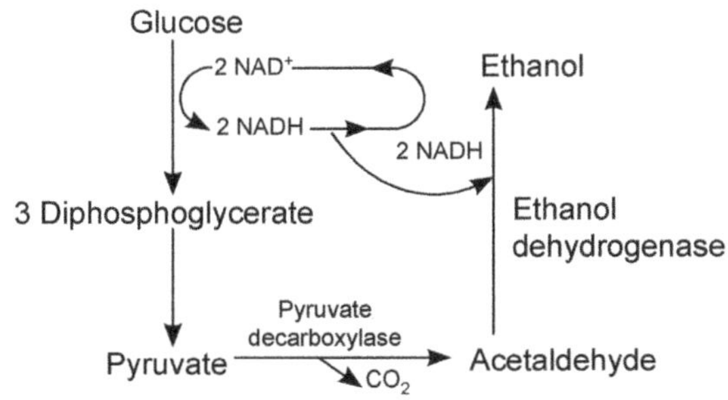

Figure 6.11: Pathway of anaerobic respiration in yeast cell.

Ethanol might accumulate in a culture medium of yeast. Alcoholic fermentation is also used in baking-industry in which CO_2 produced causes doughing of wheat flour. It occurs in the presence of baking yeast.

(b) Lactic Acid Fermentation.

It is found in muscle cells of animals (where anaerobiosis conditions often prevail especially when oxygen utilization is more rapid than its input, called oxygen-debt) or in lactic acid bacteria (useful in milk-industry). It is a one-stepped process in which pyruvic acid (produced during glycolysis) is reduced to lactic acid in the presence of enzyme lactate dehydrogenase (LADHase) and reduced co-enzyme NADH, which gets oxidised to NAD. It is an exergonic reaction ($\Delta G°= -6.0$ kcal/mole). Lactic acid is excerted out so large amount of energy of glucose is lost.

Lactic acid fermentation is of 2 types on the basis of nature of end products :

1. **Homolactic fermentation (Figure 6.12):** It is found in homo fermentative strains of bacteria like Streptococcus, Leuconostoc, Pediococcus and Lactobacillus. It is also found in the muscle cells of animals. In this, monosaccharides like D-glucose, D-mannose and many others; and disaccharides like sucrose, maltose and lactose are fermented and the predominant product is lactic acid while other products are found in traces. It yields two ATP molecules. Formation of lactic acid from glucose during anaerobic muscle contraction was reported by Fletcher and Hopkins.

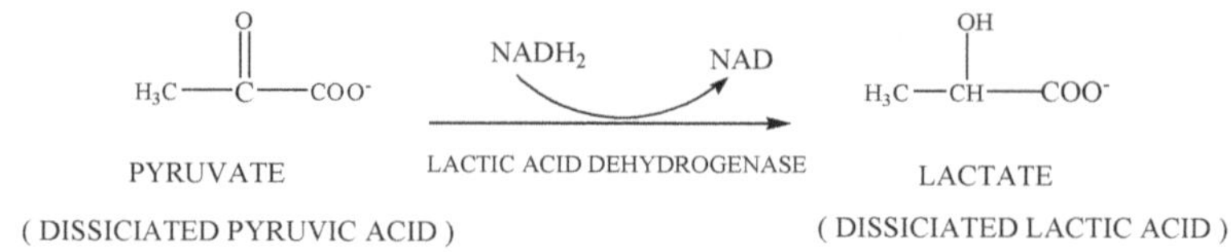

Figure 6.12: Lactic acid fermentation

2. **Heterolactic fermentation (Figure 6.13):** It is present in hetero fermentive strains of bacteria like Leuconostoc, Lactobacillus, Peptostreptococcus, Eubacterium, Ramibacterium, Bifidobacterium and Catenabacterium. In this, in addition to small amount of lactic acid (about 1.8 moles of lactic acid per mole of glucose), some other products are also formed e.g. lactic acid, acetate, ethanol and glycerol in Leuconostoc bacteria; lactic acid and propionic acid in Peptostreptococcus bacteria;lactic acid and butyric acid in different species of Clostridium like C. butyricum and C. acetobutylicum.

In *Lactobacillus brevis and Leuconostoc*, glucose undergoes phosphoketolase pathway to produce pyruvic acid and acetic acid. Pyruvic acid is then reduced to lactic acid while acetic acid to ethanol through acetaldehyde .

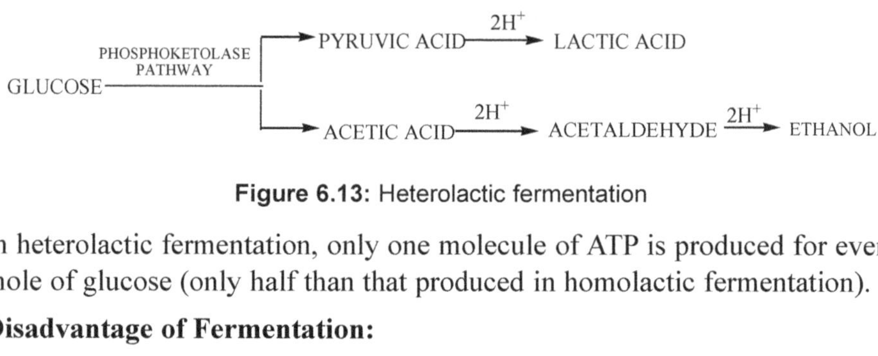

Figure 6.13: Heterolactic fermentation

In heterolactic fermentation, only one molecule of ATP is produced for every mole of glucose (only half than that produced in homolactic fermentation).

Disadvantage of Fermentation:

(a) It is a low energy producing process e.g. only 50 kcal of energy is formed in anaerobic respiration while 686 kcal of energy is produced in aerobic respiration. It is so that electrons are not transported to oxygen.

(b) Accumulation of lactic acid reasons for the muscle fatigue.

(c) There is loss of considerable amount of energy in its end products.

Advantage of Fermentation:

(a) It supplements the aerobic respiration.

(b) It has necessary role in brewing, baking and milk industries.

(c) It regenerates NAD to trap more H-atoms. If this does not happen, glycolysis might stop resulting the death of the organism.

6.8 Krebs Cycle Or Tricarboxylic Acid Cycle (TCA)

Acetyl Co.A (also called active acetate) is a 2-carbon unit derived from maximum of the major energy-generating metabolic pathways of the cell.

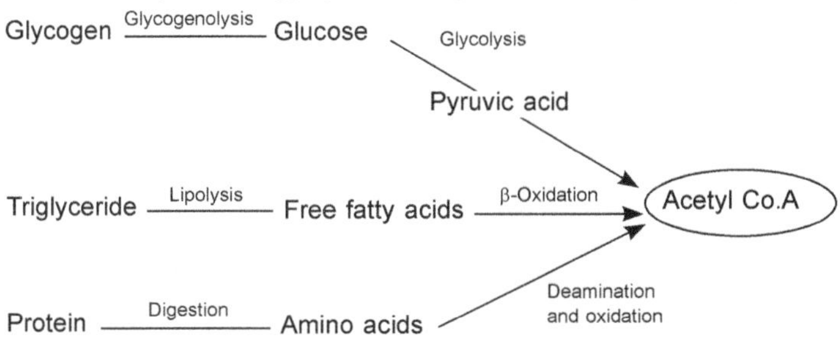

Acetyl Co. A could further be metabolized in different metabolic pathways depending upon the tissue-types and metabolic state of that tissue:

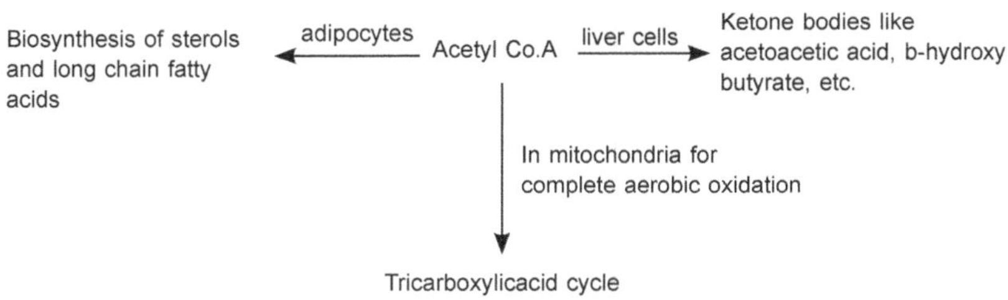

Discovery. T, Thunberg, *et al* (1910-20) showed that anaerobic suspension of minced animal tissue catalyze the transfer of H-atoms from certain organic acids like succinic, malic and citric acids, In 1935, Szent-Gyorgyi concluded that these acids stimulate the oxidation of some endogenous substrate, especially glycogen, producing one molecule of malate or oxaloacetate promoting the oxidation of many molecules of the endogenous substrate. Later, C Maltius and F. Knoop reported that citrate is enzymatically oxidized to a ketoglatarate and finally to succinate. The details of Krebs cycle were given by Sir Hans Krebs (1937) and was initially named "citric acid cycle" (first product formed is citric acid) It is now called "Tricarboxylic Acid Cycle" (TCA Cycle as citric acid has 3-COOH groups-a tri carboxylic acid).

Krebs cycle is an enzymatically controlled stepped process. The enzymatic reaction of TCA-cycle occurs within the mitochondrial matrix of inner chamber of mitochondria. It was confirmed by E.P. Kennedy and A.L Lehninger (1948) by catalyzing the oxidation of all the intermediates of TCA cycle and pyruvate at the cost of molecular oxygen and isolated mitochondria, from rat-liver homogenate.

Mechanism (Figure 6.14). TCA-cycle Involves Following Steps:

1. **Condensation:** It involves the joining of acetyl group (2C) of acety1 Co.A and oxaloacetate (dicarboxylic acid-4C) to form citrate (6C) in the existence of a condensing enzyme, citrate synthetase (mol. wt. - 100,000 and reported by S. Ochoa). The reaction involves the aldol condensation between the methyl groups of acetyl Co. A and carbonyl group of OAA. An unstable intermediate compound citroyl Co.A is formed which on hydrolysis yields citrate and Co.A. It is an exergonic reaction and ΔG is -7.7 kcal/mole. This reaction is the primary pacemaker step of TCA cycle and the rate depends upon the availability of acetyl-Co.A and OAA; and inhibited by high concentration of succinyl Co.A (competitive inhibition with acetyl Co.A for citrate syrthetase), ATP, NADH and long chain acyl Co.A.

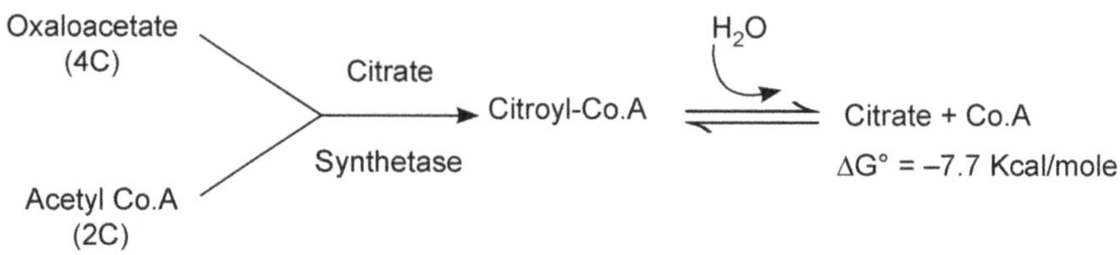

2. **Isomerization:** It involves the reversible interconversion of citrate and iso-citrate through enzyme-bound intermediate cis-aconitase in the presence of a Fe" consist of enzyme aconitase hydrolase (commonly called aconitase). The equilibrium mixture is formed of 93% citrate and 7% so- -citrate, the later is used in metabolism.

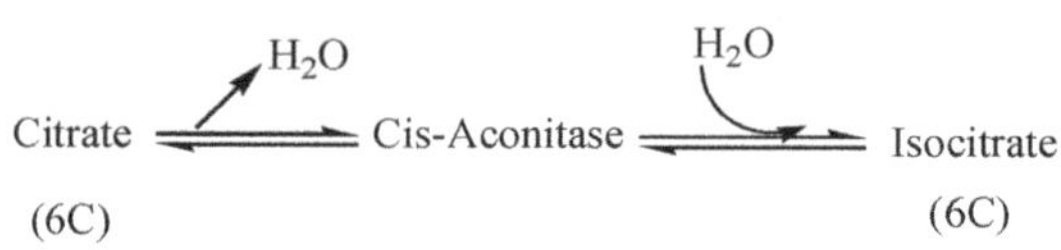

3. **Dehydrogenation-I:** Isocitrate undergoes dehydrogenation to produce oxalosuccinate (6C) in the attendance of NAD-linked or NADP linked (Mg^2 cofactor) isocitrate dehydrogenase enzyme out of which main enzyme in mitochondrial matrix is NAD-linked isocitrate dehydrogenase (Mg^{2+} cofactor and with mol. wt.-380,000). This enzyme is stimulated by ADP while inhibited by ATP and NADH.

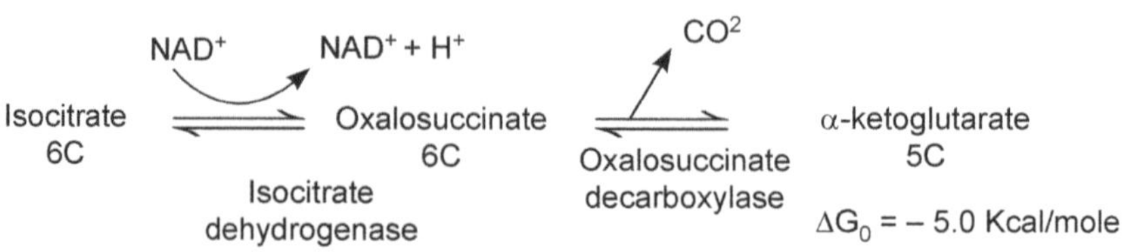

4. **Decarboxylation-I:** Oxalosuccinate undergoes first decarboxylation to form a 5-C compound, α-ketoglutarate in the presence of enzyme oxalosuccinate decarboxylase.

5. **Oxidative Decarboxylation:** In this α-glutarate undergoes dehydrogenation-II and decarboxylation-II in the presence of α-ketoglutarate-dehydrogenase multienzyme complex and changes into succinyl-Co.A (4C) which is a high-energy thioester of one carboxyl group of succinic acid. This reaction is analogous to the oxidative decarboxylation of pyruvate to acetyl Co.A and CO_2. This complex is very similar to pyruvate-dehydrogenase complex. It is formed of 3 enzymes and 5 coenzymes.

   The enzymes are:

   (i) α-ketoglutarate dehydrogenase

   (ii) Dihydrolipoyl dehydrogenase

   (iii) Dihydrolipoyl transuccinylase

   The conenzymes are :

   (a) TPP

   (b) Lipoic acid

   (c) NAD^+

   (d) FAD

   (e) Coenzyme-A

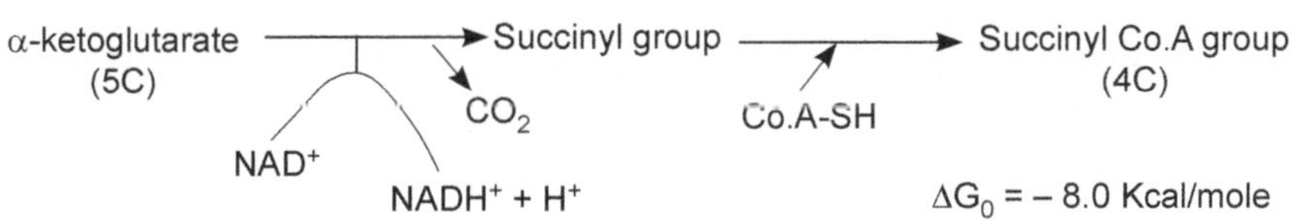

This reaction produces second molecule of carbon dioxide /CO_2 and the second reducing equivalent which reduces corresponding (i.e $NADH^+$, H^+). Succinyl Co. A has an energy-rich thioester linkage which is used in the phosphorylation of GDP (guanosine diphosphate) to GTP in the existence of another enzyme succinate thiokinase. So succinyl Co. A changes to succinic acid (4C) and coenzyme-A is released. This reaction is an example of formation of the high energy phosphate at the substrate level. So it is called substrate-level phosphorylation.

$$\text{Succinyl Co.A} + \text{Pi} + \text{GDP (C4)} \underset{\text{thiokinase}}{\overset{\text{Succinate}}{\rightleftharpoons}} \text{Succinate} + \text{GTP} + \text{Co.A} \sim \text{SH (4C)}$$

GTP might be used in the formation of ATP as shown below as both GTP and ATP have almost same standard free energy of hydrolysis.

$$\text{GDP} + \text{ADP} \xrightleftharpoons{\text{Phosphokinase}} \text{GDP} + \text{ATP}$$

(6) Dehydrogenation-III: In this reaction, succinate loses 2H-atom in the presence of inner membrane-bound enzyme succinate dehydrogenase (mol wt -100,000) which is associated with one molecule of FAD (Flavin adenine dinucleotide), eight atoms of non-haeme iron and eight labile sulphur atoms. Succinate changes to fumarate while FAD is reduced to $FADH_2$.

$$\underset{(4C)}{\text{Succinate}} \xrightleftharpoons[\text{Succinate Dehydrogenase}]{\text{FAD} \;\curvearrowright\; \text{FADH}_2} \underset{(4C)}{\text{Fumarate}}$$

$FADH_2$ passes the pair of electrons to ETC to yield 2 ATP molecules Succinate dehydrogenase is stimulated by succinate and ATP while it is inhibited by malonate and oxaloacetate.

(7) Hydration: It involves the addition of water molecule in the fumarate in the presence of enzyme fumarase (a tetrameric protein, molecular weight of 200,000).

After hydration, fumarate changes to L-malate

$$\text{Fumarate} + H_2O \xrightleftharpoons{\text{Fumarase}} \text{L-Malate} \quad \Delta G_0 = 0$$

(8) Dehydrogenation-IV: In this, L-malate loses 2H-atoms to an H-acceptor NAD^+ (which gets reduced $NADH + H^+$) and in the presence of an enzyme malate dehydrogenase changes into oxaloacetate. It is a reversible and endothermic reaction.

$$\underset{(4C)}{\text{L-Malate}} \xrightleftharpoons[\text{Malate Dehydrogenase}]{\text{NAD}^+ \;\curvearrowright\; \text{NADH+H}^+} \underset{(4C)}{\text{Oxaloacetate}}$$

$$\Delta G_0 = +7.0 \text{ Kcal/mol}$$

$NADH + H^+$ pass their electrons to ETC and produce 3ATP molecules.

Various steps of Krebs cycle have been summarized in Table-4 and shown diagrammatically in **Figure 6.14**.

Oxaloacetic acid again combines with acetyl Co.A to enter Krebs cycle and repeat the process. So in Krebs cycle, one molecule of acetyl Co.A undergoes two decarboxylation and four dehydrogenations. The four pairs of hydrogen by enzymatic dehydrogenation catalysed by dehydrogenases and oxido-reductases are removed from the intermediate compounds. These three pairs of hydrogen are utilized to lessen three molecules of NAD^+ to NADH and H^+, while one pair is used to reduce FAD to $FADH_2$ (Table 6.3).

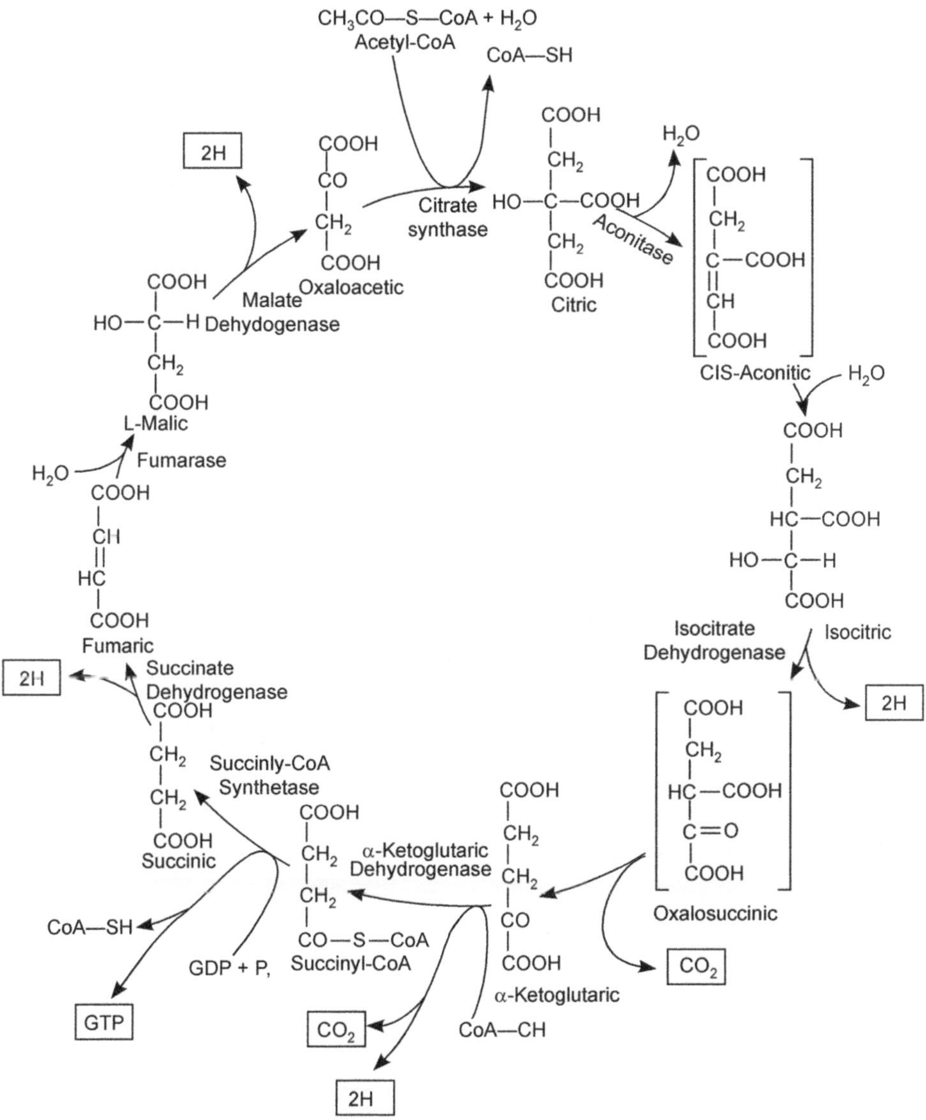

Figure 6.14: TCA (Tricarboxylic acid cycle) Image source (Google images)

Table 6.3: Various stages of TCA cycle

	Steps	Substrates	Enzyme	End-Products
1.	Condensation	Acetyl-Co. A+Oxaloacetic acid $+H_2O$	Citric acid synthetase	Citric acid + Coenzyme-A (6C)
2.	Isomerization	It is again completed in two steps: (a) Citric acid (b) Cis-Aconitic acid $+ H_2O$	Aconitase Aconitase	Cis-Aconitic acid + H_2O Iso-citric acid (6C)
3.	Dehydrogenation-I (–2H)	Isocitric acid + NAD^+ (H-acceptor)	Isocitrate dehydrogenase	Oxalosuccinate + NADH+ H^+
4.	Decarboxylation ($-CO_2$)	Oxaloscucinate	Oxalosuccinate decarboxylase	α-Ketoglutaric acid + CO_2 (5C)
5.	**Oxidative Decarboxylation** It is also completed in two steps: (–2H) ($-CO_2$)	(a) α -Ketoglutarate + H_2O + NAD^+ + Coenzyme-A (b) Succinyl-CoA + GDP + Pi	α-Ketoglutarate dehydrogenase complex Succinate thiokinase	Succinyl-Co.A + NADH + H+ (4C) + CO_2 Succinate + Coenzyme-A (4C) + GTP
6.	**Dehydrogenation-III** (–2H)	Sucinate + FAD (H-acceptor) (FAD = Flavin Adenine Dinucleotide)	Succinate dehydrogenase	Fumarate + $FADH_2$ (4C)
7.	**Hydration** ($+H_2O$)	Fumarate + H_2O	Fumarase	Malate (4C)
8.	**Dehydrogenation-IV**	Malate + NAD^+ (H-acceptor)	Malate dehydrogenase	Oxaloacetic acid + NADH + H^+ (4C)

As two molecules (–2H) of acetyl Co.A are formed from one oxidized glucose molecule, Krebs cycle must rotate twice for every molecule of glucose to respire. So net gain of Krebs cycle is:

$$2\ \text{Acetyl Co.A} + 8\ H_2O + 6\ NAD^+ + 2\ FAD + 2\ GDP + 2\ Pi$$

$$\downarrow$$

$$4CO_2 + 2H_2O + +6NADH + 6H^+ + 2\ FADH_2 + 2GTP + 2\ \text{Coenzyme A}$$

So Krebs cycle is also a low energy producing process as only 2 GTP (or ATP) molecules are produced. NADH + H^+ and $FADH_2$ could release their energy only in Electron Transport Chain (ETC). The 6 $NADH_2$ and 2 $FADH_2$ formed in Krebs cycle provide 22 molecules of ATP in ETC.

The overall reaction of glycolysis, oxidative decarboxylation and Krebs cycle is (Table 6.4):

$$C_6H_{12}O + 6H_2O \longrightarrow 6CO_2 + 4ATP + 12\ H_2$$

Table 6.4: Differences between glycolysis and krebs cycle

S. No.	Character	Glycolysis	Krebs Cycle
1.	Occurrence	In the cytoplasm outside the mitochondrion	Inside the matrix of mitochondrion.
2.	Conditions	It exist in both the aerobic and anaerobic conditions.	It occurs only in aerobic conditions.
3.	Nature of pathway.	Linear pathway.	Cyclic pathway.
4.	Energy gain	It produces 2ATP molecules per glucose molecule.	It produces 2 GTP molecules from 2 molecules of acetyl-CoA
5.	Reduced Co-enzymes	Produces $2NADH_2$ molecules per glucose molecule.	Two molecules of acetyl-CoA produce $6NADH_2$ and 2 $FADH_2$ molecules.
6.	End-products.	Two molecules of pyruvic acid from one glucose.	Acetyl CoA produces CO_2 and water molecules.
7.	Carbon-dioxide	No CO_2 is produced.	CO_2 is evolved.

Significance

(a) It is a common pathway of oxidative breakdown of carbohydrate, fatty acids and amino acids.

(b) Primary function of TCA-cycle is to produce reducing equivalents which yield ATP molecules in the ETC-oxidative phosphorylation sequence.

Krebs cycle produces

(1) $3NADH + H^+$ molecules which generate 9 molecules of ATP along route-I (every $NADH + H^+$ produces 3ATP molecules).

(2) One $FADH_2$, molecule which generates 2 molecules of ATP along route-II

(3) One GTP molecule which could produce one ATP molecule by transphosphorylation.

Thus in a Krebs cycle proper, one acetyl Co.A produces 12 ATP molecules.

(c) Acetyl Co.A and intermediate products of Krebs cycle could be used in different metabolic pathways e.g.

(1) Citrate could be used as a carbon source for cytosolic biosynthesis of fatty acids, steroids, etc.

(2) Isocitrate could be used in the biosynthesis of aspartate.

(3) α -ketoglutarąte could be converted to glutamate in the presence of NADPH and ammonia. Converted Glutamate could be used in the biosynthesis of aspartate amino acid by transamination.

$$\alpha\text{-ketoglutarate + Alanine} \rightleftharpoons \text{Glutamate + Oxaloacetate!}$$

$$\text{Glutamate + Pyruvate} \rightleftharpoons \text{Aspartate + }\alpha\text{-Ketoglutarate}$$

(4) Succinyl Co.A could be utilized in the synthesis of porphyrin.

So Krebs cycle is called an amphibolic pathway and involves not just catabolism but also generates intermediate precursors for anabolic pathways.

6.9 Interrelationship Between Various Metabolic Pathways of Proteins, Fats and Carbohydrates

Some Metabolic Interrelationships between Carbohydrate, Fat and Protein depicts in **Figure 6.15**. The significance of this irreversibility is that acetyl-coA couldn't act as a precursor for glucose or glycogen, while many of the TCA cycle intermediates are effective precursors of glucose. Atoms are lost as carbon dioxide (CO_2) before a carbohydrate precursor is formed, thus there could be no net synthesis of carbohydrate from fatty acids in animals. Glycerol moiety of fats might, however, be converted to glucose. Because some organs such as the brain require blood glucose for energy, it is of maximum importance that a certain circulatory level of glucose be maintained even when carbohydrate is not being absorbed and the supply of glycogen is nearly depleted. The glucogenic amino acids are of maximum importance in this regard. The majority of the amino acids are glucogenic as illustrated in Fig. 13 (glutamic acid, aspartic acid, glycine, alanine, serine, valine and a number of others), also apparent from Figure 15 carbohydrates and the carbon chain from amino acids are effective precursors of body fat.

The central role of acetyl-coA in the oxidation of fats and carbohydrates has been mentioned. Acetyl-coA serves as a precursor of several other products, including ketone bodies and cholesterol. When large amounts of fat are being metabolized with very little carbohydrate, such as in starvation and particularly in diabetes, complications of ketosis arise. Acetyl-coA is apparently formed at a more rapid rate than it could be handled by the TCA cycle and ketone bodies accumulate. While the many organs and tissues (i.e. muscle) of the body could metabolize the ketone bodies for energy, the liver could not do so. The excess ketone bodies are excreted and because two of these are acidic they pull sodium ions from the body. This lowers the base reserve and could result in acidosis. A number of explanation for failure of the TCA cycle to handle the acetyl-coA of fat metabolism have been put forward, such as a relative lack of TCA cycle intermediates (oxaloacetate) and excessive fat breakdown by the liver, but none has been universally accepted. One current idea is that triglyceride in the adipose tissue is being continuously broken down to fatty acids and glycerol and also is being synthesized. Resynthesis of triclyceride from glycerol apparently could not occur in adipose tissue as because of lacking the enzyme, glycercl kinase, which produces the essential intermediate, a glycerol phosphate down of glucose via dihydrozy-acetone-phosphate of fat metabolism depends on glucose metabolism by the adipose tissue. Presumably, if glucose is in short supply, the catabolism "overshoots" and too much fat is mobilized resulting in ketosis. In addition to entering the TCA cycle, labile many organs and tissues.

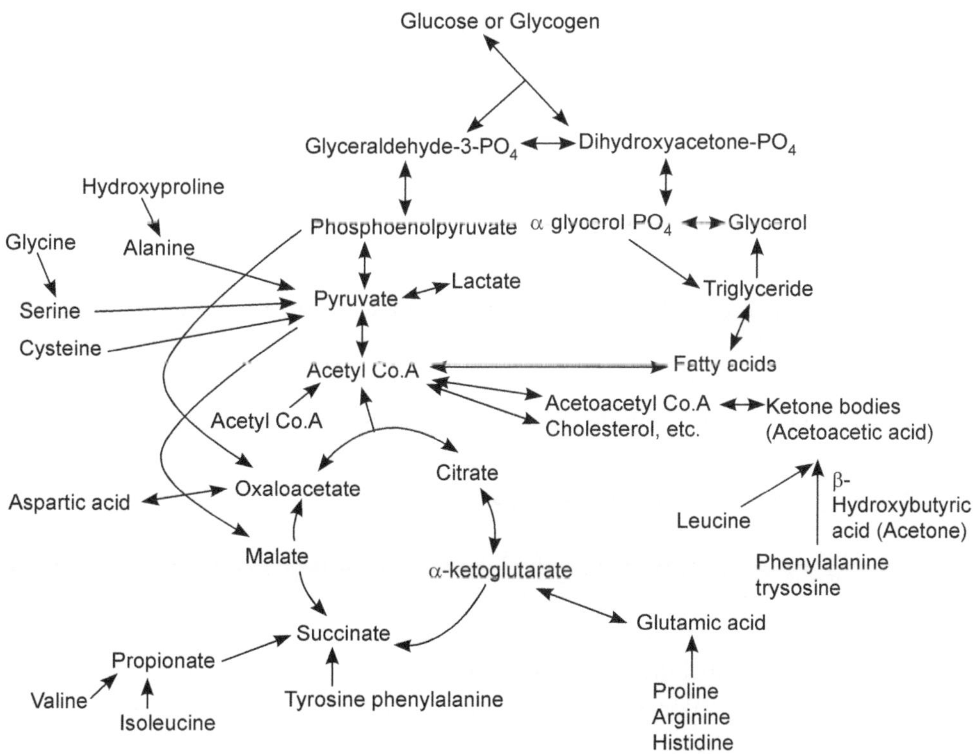

Figure 6.15: Example of interrelationship between carbohydrate, fats and protein metabolism

Some of the amino acids are also ketogenic and if supplied singly in fairly large amounts they will cause a ketosis. Particularly leucine and to some extent phenylalanine, tyrosine and isoleucine are ketogenic as is apparent in Figure.

Some amino acids contributed some of its carbon to both glucose and ketone bodies, and there are not really either glucogenic or ketogenic.

Fat mobilization from adipose tissues is a subject which only recently has yielded to investigation. Mobilized fatty acids circulate in the blood as albumin complexes, usually occur in inverse proportion to the level of blood glucose, and serve as effective sources of energy for many tissues of the body. Their turnover rate is exceedingly rapid and although they make up only a small portion of the total lips of the blood they are quantitably very necessary energy sourced. Their release from adipose is at least in part, under hormonal control.

Another interrelationship of current interest between carbohydrate metabolism and biosynthesis of fat is so-called hexose-monophosphate shunt.

Although this metabolic scheme will not be discussed in detail, it explains the conversion of glucose to pentoses and a number of other sugars and is an alternate route of glucose metabolism. Significance of the fact that an early reaction of the shut generates NADPH rather than NADH as in glycolysis and the TCA cycle. NADPH is the reduced coenzyme involved in many reductive syntheses such as the synthesis of fatty acids. It is believed hence that when a major share of the glucose in a tissue is metabolized by the shunt mechanism, as opposed to glycolysis, fatty acid synthesis and other biosynthesis are favored. Perhaps this is one mechanism whereby the body controls the extent of synthesis versus the extent of complete oxidation.

6.10 Calorific Value of The Food

The energy gained by the human body through metabolism is shown by the energy value of a food indicated in kilojoules (kJ)/100 g or 100 ml. Additionally, the specification is in kilocalories (kcal). Carbohydrates, fats and proteins, contribute diets 90% dry weight and 100% energy. These three offers energy (in calories), however the amount of energy delivered in 1 gram (1/28 ounce) varies:

- 4 calorie/gram of carbohydrate/protein
- Whereas 9 calories/gram of the fat

The above mentioned nutrients vary in how fast they source energy too. The carbohydrates being the fastest and fats the slowest one.

All food stuffs get digested in the intestine, and further breakdown into the basic units as mentioned:

- The Carbohydrates broke down to sugars
- The Proteins into amino acids
- The Fats in the fatty acids and glycerol

The human body from these basic units builds the ingredients which are necessary for the growth, maintenance and various activities of body.

Carbohydrates

Carbohydrates, might be simple (like glucose, sucrose)or complex (like starches and fibers) depending on the size of the molecule,. Carbohydrates might be unrefined and refined. The refined being the food that is extremely processed, the body processes the carbohydrate molecules rapidly, and they offer little nutrition even though they comprise of almost the same amount of calories. The refined products are so often supplemented, means vitamins and minerals been additional for enhancing their nutritional energy value.

Maximum authorities endorse that about 50-55% of the total daily calories must comprise of carbohydrates. Less than 10% of total daily calories must come from added sugars.

Proteins

A Protein is made up of the complex structure formed from small entities termed as amino acids. Since proteins are complex molecules, the breakdown into body takes longer time. Consequently, they are a considerable slower and longer-lasting energy source in comparison to carbohydrates. The body uses almost 100% of egg protein and a great amount present in eat and milk. The body can utilizes less than half of the protein in the cereals and vegetables. The body requires protein for maintaining and replacing tissues and for functional growth. Protein is not generally utilized for energy. But in case the body is not getting adequate calories from the fat stored or the other nutrients the protein is the source of energy.

During excess protein production by the body, the protein is fragmented down and fat is stored as the component.

The proteins are main building blocks of the body and are the chief component of the most cells like connective tissue, skin cells and muscles.

An adult must consume 60 grams of protein/day (0.8 grams/Kg of body weight or some 10-15% of the total calories). The people involved in body building and the growing children require extra protein in the diet. Those limiting their calories for weight loss normally need more protein for averting the muscle loss during weight loss.

Fats

Fats required for the energy and growth of the body are the complex molecules that comprising the fatty acids and glycerol. It also utilizes them for synthesizing hormones and the other substances which are required for the different activities (like prostaglandins and others) of body. Fats are the slowest among the all energy source but are also the most energy-resourceful form. 9 calories is delivered by every gram of fat, being more than double of carbohydrates and proteins. The body stores the additional fat (as subcutaneous fats) beneath the skin and the abdomen for the time when more energy is needed. The excess fat can also be stored in the blood vessels and the organs blocking the blood and damage the organs, causing serious disorders. The various kinds of fat in the nature are:

(i) Saturated fats

(ii) Monosaturated fats

(iii) Polysaturated

The saturated fats rise the risk of atherosclerosis by increasing the body cholesterol level. The food items from the animals consist of saturated fats having the solidifying tendency at room temperature (R.T). On the other hand the plant fat mostly consists of mono or the polyunsaturated fatty acid that liquefies at R.T/ room temperature with exception of coconut and palm oil which comprises more saturated fats than any other plant oils.

The fats necessary to be consumed $<$ 28% of daily calories (90 gm. /day) and the saturated fats to be less than 8%.

Questions

1. Give the structure of ATP and write the reactions involved in the complete hydrolysis of ATP.
2. What is glycolysis and where does this process occur in the cell?

3. Give the significance of Glycolysis.
4. Differentiate the preparatory phase and pay off phase of glycolysis
5. Give the energy balance of glycolysis.
6. Define fermentation. What are the advantages and disadvantages of fermentation?
7. Compare the efficiency of aerobic and anaerobic respiration.
8. Differentiate glycolysis and krebs cycle.
9. Enumerate the mechanism of krebs cycle. What is its significance
10. Outline the complete glycolytic pathway giving structure of intermediates and enzymes involved.
11. How is pyruvate converted into lactic acid and ethanol under aerobic condition? What are these reactions called?
12. How many types of reactions involved in metabolism? Discuss with an example of each type.
13. Define calorific value of food. Fats have more calorific value than carbohydrates, explain.
14. Explain the interrelationship in the metabolic pathways of carbohydrates, proteins and fats.
15. What is metabolism? Explain briefly each type of metabolism.
16. What ids the fate of pyruvate in the biological system.

❑❑❑

Index

For Product Safety Concerns and Information please contact our EU
representative GPSR@taylorandfrancis.com
Taylor & Francis Verlag GmbH, Kaufingerstraße 24, 80331 München, Germany

www.ingramcontent.com/pod-product-compliance
Ingram Content Group UK Ltd.
Pitfield, Milton Keynes, MK11 3LW, UK
UKHW020143200926
475702UK00005B/896
9781041206828